Inspiration and Design for Bio-Inspired Surfaces in Tribology:

Emerging Research and Opportunities

Hisham Abdel-Aal
Drexel University, USA

A volume in the Advances in Chemical and Materials Engineering (ACME) Book Series

IGI Global
PUBLISHER of TIMELY KNOWLEDGE

Published in the United States of America by
IGI Global
Engineering Science Reference (an imprint of IGI Global)
701 E. Chocolate Avenue
Hershey PA, USA 17033
Tel: 717-533-8845
Fax: 717-533-8661
E-mail: cust@igi-global.com
Web site: http://www.igi-global.com

Library of Congress Cataloging-in-Publication Data

Names: Abdel-Aal, Hisham, 1960- author.
Title: Inspiration and design for bio-inspired surfaces in tribology : emerging research and opportunities / by Hisham Abdel-Aal.
Description: Hershey, PA : Engineering Science Reference, an imprint of IGI Global, 2019. | Includes bibliographical references and index. | Summary: "This book examines developments in bio-inspired engineered surfaces for tribological applications"-- Provided by publisher.
Identifiers: LCCN 2019031690 (print) | LCCN 2019031691 (ebook) | ISBN 9781799816478 (hardcover) | ISBN 9781799816485 (paperback) | ISBN 9781799816492 (ebook)
Subjects: LCSH: Surfaces (Technology)--Design. | Tribology. | Biomimetics. | Contact mechanics.
Classification: LCC TA418.72 . A23 2019 (print) | LCC TA418.72 (ebook) | DDC 620/.44--dc23
LC record available at https://lccn.loc.gov/2019031690
LC ebook record available at https://lccn.loc.gov/2019031691

This book is published in the IGI Global book series Advances in Chemical and Materials Engineering (ACME) (ISSN: 2327-5448; eISSN: 2327-5456)

British Cataloguing in Publication Data
A Cataloguing in Publication record for this book is available from the British Library.

Advances in Chemical and Materials Engineering (ACME) Book Series

ISSN:2327-5448
EISSN:2327-5456

Editor-in-Chief: J. Paulo Davim, University of Aveiro, Portugal

MISSION

The cross disciplinary approach of chemical and materials engineering is rapidly growing as it applies to the study of educational, scientific and industrial research activities by solving complex chemical problems using computational techniques and statistical methods.

The **Advances in Chemical and Materials Engineering (ACME) Book Series** provides research on the recent advances throughout computational and statistical methods of analysis and modeling. This series brings together collaboration between chemists, engineers, statisticians, and computer scientists and offers a wealth of knowledge and useful tools to academics, practitioners, and professionals through high quality publications.

COVERAGE

- Heat Treatments
- Coatings and surface treatments
- Artificial Intelligence Methods
- Ductility and Crack-Resistance
- Chemical Engineering
- Multifuncional and Smart Materials
- Computational methods
- Materials to Renewable Energies
- Ceramics and glasses
- Thermo-Chemical Treatments

IGI Global is currently accepting manuscripts for publication within this series. To submit a proposal for a volume in this series, please contact our Acquisition Editors at Acquisitions@igi-global.com or visit: http://www.igi-global.com/publish/.

The Advances in Chemical and Materials Engineering (ACME) Book Series (ISSN 2327-5448) is published by IGI Global, 701 E. Chocolate Avenue, Hershey, PA 17033-1240, USA, www.igi-global.com. This series is composed of titles available for purchase individually; each title is edited to be contextually exclusive from any other title within the series. For pricing and ordering information please visit http://www.igi-global.com/book-series/advances-chemical-materials-engineering/73687. Postmaster: Send all address changes to above address.

Titles in this Series

For a list of additional titles in this series, please visit:
https://www.igi-global.com/book-series/advances-chemical-materials-engineering/73687

Natural Fiber-Reinforced Geopolymer Composites Synthesis, Properties, and Applications
It-Meng (Jim) Low (Curtin University, Australia) Thamer Alomayri (Umm Al-Qura Univeristy, Saudi Arabia) and Hasan Assaedi (Umm Al-Qura Univeristy, Saudi Arabia)
Engineering Science Reference • © 2020 • 230pp • H/C (ISBN: 9781799834595) • US $185.00

Nanotechnologies and Clusters in the Spaces of Higher Dimension Emerging Research and Opportunities
Gennadiy Vladimirovich Zhizhin (Independent Researcher, Russia)
Engineering Science Reference • © 2020 • 240pp • H/C (ISBN: 9781799837848) • US $195.00

Processing, Structure, and Applications of High-Temperature Materials
Brahma Raju Golla (National Institute of Technology Warangal, India)
Engineering Science Reference • © 2020 • 300pp • H/C (ISBN: 9781799831792) • US $195.00

Metaheuristic Approaches for Optimum Design of Reinforced Concrete Structures Emerging Research and Opportunities
Aylin Ece Kayabekir (Istanbul University-Cerrahpaşa, Turkey) Gebrail Bekdaş (Istanbul University-Cerrahpaşa, Turkey) and Sinan Melih Nigdeli (Istanbul University-Cerrahpaşa, Turkey)
Engineering Science Reference • © 2020 • 229pp • H/C (ISBN: 9781799826644) • US $185.00

Properties, Techniques, and Applications of Polyaniline (PANI) Thin Films Emerging Research and Opportunities
Subhash Chander (Indian Institute of Science Education and Research, Mohali, India) and Nirmala Kumari Jangid (Banasthali University, Tonk, India)
Engineering Science Reference • © 2020 • 236pp • H/C (ISBN: 9781522598961) • US $165.00

IGI Global
PUBLISHER of TIMELY KNOWLEDGE

701 East Chocolate Avenue, Hershey, PA 17033, USA
Tel: 717-533-8845 x100 • Fax: 717-533-8661
E-Mail: cust@igi-global.com • www.igi-global.com

Table of Contents

Preface

The journey of the human engineer in the world of surfaces is indeed intriguing. It is an adventurous endeavor characterized by challenges, riddled with misunderstandings, and enabled through ingenuity. The most interesting feature of that journey is that it started from interaction with nature, and perhaps its final destination is heading towards mimicking nature. From departure to destination the journey traces a complex irregular path that offers questions more than it provides answers.

Although our encounters with surfaces are rooted in antiquity, our understanding of their real impact on system performance is relatively recent. Surfaces, for a long time, were regarded as covers that do not influence the function of whatever system they enclose. Occasionally, there were efforts to depart from that prevailing perception through attempts of engineering a surface to obtain a target performance. These efforts, however, were individual attempts that stood on empirical foundations. They did not impact the collective comprehension of how surfaces function. A familiar example is that of samurai swords and later on Damascus steel. Here, surface engineering was performed through imposing functional property gradients. The objective was to increase durability on impact. However, surface engineering in this case was imposed based on empirical rules mainly acquired through trial and error. This led to nonuniform performance across processed material patches (i.e., no quality control by modern standards). In tribology, the subject of this work, increase of "grip" was a requirement for many applications. This was achieved by introducing a "rough surface". The question of "how rough is rough" remained a question with no definite answer since the role of the roughness features was not appreciated nor fully understood.

At the dawn of modern industrialization, real appreciation of the function of surfaces still eluded engineers. However, encountered material and equipment failures prompted rethinking surfaces. Advances in mathematics, analytical tools, and modeling skills assisted the emergence of new ideas about surfaces and their function. It was late in the 19th century when the effort of rethinking surfaces acquired a humble momentum (perhaps because of the introduction of the automobile). Despite this fresh look at surfaces, the newly forming view still was problematic. In the early 20th century, the surface-cover view was practically replaced by a "matter-of-fact" paradigm. Common convention, dominant at the time among surface engineers, still promoted a modernized version of surface-cover paradigm with a slight twist. The topography of surfaces was a function of the manufacturing method. Therefore, any change in roughness or smoothness would take place by adopting a different manufacturing route. Moreover, the answer to how "smooth" or "rough" a surface is remained subjective. It took the industrialized world several disasters to recognize that surfaces are vital elements of any rubbing system and that they are the gateway to structural health and optimized function of engineering systems.

A significant milestone in the efforts to understand the role of surfaces in tribological behavior was the ability to statistically describe topography. The efforts practically started in the 1930s with the seminal work of Abbott and Firestone who introduced the concept of the "material curve". The introduction of this concept ushered the quest for controlling surface behavior into a meaningful phase. It initiated efforts for systematic and consistent description of the elements of topography within a surface. The description evolved into an elaborate, albeit massive, system of parameters that enabled further in-depth understanding of how surface topography contributes to tribological processes.

Towards the end of the 20th century a major paradigm shift in surface design took place. That shift was motivated by the realization that topography of surfaces, which result from conventional manufacturing processes, are random in essence and therefore are difficult to control. Thus, because of randomness, description of surfaces would lead to probabilistic predictions of system performance. Predictions of such a nature may fit with systems of macro-scale. However, they do not offer satisfactory control of performance for miniature systems. This realization motivated attempts to advance control of roughness features and led to the eventual introduction of the so-called structured, then later on, functional-deterministic surfaces.

Controlling the topography, shape, and size of roughness features practically shifts surfaces into an entirely unchartered technical domain. It mandates a shift from randomness to order. Passing from the random to the structured is demanding from a design-optimization perspective. It is crucial to develop the tools needed for inception and implementation of a deterministic-structured surface design. Tools of design, on the other hand, originate from a clear vision of the intended use of the surface. Building a viable structured surface requires, first, clear definition of the traits that should be available within an ideal construct; and second, defining a path to realize the design. Both of these steps are currently underdeveloped within the realm of tribology. A logical first step towards gaining this new insight is to define the traits of an ideal tribo-surface.

The question of what is an ideal tribological surface, although straight forward, has no simple answer because of the complexity of the factors involved. The answer is further complicated by the view of a surface as a smart system, which not only optimally functions at an initial point, but also is capable of self-monitoring and of dynamically adapting its performance. Keeping in mind the limitations of any generalized definition, one may compose a "generic list" that highlights the major attributes favored to exist within a modern tribological surface. These are: minimized damage, control of tribological response, maintaining a preset damage limit, adaptability to changes in external rubbing conditions, self-healing-ability, and self-condition monitoring. More important an ideal surface must maintain optimized performance within a domain of conditions (e.g. range of loads and speeds). These attributes reflect a complex interaction between material choice, geometry and manufacturing technique. In this capacity, the ideal surface is a smart multi-tasking complex construct. Successful design and implementation of a surface of such nature requires advanced design paradigms that emphasize management of the complexity that arises due to the interaction between structural attributes and performance constraints. Design paradigms in human engineering do not address surfaces of such nature. Traditionally, human engineering aims at peak performance at a point within a potential application domain and a single dominant task. Developing a viable design for a surface in the envisioned future sense, under the current paradigms, is challenging. Many practitioners, therefore, look for inspiration in our encountered surroundings especially within biological constructs.

In nature, a surface is an integral part of the host biological system. Many biological surfaces reflect designs that maintain structural integrity, and in case of tribological encounters, can mitigate friction and friction-induced damage. They also reflect structural traits that induce functional adaptation to their local tribological habitat. This system-specific surface design represents a design dilemma for the human engineer. First, the essence of optimizing a surface of interest has to be understood from the point of view of the constraints on the biological analogue. The philosophy of design generation in nature centers around function. Underpinning this paradigm is the general principle of "separation of function". Separation of function enhances survivability of species in a fiercely competitive world that promotes economy of living material. Natural designs, also, portray an integrative relation between form and function. The pursuit of a single purpose requires careful considerations of the imperatives of balancing economy of design (with regards to materials and effort) and efficiency of function. Following understanding of constraints, a decision of whether to mimic the natural design or to use it as an inspiration has to be made. Each of these decisions has a separate flowchart. Moreover, decisions have to be made in light of the technological constraint that apply to the target surface. This raises the interesting question of how to find analogies between technological constraints and biological (i.e., environmental-habitual) constraints. The answer to this question, defines a viable roadmap to successful bio-inspired surface engineering.

There are many challenges in the field of bio-inspired tribological surfaces. These originate from the multidisciplinary nature of the topic and what information is useful to different stakeholders. The term" bio" indicates the involvement of biologists. Biologists provide useful information about species, their habitats, and their functional niche in the natural world. However, improving the clarity of information exchange between biologists and surface engineers is tasking. Biologists "describe" behaviors mainly through statistics or through generalized models. Surface engineers, however, use a different set of numbers, parameters, and metrics to describe a surface and to characterize performance. This situation, by default, manifests a challenge in conveying important information and streamlining technical feedback throughout the design process. This book, therefore, is an attempt to address part of the communication demands.

In the past ten years, or so, several works with the general topic of biomimetics and bio-inspired engineering appeared. Almost all of these works were of a general nature. They provided an overview of the efforts within the field of bio-inspired surfaces. However, there was no in-depth exposure of the building blocks of the bio-inspiration process whether applied to surfaces or to general mechanisms. An alternative approach, among other works, was to focus on the chemical processes involved in producing textures and the use of these methods to impose bio-inspired roughness features on a given surface. The field, however, has grown considerably in the past few years. Many new applications, approaches, and directions have emerged. To this end, writing about the field in the same generalized manner would only contribute towards widening the gap between the main stakeholders within the discipline. Indeed, the growth of the discipline have revealed theoretical and practice differences related to target applications. As a result, one cannot consider a generalized approach that satisfies the various design needs across the board. In surfaces for example, the theoretical foundations and manufacturing methods relating to hydrophobic or self-cleaning surfaces differ from those necessary to design tribological surfaces. As such, there is a need for specialization upon writing. This book, therefore, focuses on one class of surfaces, namely, tribological surfaces.

The emphasis throughout the chapters of this book is the fundamentals of developing a tribological surface. Therefore, the manuscript provides the fundamental information needed to deduce innovative solutions for problems dominant in the tribological domain from natural design constructs. The work does not promote biological surface designs as the "gold standard" to be achieved in the technological realm. Rather, the premise of the presentation is advocating biological constructs as potential candidate solutions. It is understood that biological constructs, while elegant in their own context, may not manifest a viable option within a constrained technological application. Alternatively, they may only point the designer towards a solution that needs further development. The manuscript, therefore, provides the necessary background information, and tools, that should allow judging the viability of a candidate natural design. Therefore, the work is viewed as a contribution towards streamlining the design process.

The book comprises nine chapters designed to initiate the reader into the world of surfaces, their evaluation, and their design. Chapter 1 reviews the origins of surface-system considerations. It highlights the fundamental role a surface plays in preserving the structural integrity of a tribological system and the crucial role of surface texture in maintaining the state of a rubbing material. Here we make the case for custom engineering of texture. It is shown that the idea of engineering textural features, while being fundamental, is not easy to implement. We discuss the complexity of the factors involved and how they render the customization process industrially demanding. The main emphasis is on one fundamental factor, namely, the absence of a texture design paradigm that caters to the multi-functionality requirement for futuristic surfaces.

Chapter 2 reviews the laws governing the friction behavior of objects. The material starts with a historical view of the evolution of friction laws and how they shaped the science of tribology. The second part of the chapter provides a generalized overview the mechanics of contact between complying solids. Major contact models are listed and formulas for calculating area of contact, contact forces, contact stresses are also developed. Finally, the chapter applies the presented information to the contact of biological species. The information presents a summary of the major rules in biological attachments, their adhesion and friction, along with their contact mechanics roots.

Chapter 3 discusses routes for the classification of surfaces and the evolution of surface metrology. There are two basic ingredients to the surface of a manufactured workpiece: the manufacturing process and the production technique. Each of the ingredients has an effect on the functionality of a rubbing part. The material provides an answering scheme for a question that often arises about how to characterize the influence of roughness features and how to utilize characterization metrics to predict, then monitor, performance.

Chapter 4 reviews both the economic and technological landscapes of surface engineering. It is shown that surface engineering aims to modify the properties of surface and subsurface layers of an object to meet a specific performance goal. The focus is on the enormous growth in the field in light of new emerging technologies that enable manipulation of matter at virtually any scale. The presentation emphasizes the diversity of the subject and its relation to desired performance of objects in tribological situations. The chapter also defines the traits of an ideal tribological surface and briefly discuses obstacles that hinder materialization the envisioned ideal surface construct.

Chapter 5 explores methods of engineering the elements of the roughness of a tribological surface and the effect of that process on performance. Texture and topography have a crucial influence on the tribological performance of surfaces. The presentation focuses on the dominant technology of Laser Texturing. We review the principles of the texturing process, its effects on friction and wear then discuss existing technical difficulties. It is shown that despite being practiced for a long time, and being promising, laser texturing still lacks standardization. To date, there is no agreement on a standardized design methodology that yields optimized textural features that meet the requirements of a prespecified application.

Chapter 6 discusses the historical origins of the concept of borrowing from nature and coining the science of Bio-mimetics. The material also surveys examples of tribological systems in nature. Generation of design in natural systems (geometry, pattern, form, and texture) is shown to be holistic in essence. It synchronizes simple interaction of design constituents and efficient performance. Such an approach yields deterministic design outputs that while conceptually simple are of minimized energy footprint. Natural engineering, it is shown, seeks trans-disciplinary technically viable alternatives to our technological practices. These alternatives, given functional constraints, require minimum effort to construct and economizes effort while functioning.

Chapter 7 discusses the impact of absence of a holistic surface-design methodology in the technological realm. We show that manifesting designs which merge function, form, and topography to achieve lean performance is currently a bottleneck in the field of tribology. The presented material shows that merging function and topography, while in essence, has not matured within the realm of man-made surfaces is advanced in natural designs especially within the scaled reptiles (Squamata). This prompts many engineers to scour biological analogues for design alternatives. However, the problem of evaluating the viability of a natural surface, and judging its suitability for a technological application is frequently encountered. The chapter addresses this problem through a detailed case study that involves extracting metrological information for snakes that engage in rectilinear locomotion.

Chapter 8 discusses the multi-scale nature of biological materials. It is shown that this characteristic motivated several design attempts within the field of tribological surfaces. These designs were not easy to implement based on lack of technological means. Until the push for nanoscale material manipulation, many designs, although conceived and conceptually verified, were not technologically possible. The leap in technologies that matured within the past decade resurrected efforts to manufacture many discarded designs on a commercially scale. The material within this chapter presents samples of existing bio-inspired tribological surfaces. The examples are either a direct replica of the bio-analogue or represent a modification of the surface through a combination of chemical and geometrical changes.

Chapter 9 discusses the potential of bio-inspiration in 3-and 4-D printing. We argue that the true potential of texturing hasn't been realized yet not because of the lack of enabling texturing technologies but because of the severe lack of detailed information about the functional details of texturing in a tribological situation. That is how surface features, their geometry, interact with the functional gradients present within the subsurface layers to control the friction profile of a structure. The material emphasizes the potential of Bio-inspired surfaces in providing a pathway for realizing true synchronization of function through a layer-by-layer customization of surface and subsurface material. In particular the chapter discusses methodologies to extract design parameters that lead to manifesting 4-D printed tribological constructs where surface and sub-surfaces respond optimally to external stimulants represented by the operation conditions of load, speed and ambient temperature.

Successful design of functional deterministic surfaces is not a product of mere bio-mimicry. Rather, it culminates probing the geometry, texture, form, and construction of the bio-analogue and linking these ingredients to the desired functional profile of the surface in the Human Engineering Domain. That is, generation of Bio-inspired functional surface designs stems from implementing design rules rather than replication of natural constructions. Deduction of design rules requires decoding the metrological features, and the analysis of surface performance, of bio-analogues using standardized engineering methods. Success in designing a bio-inspired surface, therefore, requires a trans-disciplinary approach that combines engineering, physics and biology. These don't combine naturally since they entail different methodologies of problem solving and investigations. It is hoped that this book would bridge the gap between the disciplines in the context of biomimetic surface design and construction. Further, it is hoped that the material would equip the reader the basic skills needed to navigate between the biological and the engineering domains.

Chapter 1
Surfaces and Function

ABSTRACT

This chapter reviews the origins of surface-system considerations. It highlights the fundamental role a surface plays in preserving the structural integrity of a tribological system and the crucial role of surface texture in maintaining the state of a rubbing material. Here the authors make the case for custom engineering of texture. It is shown that the idea of engineering textural features, while being fundamental, is not easy to implement. They discuss the complexity of the factors involved and how they render the customization process industrially demanding. The main emphasis is on one fundamental factor, namely, the absence of a texture design paradigm that caters to the multi-functionality requirement for futuristic surfaces.

THERMODYNAMIC PERSPECTIVE

Everything that is interesting in nature happens at the boundaries: the surface of the earth, the membrane of a cell. The moment of catastrophe, the start and finish of a life. The first and last pages of a book are the most difficult to write. (Humphrey 1999)

Matter is a general term that describes any physical substance. There are several approaches that one can adopt to define matter. We can describe matter in terms of its physical and chemical structure (atoms and subatomic particles). We can also use its behavior for description. Alternatively, moreover, we may base our definition on defining characteristics or traits.

DOI: 10.4018/978-1-7998-1647-8.ch001

In bulk, matter can exist in different forms, or states of aggregation, known as phases. These depend on ambient conditions (pressure, temperature and volume). A phase is a form of matter that has a relatively uniform chemical composition and physical properties (such as density, specific heat, refractive index, and so forth). These phases comprise the three familiar ones (solids, liquids, and gases), in addition to the rather exotic states (such as plasmas, super-fluids, super-solids, and the so-called Bose–Einstein condensates, etc.).

In free-form, phases of matter exist under equilibrium with their surroundings. This equilibrium is defined by the intensity levels of a so-called fundamental energy potentials (mechanical, thermal, chemical, etc.). Shaping of matter into a form suitable for a technological application (e.g. crankshaft, surrendered, etc.) involves deliberate upset of the original equilibrium state. Any shaping process includes a change of phases, volumes, and internal structures. After shaping is completed, the new object settles into a new equilibrium state. The state meanwhile may undergo disturbances due to operation of the object. Changing the original equilibrium state of matter, along with frequent disturbances while usage, upsets the balance between the energy potentials. This, in turn, triggers an exchange between the newly formed object and its surroundings. The goal of that exchange is to reestablish a stable equilibrium state suitable to the new environment induced by shaping. Depending on the state of the object the magnitude, intensity, and direction of the interaction with the surroundings may change continuously. This change manifests itself in terms of forces and changes in properties or structure of the object. That change may extend to affect the surroundings as well. In all, inducing a change in matter from a free-form to that of an object triggers a communication between the energy potentials within the object and its surroundings. This communication reflects changes in the gradients of those energy potentials due to the activity of an object. Our notion of the dynamics of that communication, traditionally, have evolved in terms of a system-surrounding paradigm. Such a paradigm is fundamental to all analysis pertaining to behavior of technological objects and the quantification of changes in their state.

The mechanistic steps describing the communication, and evolution of system changes, involve energy transformations. Energy transformations, in turn, are governed by conservation principles. Conservation, meanwhile, is fundamental to thermodynamics. It is not a surprise, therefore, that our motion of the system and its components has strong thermodynamic roots. To this end, it is beneficial to first review the basic definitions of the system-surroundings paradigm and implied information exchange from a thermodynamic perspective.

Systems and Boundaries

A system is a simply a domain of interest. It is the entity being investigated or subjected to analysis. Formally, the collection of matter or region in space that is focused upon constitutes a "thermodynamic system". A system is differentiated from the surroundings by a "boundary" or a wall.

Figure 1. Schematic of a generalized thermodynamic system

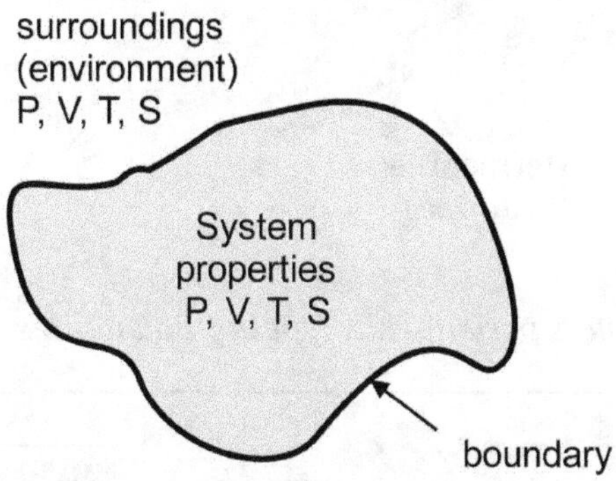

Boundaries can be actual or theoretical. The boundary is of zero mathematical thickness and does not affect the properties of the matter within, nor that outside of, the system it encloses. Conserved or non-conserved quantities can pass into and out of the system across the boundaries (Figure 1). The space outside the thermodynamic system is known as the surroundings, a reservoir, or the environment.

The boundary is also shared between system and surroundings because they are at contact at all common points. Moreover, properties vary continuously across the surface of the boundary. Such a condition is called the no-jump condition, which implies that there are no sharp jumps in properties upon passing from system to surroundings across the boundary nor upon passing from surroundings to system. In thermodynamics it is the interaction between system and surroundings that is the focus in most of the cases encountered.

The properties of the boundaries determine what transfers can occur. A boundary (wall) that allows transfer of a quantity is said to be permeable to it. The thermodynamic system is classified by the permeabilities of its several walls (Table 1). A transfer between system and surroundings can arise by contact, such as conduction of heat, or by long-range forces such as an electric field in the surroundings.

Figure 2. A generalized view of the exchange of matter and energy across a permeable boundary

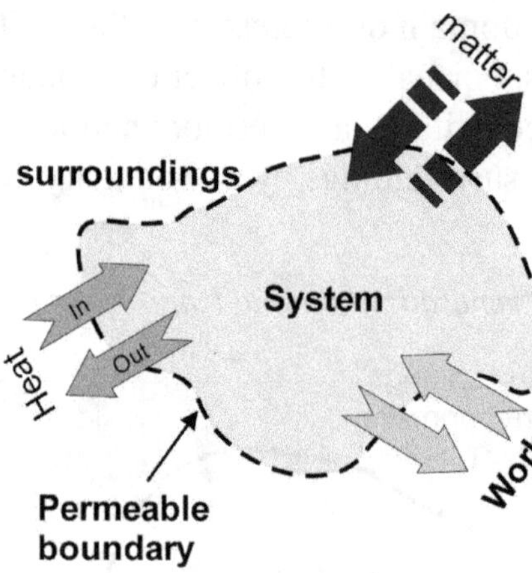

Table 1. Types of transfers permitted by types of Boundaries

Type of Permeability	Type of Transfer		
	Matter	Work	Heat
permeable to matter	Y	N	N
permeable to energy-impermeable to matter	N	Y	Y
adiabatic	N	Y	N
dynamic and impermeable to matter	N	N	Y
isolating	N	N	N

What Is a System?

Motion, in general, is a characteristic trait of matter. It is inherent to its very nature. Motion of matter takes many forms. Quantifying all these forms is possible through the unifying concept of energy. That is, energy is the vehicle that allows accounting of the various forms of motion, and the amount of resources these forms consume or produce. This accountability is possible due to the nature of energy which allows its sum to remain constant in all situations (even when various forms of motion are simultaneously encountered). Motion of matter, and energy corresponding to motion are coupled. That is, for energy to change form, or motion to take place, an interaction between the system and its surrounding has to take place as well. The direct transfer of any form of motion, along with its corresponding quantitative measure, energy, gauges all system-environment interactions.

Typically, either energy or motion transfer upon system interaction with surroundings. An exception takes place when matter, from which motion is inseparable, is exchanged between system and surroundings. In this case, momentum, is transferred along with matter and energy. Systems are therefore classified based on the nature of that interaction with their surroundings. The classification of the system, in thermodynamics, considers if motion, and its quantifying energy, only has been transferred to, or from, a system in and interaction. In addition, classification of systems considers whether matter has transferred along with motion, and it's quantifying energy, or not. In all, thermodynamics adopts a classification of systems based on how energy crosses from system to surroundings. Thus, energy transferred in association with mass is distinct from that which transfers independent of mass transfer. It is that distinction that brings about the classification of thermodynamic systems.

Isolated System

The transfer of energy from and to a system, whether accompanied by transfer of matter, or not, brings about an ability to inflict change on the surroundings by the system (or in the reverse inflict change on the system itself). The absence of communication between system and environment, indicates that the zone chosen for study, is isolated from its surroundings. Therefore, it cannot inflict, or communicate, any changes. Further it cannot accept any transfer interaction that causes it to change. The boundary in such systems is constrained. It does not move nor the form (i.e. it displays rigid behavior). Such a system is termed an isolated system (Figure 3-a) in thermodynamics.

Adiabatic System

Any system which does not maintain energy interaction, in terms of heat, with the surrounding is classified as an adiabatic system (Figure 3 b) All work interactions in such systems are done at the cost of the internal energy of the system.

Closed System

A closed system is the system that has a boundary that is permeable only to heat or work. Thus, this system cannot exchange matter (i.e. mass) with the surroundings.

Open System

An open system is one that can exchange both matter and energy across its boundary. Its boundary is permeable. It is also known as a flow system since matter and energy are permitted to flow in and out. This system can have multiple inlets and multiple outlets. Analysis of open systems typically is associated with the concept of a control volume.

Control volume is a mathematical abstraction that facilitates monitoring the flow of matter and energy in and out of the boundary of an open system. The concept of a control volume considers whether time is a factor in quantifying the amount of matter and energy exchange between an open system and its surroundings. If time is effective, i.e. at equal time intervals of observation the rate of flow of matter and energy in and out displays considerable fluctuations, then we speak of non-steady open system. Alternatively, if there is no fluctuation in the rate of flow or if these fluctuations are negligible then we speak of steady flow within an open system.

Surfaces in Technological Systems

A physical boundary of a technological object encloses a system. It represents the contour that defines the collection of matter shaped. This control is commonly known as a "surface". The functional definition of the surface, it would be noticed, is consistent with the thermodynamic definition of a "boundary". In contrast to thermodynamics, however, surfaces reflect different traits.

Physical surfaces manifests in nature that is different than that the deduced from a thermodynamic description. A boundary of a thermodynamic system is of zero thickness. A physical surface, in contrast, has thickness. No matter how thin it still contains several layers of matter. The method contained does not necessarily have to be homogeneous nor it must be the same as that comprising the system. In addition, even if nominally the surface is of the same material, as that of the enclosed system, it does not have to be of the

Figure 3. General classification of systems in thermodynamics. The classification considers the exchange of two types of energy between system and surroundings. The first is energy that is not associated with mass (e.g. heat and work), whereas the second is associated with mass (e.g kinetic and potential energies). The distinction between open and closed depends on which type of energy is permitted across the boundary.

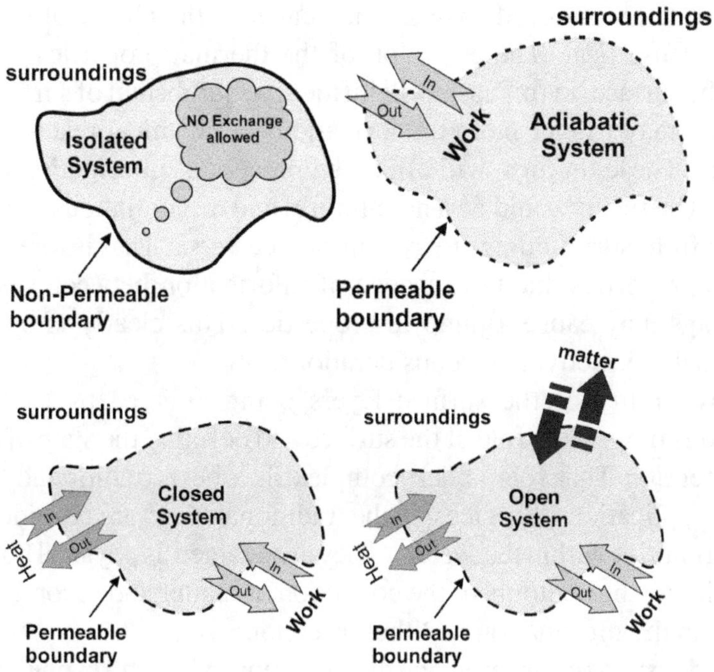

same microstructure as that of the bulk. In fact, many surfaces reflect different micro-structure than the bulk of the system despite being nominally made of the same material (metal or polymer). This takes place due to the nature of manufacturing processes used in producing an object. It follows that real surfaces may reflect their own physical, chemical, electrical, and mechanical properties. These properties may, or may not, be the same as those of the bulk.

Consistent with thermodynamic considerations, surfaces in technological systems are permeable to information. They permit the exchange of mechanical, thermal, chemical information between system and surroundings. However, because real surfaces contain layers of matter, and have thicknesses, they may induce changes in the magnitude and intensity of the information being transferred. The material of the surface, although permitting exchange of information, may also be affected by the transfer process itself. And mechanical systems, for example, the exchange of mechanical and thermal information may be intense. Transfer of forces causes deformation within the structure of the surface layers. Deformation, in turn, may alter the transport properties of the layers. Such a change reflects on the amount of heat, for example, crossing the surface in either direction. Depending on the evolution of the properties of the surface layers, the system may accept, reject, or even store heat. Deformation induced changes may enhance the ability of the surface layers to diffuse heat. The evolution of the thermal properties within the layers of the surface, in this case, will influence equilibrium of surface layers, and thereby may trigger more the formations or chemical reactions within the surface. These, in turn, will affect the process of information exchange and so on. The result would be a nonuniform and discontinuous transition of properties from surroundings to system or vice versa. That is, the evolution of surface properties due to exchange of information between system and surroundings may cause a jump in properties. This clearly is opposite to conventional thermodynamic considerations.

The probability of the surface layers being affected by exchange of information points at the role of the surface and defining the state of a system and its integrity. That role, albeit complex, has been traditionally ignored and many applications. Surfaces, in the traditional view, are considered non-functional entities within the system. They are deemed as physical boundaries that first define the contours of the collection of matter and second represent a gateway to the information in either direction.

The modern view of surfaces, and their function, while still evolving, regards surfaces as an integral functional entity within the system. Therefore, the design of surfaces is emphasized. It follows that the modern view of surfaces promotes a surface-by-design paradigm. For this paradigm to materialize there is a fundamental need for understanding the functional requirements of systems on a case-by-case basis.

Friction (Tribological) Systems

An important class of technological systems is the so-called tribo-systems. These are mechanical systems of which one or more of its components engage in relative sliding against a moving or a stationary object (e.g., pistons in engines, brakes, cutting tools). In these systems, part of the surfaces engages in sliding relative to a contacting counterpart. Sliding may take place in the presence, or lack of, a lubricating medium. The sliding environment may be hostile in the sense that motion takes place subject to one or more of high stress, severe heat, intense electrical potential, or chemical environment.

By nature of their function, tribo-systems are energy driven. They sustain intense exchange of information. A surface in a rubbing situation, not only has to efficiently exchange information with its surroundings, but also must maintain its structural integrity throughout its lifespan. Failure of surfaces in this case expose the enclosed bulk components and thereby induces failure of the system as a unit.

Rubbing materials develop characteristic subsurface regions during sliding. These zones are of interest because their analysis leads to important findings concerning the types of operative tribo-mechanisms, and to the reasons for observed wear behaviors. The composition and depth of these characteristic zones may not differ if test conditions are identical for two specimens. In contrast, for different materials subject to different test parameters the same zones may still be identified, however, with relatively different details with respect to depth, morphology and, dimension. In sliding, both thermal and mechanical events are interlaced. Thermal events may change the mechanical properties of the rubbing pair (e.g., hardness yield point, etc.). This change, in turn, affects the rate of heat release at the surface thereby altering the thermal and mechanical properties of surface and near surface layers and so on. A generalized description of the zones developed in a rubbing material is discussed in reference to Figure 4. Where, the different zones are identified with respect to a steel tribo-specimen that have undergone continuous sliding for some 240 seconds and reached the quasi-equilibrium wear state for the particular test conditions.

Three distinct zones may be identified. Zone 1, compositional mix, is the region closest to the actual sliding interface. The composition of zone 1 is different from the underlying regions in that it consists of the original specimen material, the counter-face material, and material from the environment (Abdel-Aal, 2005). Thermally, however, zone 1 constitutes a thermal skin that envelope a thin sub-layer from which prospective asperities are generated. Severe heating, in the form of strong temperature gradients, is characteristic to this thermal skin, note the severe deformation and orientation of the material grains at the interface.

Zone 2, which is characterized by an accumulation of plastic strain, is separated from zone 1 by a compositional interface. Within itself, this region consists entirely of original specimen material. Temperature rises are augmented by the plastic deformation (internal heat generation). The temperature gradients, however, are not as pronounced as those in zone 1,

Figure 4. Conceptual illustration of the analogy between thermal and mechanical zones developing in a dry sliding sample of a mild steel (AISI 1020) pin (6 mm×1 mm, sliding in the 1 mm direction) after sliding for 240 s at a speed of 0.5 m/s. Schematic illustration of the various thermal zones encountered in sliding (left hand side). Schematic illustration of the various mechanical zones encountered during sliding (right hand side).

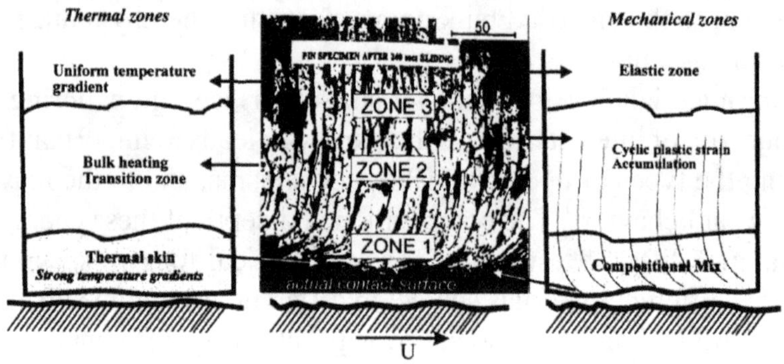

compare the deformation and relative grain orientation in this zone compared to those at the interface. With temperature gradients decaying away from the interface with zone 1, zone 2 represents a transition between the severe temperature rise at the contact interface and the thermally inactive bulk of the material. A region which undergoes elastic strains, zone 3 in figure 1 follows zone 2. An elastic/ plastic interface defines the boundary between these two regions. Like zone 2, region 3, consists entirely of original specimen material. Thermally, however, this region may be considered relatively inactive. It is seen that the surface zone is the most active layer in the rubbing material. Further, it is the layer that is dynamic in terms of information exchange. This exchange is a consequence of the friction work introduced to the material as a result of the relative motion and compliance.

Friction Work and Energy Conversion

When two materials in contact, rub against each other, a tangential force, representing the frictional tractions, develop at the interface. This force works opposite to the direction of motion. Thus, it works to resist the relative motion between the contacting surfaces. In doing so, it performs mechanical work on both contacting solids. The work performed on the rubbing surfaces will provide the energy necessary for the formation of new phases within the contacting layers, initiating deformations. It also provides energy necessary to initiate secondary processes (both interfacial and within the sub-layers) and for generation of damage (wear particles). The major portion of the frictional work, however, will be transformed into thermal energy (heat). This heat, eventually, will be dissipated into the rubbing pair and transported from the surface layer toward the bulk of each of the rubbing materials.

Conversion of the work of friction is a thermodynamic process that entails both activation and passivation (Abdel-Aal 2003). In this process, part of the mechanical work dissipates within the tribo-system to cause an increase in the internal free energy. The remaining part is converted into heat. The particulars of the conversion process, i.e., exact proportion of frictional work that dissipates as free energy and that which converts to heat, depends on several factors. These may include geometry of the tribo-system, structure and composition of the rubbing materials, physical and chemical properties, and kinetics of the interaction between the rubbing materials and the environment.

The increase in free energy due to dissipation may disturb the thermodynamic equilibrium of the affected layers of the rubbing material (Abdel-Aal 2003). The response of the material will be an attempt to dissipate the disturbance in order to re-establish equilibrium or at least to attain a steady state. The steady state, which enables the system to maintain relative motion between the rubbing surfaces at minimum energy expenditure, represents the shift in the position of equilibrium. The dissipation process is in general a manifestation of the Le-Chattelier-Prigogine principle (Abdel-Aal 2002), which states "a change in one of the variables that describe a system at equilibrium produces a shift in the position of the equilibrium that counteracts the effect of this change". Such a shift is facilitated through maximizing the rate of entropy generation within the contacting layers. For maximizing entropy generation, meanwhile, the response of the system to any disturbance will be irreversible.

The "activation" of many surface and sub-surface material responses is a consequence of conversion of frictional work. Activation responses start at the interface. They may assume many forms depending on the amount of available energy and rate of work conversion Figure 5. Possible responses include oxidation, structural transformations, hardening, crack formation, and thermal checking among others. Thermal energy within this view is a "passivation" influence since transport of heat, away from the surface, will reduce the amount of energy available to trigger any of the dissipative responses. The process of mechanical energy conversion is rather complex. To be emphasized, however, is that conversion of mechanical work into different forms of energy influences. In fact, focusing on the conversion process yields a clear idea about the nature of information exchanged between system and surroundings.

The information exchanged emanates from the so-called enthalpy of a strained solid (h). This quantity, in general, represents the total energy within a solid at any given situation. In friction there is a time dependency because of motion, thus, we consider the rate of enthalpy generation in a solid that is undergoing mechanical deformation. For a deformable body, the specific enthalpy is a function of both temperature and "volume increase" of the body Viz,

$$h = h(T,e) \tag{1.1}$$

where e is the change in volume given by,

$$e = e_{(11)} + e_{(22)} + e_{(33)} \cong \frac{\Delta V}{V} \tag{1.2}$$

$e_{(11)}$, $e_{(22)}$, and $e_{(33)}$ are the so-called "Lagrangian Principal Strain" values given by the roots of the equation,

$$e^3 - I_L e^2 + II_L e + III_L = 0,$$

with

$$I_L = e_{ii}, II_L = \tfrac{1}{2}\left(e_{ii}e_{jj} - e_{ij}e_{ij}\right), III_L = \left|e_{ij}\right|$$

The time rate of change of the specific enthalpy is thus written as,

$$\rho h^o = \rho \left[\frac{\partial h}{\partial T}\right]_e T^o + \rho \left[\frac{\partial h}{\partial e}\right]_T T e^o \tag{1.3}$$

The first term ρ(∂h/∂T)e is recognized as the volumetric specific heat, Cp, under constant pressure. The second term (∂h/∂e)T ≡ Ck can be defined as another type of "specific heat" that measures the "isothermal" change in volume per unit volume of the continuum body. Since the volume change in a continuum body is a result of mechanical straining, the heat capacity Ck serves as a bridge to thermo-mechanical deformation.

Figure 5. Schematic comparison of energy requirements that are necessary for processes taking place during sliding. The work of the friction force initiates many processes at the atomic level. These, in turn, dissipate into heat. The amount of heat resulting is around 80–90%

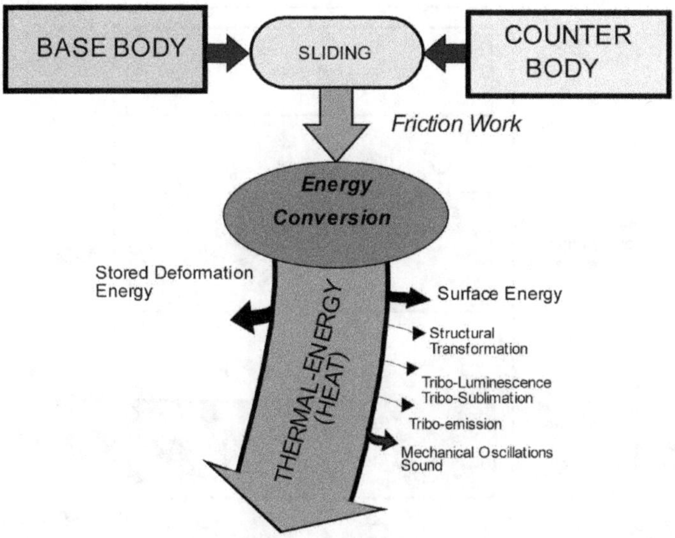

Equations (1.1-1.3) highlight the importance of events within the contacting layers of the rubbing material. The layers of the complying-sliding material, practically, manifests three sub zones. Each of these zones, may be considered as a subsystem (Figure 6) that communicates with other parts of the rubbing solid. All subsystems, meanwhile, respond to the environment as a unit. The most active zone is that which contains the contacting layers (surface layer in Figure 7). This layer entertains both severe time-dependent strains, and strong settlement effects. Meanwhile a strong thermomechanical coupling exists within the layers of the material and that zone. This layer is the so-called thermal skin which contains compositional mix in Figure 4. The enthalpy of the solid and his own is the sum of the two components on the right-hand side of equation 1.3. Each of these components is of the same order of magnitude. A significant contribution to the enthalpy derives from the term Teo which denotes thermomechanical coupling. In this zone, therefore, the surface entertains both mechanical and thermal information with each of the contributions being affected by the evolution of the structure of the zone itself. Note also that the interface (i.e. the layer that includes the micro-roughness) is the initial active layer within this zone. It is the gateway for any thermal or mechanical information to pass from and to the rubbing material.

Figure 6. A generalized illustrative model of a rubbing system and its active subsystems. The zone with intense activity, the thermal skin, entertains considerable time dependent thermal and deformation gradients as well as coupling between the thermal and the mechanical states.

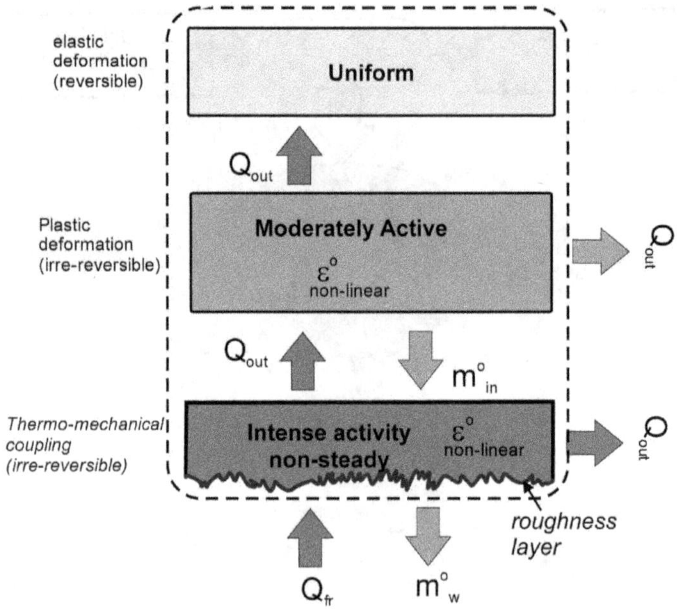

The following zone is semi-active, with higher temperatures than ambient, but linear strain rate gradients. As such most of the contribution to the enthalpy within this zone comes from several events. The upper zone, uniform zone in the figure, is virtually an active as entertains weak strain rate gradients accompanied by sly temperature rise compared to the ambient environment. It is seen, therefore, that the very active layer among the rubbing solid is that which contains the interface and a very thin layer above it. Almost all changes and exchange of information, mechanical and thermal, originates at this layer. It is no surprise, therefore, the damage and degradation in rubbing elements initiates at this level. The nature, and intensity, of activity within this layer is the catalyst for deliberate engineering of roughness. This is because roughness manifests the gateway for any meaningful information exchange between the rubbing material and its environment. The task of deliberate engineering of the roughness layer, although fundamental, is not easy to achieve because of the inherent complexity in rubbing.

Complexity in Rubbing Systems

A mechanical system that encounters friction is by default complex. This complexity has two contributing factors, the first is inherent whereas the second involves with operation. The inherited contribution to complexity has its origins in the manufacturing phase of the rubbing component itself. Due to the shaping, and finishing, steps that a material undergoes during manufacturing causes the existence of layers of different microstructure within the resulting object. The differences in microstructure manifests in either different chemistry or different scales within the various layers comprising the object. In other words, in manufactured components, no matter how precise the manufacturing is, which contain inhomogeneities. Inhomogeneities in turn render what seems to be a whole system, practically, a group of different subsystems. These subsystems, upon undergoing friction, would evolve according to their initial conditions and to the local conditions of loading both mechanical and thermal.

So, in effect, each layer with similar or identical microstructure will act as separate sub-components. Each of the subcomponents would communicate with other subcomponents (or sub layers), and depending on how streamlined that communication is, the system as a unit, may, or may not, be efficient. During sliding, moreover, the sublayers will react to the parameters of operation and to the energy release due to friction.

The frictional behavior of a sliding material, thus, will be influenced by the forces developing at the interface of the contact with the mating counterpart. Focusing on the surface layer, the behavior of the material within this layer will evolve according to the physical, and chemical properties of the real area of contact. This area depends on the geometry of the roughness (or texture) of both rubbing bodies. As such geometry of the roughness features will contribute to the complexity of the rubbing materials. It to be noted, that the properties of the interface need not be the same as those of the bulk of the mating materials. Moreover, the mechanical response of the interface (e.g., strain hardening or strain rate response) may also differ from that of the material bulk. In many cases, especially in sliding of metals, the state of cleanliness of the interface (i.e., the presence or absence of contaminants or oxides), plays a significant role in determining the path of the friction process and the level of damage that may be incurred upon either of the sliding materials.

Two features of surfaces are important in analysis because of their direct connection to the state of the real area of contact. These are the roughness and the state of contamination of the surface. In metals for example, the roughness will determine the size of the area of true contact and the actual loading (true contact stress). Contamination, on the other hand, will determine the likelihood of true metal-to-metal rubbing. In the presence of contaminants, especially a very thin oxide layer, in ambient air, metal to metal contact may not occur, except in cases where severe damage is sustained and when sliding takes place in vacuum. Presence of an oxide layer also prevents cold welding of metallic surfaces upon contact and facilitates relative motion.

Surfaces are never absolutely clean. A layer, however thin, will always exist on the original material. In metals, Figure 7 layers have either a mechanical origin or may result from chemical activity catalyzed by the release of thermal energy due to sliding. Cold working of the metal grains (cold worked layer in figure) for example may be a result of residual stresses due to the method of surface preparation (turning, milling, drilling, forging, etc.,). It also may result from the response of the contact layer to the strains imposed by the friction tractions at the interface.

Figure 7. Schematic showing typical surface and subsurface microstructures present in rubbing metallic solids.

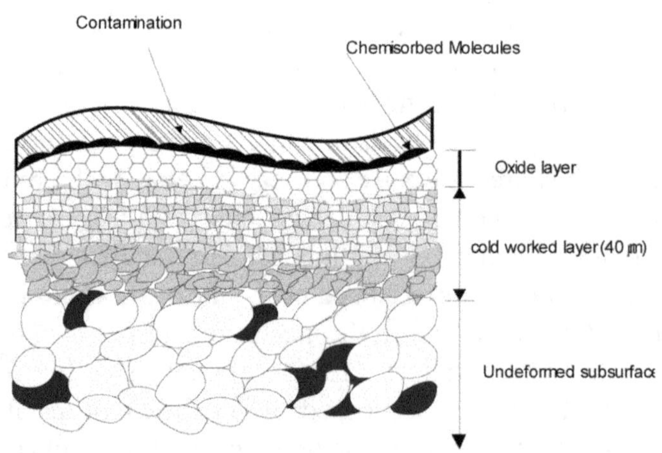

Transfer layers, Figure 8, may result from mechanical mixing, affinity of the surfaces. In some cases, such as in the case of formation of a build-up layer on cutting tools, it may result from the contribution of coupled thermo-electrical activity at the interface Chemisorption layers are due to accumulation of adsorbed species, such as water molecules from atmospheric moisture, carbon from atmosphere or from lubricant layers. Adsorbed layers may also originate from human handling (e.g., the transfer of salt from human sweet).

Figure 8. SEM micrograph of a transfer layer of Commercially Pure Titanium, CPT, (moving specimen) that formed on a stationary WC-CO pin during dry reciprocating sliding

Oxidation is due to the reaction of the metal with oxygen. In the presence of heat, released from friction, protective oxides may form. These may shield the substrate material from damage by preventing metal-to-metal contact. The formation of a protective oxide layer, and the eventual destruction of it, is process that is aided by the accumulation of friction-induced thermal energy. The presence of that heat allows oxygen atoms to penetrate the surface and react with the substrate material to form the protective oxide which grows at a rate controlled by the dilution of thermal energy. When the oxide layer reaches a critical thickness the difference in temperature between both sides of the layer, the substrate side and the ambient surroundings side, will initiate cracking of the oxide and whence its' destruction.

The different layers, or subsystems, will manifest a distribution in stresses, strains, and the rate of energy dissipation. Consequently, the initial degree of inhomogeneity in microstructure, the origin of the function layers, would increase. That increase adds to the complexity of the system and in some cases would disrupt the transmission of information that crosses the surface toward the bulk of the material from the surroundings and vice versa. This highlights the important role of streamlining, and integrating, surface and subsurface design in the rubbing system. Moreover, for optimization of performance with respect to friction-induced energy dissipation, and structural integrity, there is a need for some measure of active control over the internal state of the sliding material. This is necessary to ensure optimal performance in rubbing, efficient multi scale management of complexity (structural and operational).

A generalized view of complexity and tribological systems considers the connection between the phenomena observed in rubbing. In particular, it considers how different phenomena initiate, then evolve, in relation to each other. This view is rather important to developing an insight the dynamic evolution of damage of a rubbing material.

As mentioned earlier, the work performed by the friction force is a destabilizing agent. Instabilities manifest in various forms (depending on the intensity of release of friction work). Any dominant damage for should be related to the ability of the material layers to dissipate or mitigate the disturbance initiated by friction work. Once relative motion between the complying surfaces initiates, the work release, and its induced effects, take place. It is expected that the material will have a different threshold of heat intensity, for example, to initiate a particular chemical reaction. Another threshold may trigger softening of the material layers, and so on. A question that is to be asked, once accepting the energy threshold proposition, which is how such effect evolve with the duration of relative motion.

A conceptual relationship between events in rubbing was developed by Abdel-aal (2001). (Figure 9). The figure zones on the surface and immediate subsurface layers (i.e. the so-called skin of the rubbing material). Mechanical events take place repeatedly: scratching and indentation. They continuously cause mechanical stresses and strains. They further dissipate part of the work of friction through the induce formations. The rest of the work, however, manifests as heat. Heat, in turn, triggers more chemical activities. Further, it induces softening, and thus leads in directly tomorrow mechanical deformation. Mechanical stress changes in the transport properties of the active zone within the material. This leads to triggering more mechanical deformation because of heat accumulation (Abdel-Aal 2001). In metals, moreover, thermal and electrical properties are coupled through the Weidman-Franz law. Thus, the change in thermal transport properties may trigger intense electrical activity that may change the polarity of the rubbing pair. Electrical activity may also increase adhesion and induce more mass loss through wear. In effect, therefore, the interdependence of events in rubbing is not simple. Events and tribological systems often display a high degree of coupling between the dynamic damage modes and evolution of material properties within the active zone. The inception of that coupling takes place at the level of the micro-roughness. The layer of the micro-roughness is also a dynamic layer. It evolves with sliding and in relation to evolution of material properties and local conditions (Abdel-Aal 2001). In all, surface texture, physically, represents the forefront of many of the detrimental phenomena taking place due to rubbing. Accordingly, considerable work within the so-called texture-by-design have taken place within the past two decades and continues with noticeable intensity (Abdel-Aal 2000) (Bogeum 2013).

In the past two decades, and due to the push for micro-nano scale engineering, many enabling technologies has come to existence (Ajay 2013)- (Choi 2004).. These technologies have opened new possibilities for developing sophisticated surface-engineering paradigms that address modern surface Functionality. Development of surface texturing paradigms, however, did not keep up with the ever-evolving functionality of surfaces (Costa 2008)-(Datta 1995). In fact, currently, the pressing needs are to develop comprehensive holistic surface engineering design philosophies that are holistic in nature to cater to the dynamic demands of surfaces. A focus of special interest is the so-called multi or switchable functionality of surfaces which is based on engineering of textural elements and their topography. Such an art is still fledgling within the realm of man engineered surfaces.

Texture in Relation to Function

The previous section introduced a general description of structural events within the rubbing material. It has been established that almost all events crucial to the integrity of a sliding material take place in the very thin layer that comprises the surface roughness. Texture and roughness of a surface,

Figure 9. A representative map of complexity within a rubbing material. The map reflects the importance of the surface in diffusing information coming from the surrounding environment. The model also emphasizes the critical effect of the major structural and thermal forces acting on the material. These induce changes on the nominal material properties and can trigger coupled phenomena which may accelerate material deterioration through wear.

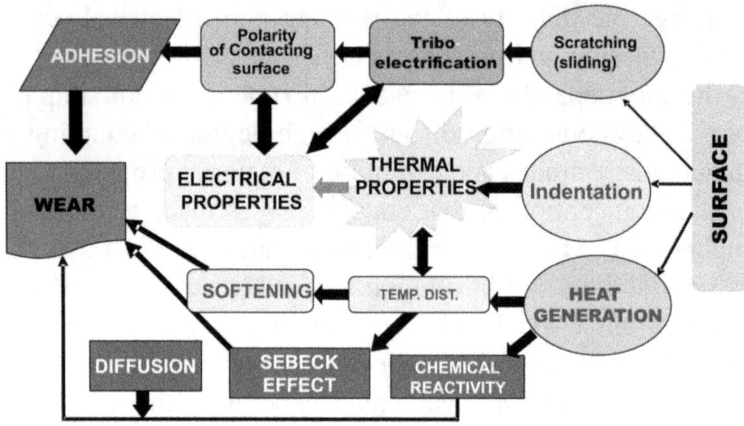

therefore, assume a primary role in the preserving the structural integrity of a rubbing material. Knowing that all changes taking place, in a material subject to friction forces, are triggered by the work of friction impacts the function of texture. This is because the gateway to the contacting layers starts at the real contact area between the rubbing solids. This area, in turn, is determined by the state of the roughness, and as will be seen in later chapters, by the geometry of the features themselves. The dependence on the roughness impacts the efficiency of performance over surface and whether a given surface fulfills its intended purpose.

In the traditional sense, performance and function of a surface are used interchangeably. In this section we would adopt the same convention. Thus, function will be taken to indicate performance over surface under a set of operation parameters. In relating surface texture to function, one is faced with the vast number of applications where surfaces are of prime importance. The diversity of applications prohibits developing a systematic approach that consistently ties textured to function. To overcome the situation the focus shifts to developing a generic analytical framework.

The case of concern in this work therefore is that of surfaces in tribological applications. To develop a generic analysis framework, we consider two aspects of function. The first is initiation of contact, then evolution of structural integrity until failure thresholds. In doing so we also consider the regime of operation. That is, we consider whether the act of rubbing itself takes place under dry or lubricated conditions regardless of the lubrication regime or type of lubricant if any (Figure 10).

Figure 10. Classification of operation regimes and influencing factors for a tribological system. The dynamic regime, actual rubbing, entails feedback from the influencing factors (e.g., roughness topography and mechanical properties).

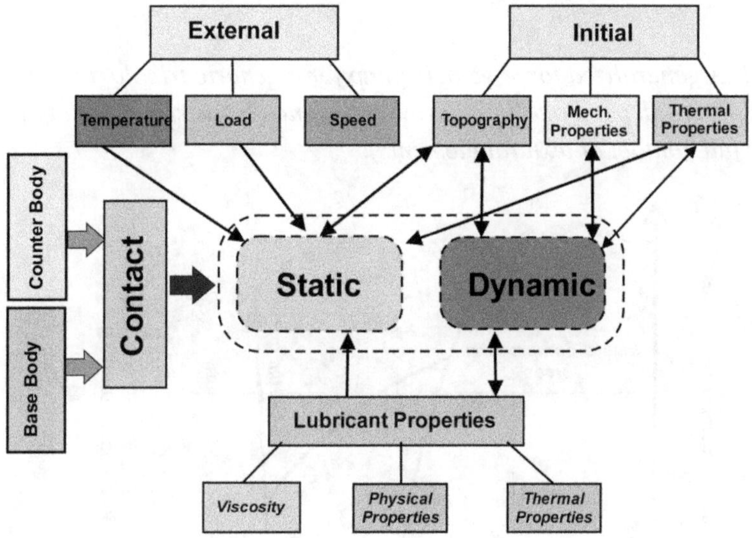

The goal of the process is to relate the texture attributes to the performance of each of the stages of the map. Figure 8 also indicates that the analysis of the relation of texture and function may be either static (i.e. at the contact stage) or could be dynamic (i.e. considering the evolution of textural attributes with operation). Naturally, the second case, dynamic analysis, is more demanding. Here, not only the evolution of the surface texture attributes is needed but the evolutionary path to the dynamic values also needs to be identified. The dynamic characterization of textural attributes, it to be noted, is not a trivial matter. Parallel to the map of Figure 10 another map that represents the forces which influence each stage of the analysis is necessary. The forces acting on the rubbing material are not confined to the frictional tractions at the interface. The interface is one of the components of a system. Although being the most active, the surface response to other forces acting on connected parts of the sliding material. Forces also depend on the scale of analysis, and the speed of motion. If force-scale map, Figure 11, may be developed to illustrate the various influences on the sliding system. Combining Figures 10 and 11 yields It is noticed in the map of Figure 11 that the number of forces, and their interference, depends on the scale of separation between two rubbing surfaces. The smaller the separation gap the higher the interference between force regimes. a clearer picture of the environment of operation for a rubbing material. Linking these the surface traits should help managing the complexity of the events within the surface layer.

Figure 11. A generalized force-velocity map for a generic tribological system. The dominant forces depend on the separation distance between the sliding pair (d=0 indicates full contact of textural elements).

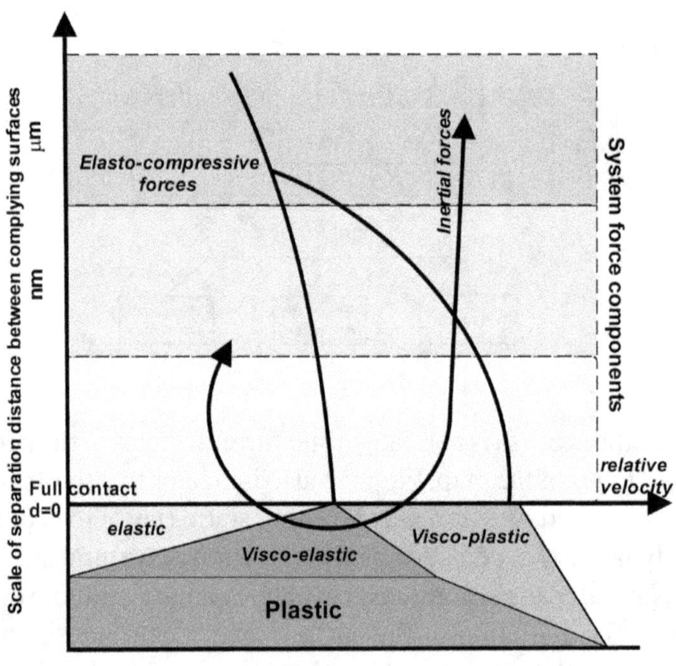

The effect of relative sliding is almost similar. Higher relative sliding speed leads to a transition in the effective force regime. Naturally, higher relative sliding speeds would cause a shift in the effect of forces due to higher deformation rates.

Why Engineer Texture?

Physical phenomena critical to the structural integrity of a rubbing material initiate at the interface between the sliding pair. The exchange of energy, it's transformation to useful or destructive forms, take place also at the surface. Signals that manifest changes in the microstructure or state of the material also initiate at the surface roughness level. In the previous sections, we discussed the degree of interconnection between surface, and interfacial, events and the state of integrity of a rubbing material. It was established that the roughness of the surface, regardless of its scale, stands at the forefront of any changes to the rubbing system (regardless of the inducing force). Such an intimate relation between the surface events and the integrity of the material in a tribological system highlights the importance of preserving surface quality.

Quality in the context of surfaces has many interpretations. In the realm of tribology, it is described as the deviation of surface texture from the through plane. This plane may be of any predefined shape (cylinder, is fear, etc.). The evaluation of quality, therefore, is gauged in terms of a relation between damage and texture. If surface deviations are insufficient, they may cause damage through wear or breaking down of a lubricant. Large deviations will be rubbed off and will act as abrasives which further scratch and damage the contacting material. Size and shape of texture is also crucial in cylinders in cylinder liners of engines. Their size can affect combustion and engine life. Thus, the main challenge in tribology is to devise a way to optimize both the shape and durability of texture. Therefore, enhancing the quality of the surface entails a geometrical dimension to ensure optimal deviation from a plane. In addition, it also entails immaterial choice dimension.

The choice of the material for a rubbing application the subject to many factors. A surface and its quality may, or may not, be essential among these factors. However, the optimization of geometry is a compromise between the life of a part and the intended functions of that part. It is often the case, especially in modern times, that the surface texture assumes many functions. These additional functions, other than minimizing friction, may include paint-ability, hydrophobicity, aesthetic considerations, etc. Thus, an optimal geometry that insures fulfilling all the required functions of the surface is always called for.

Optimization of textural elements while being a fundamental requirement, is not necessarily a straightforward matter. There are several constraints that limit an efficient optimization process. This is especially the case in conventional and fracturing processes (e.g. turning, drilling, peeling, etc.). Here the surface is a matter-of-fact. It is a consequence of the production techniques and finishing sequence of operations. Analysis of the features resulting from these processes becomes a tedious process as well. Analysis to assess the quality of textural features sustains many inaccuracies that depend on the scale of observation. Difficulties in analysis manifest into stochastic procedures which are difficult to reverse in order to control the quality of the resulting surface (Figure 12).

Figure 12. Schematic of the optimization process in surfaces. The figure highlights the feedback between generation of texture at the initial production stage and optimization through testing for enhancing performance.

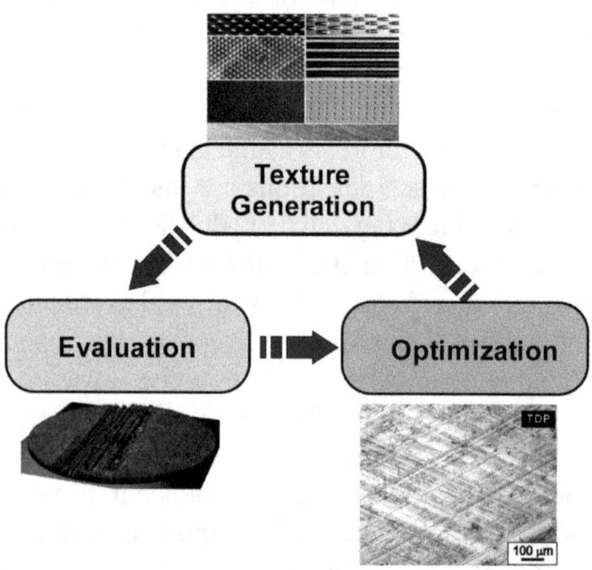

In modern times, around the late 90s of the 20th century, the need for multitasking-multi feature surfaces started to emerge. Performance of many products depended on the existence of a surface that not only is durable, with respect to tribology, but also has embedded sensory, aesthetic dimension, among other requirements. In all, the demand of multi-functioning texture became a must. These demands heavily affected the quality of texture. More precise texture geometry became essential. The advent of more precise textural engineering methods, e.g. laser texturing and lithography, enable additional requirements to emerge. However, despite all the progress in material manipulation a fundamental question still lingers. That question is: what is the optimal geometry of textural elements that fit preconceived application? Further, to date, and enabling paradigm which, given a set of functional constraints, would yield an optimal textural geometry does not exist.

Economic Impact of Surface Engineering

The fundamental challenge in designer-texturing for rubbing applications is to identify a "suitable" texturing feature. Suitability here indicates basic shape and geometry of the texturing element, its dimension, and topology on the surface (ratio of area occupied by the features to free area of the surface, distances and general shape of columns and rows of the textured array). A generalized strategy capable of yielding viable application-driven suitable texturing is non-existent so far (despite more than two decades of work). One often discussed reason is economics (i.e., cost of fabrication features versus benefits gained).

The economic landscape, however, drastically changed within the past two decades or so. In particular, surface engineering has emerged to the forefront due to vested interest of various stakeholders. In addition, redefined global priorities (e.g. sustainability, energy savings, natural resource management, etc.) have driven market needs. This resulted in enormous growth of global market needs. While no exact statistics concerning the actual size of the global surface engineering market exist, it is possible to envision the extent of that market through several financial estimates of selected constituents.

An important growth driver of growth in the field is maintaining the structural health (corrosion, and erosion) of infrastructures (especially within the energy transport sector). Market watch (Abakan) estimates the size of global market for advanced metal coatings and formulations, that addresses infrastructure demands, to be in the vicinity of 150 billion U.S dollars. Figure 13 [Surface engineering 2013) illustrates the financial building blocks of the thermal spray and coatings addressed by one company (Sulzer). As shown in the figure, the addressable market totals some staggering 7 billion U.S dollars (with the actual application occupying the biggest chunk with almost 4 billion U. S dollars). The value of the surface engineering and coating industry in the United Kingdom is worth over 11 billion pounds and affects products worth 140 billion pounds (Time for strategic change 2014). Demands for innovation in the field of surface engineering are vigorous. This is due to the pressing needs of potential stakeholders (the number of which is growing steadily).

Innovation entails developing new methods for manipulation of materials at different scales and validating the feasibility (both technical and economic) of the new technologies. Currently there exist many methods and technologies to implement textures from nano to micron size on virtually any material. These entail direct physical fabrication methods, such as sand blast (Guan 2005), (Linchun 2003) and laser ablation (F.Z. Dai 2015), or pattern transferring from templates e.g. chemical etching (Guo-Sheng 2015)-(Defourny 1992). To date, however, there is no agreement on the optimal values to be implemented given a particular surface. More importantly, a methodology for generation of specialized functional textures for tribological surfaces is virtually non-existent. Design of texture is still viewed as "black art" (Stout 2001) to the effect that currently there exists a gap between available enabling texturing technologies and conceptual texture-design paradigms. Essentially absent is a holistic methodology that merges form and topology of texture subject to function constraints. Such a paradigm while not mature within the realm of human surface engineering, is advanced in natural surface constructs. Consequently, many researchers currently look at natural systems for mimicry or inspiration. Economic outlook consolidates the potential of bio-inspired approaches for surface engineering. Future predictions for global market share of bio-inspired surface engineering technologies anticipate a five-year Compound Annual Growth Rate (CAGR) of 27.7% within the period 2012-2017 (Surface engineering 2013).

Figure 13. Financial building blocks of the surface engineering by coating and thermal spray global market, along with addressable market share.

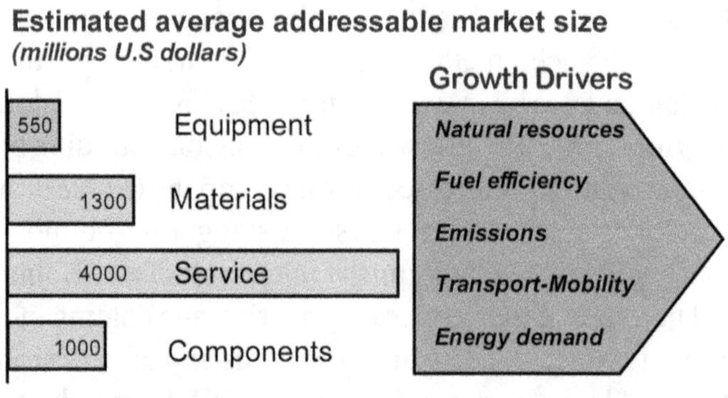

Naturally, the predicted numbers include technology that addresses tribological issues (i.e., friction and wear mitigation) among other technical challenges. However, judging by the current pressing demands for life cycle enhancement of moving parts, and sustainability considerations, it is anticipated that the market share addressing tribological issues will be sizeable.

The Case for Mimicking Biology

Biological species are subject to the same constraints that bind the performance of man-engineered systems. They must contend with inertia, gravity, magnetism, momentum, etc. Further, similar to tribological systems, the locomotion effort the subject to the resistance of ground forces and friction forces. One advantage engineered systems have over biological species is maintenance. A man engineered system may undergo periodic maintenance were damaged parts may be replaced or refurbished. Thus, the service life of a component within a man-engineered system, while important, is not, in principle of paramount concern. In biology, however, species do not have the advantage of maintaining their body parts (especially their surfaces). Thus, in biology moving parts, including those which are subject to frictional interaction, stay with the species for its entire life duration. Further, biological frictional surfaces do not have the advantage of a so-called running-in through which textural roughness optimize the shape and adjust to the working parameters. Rather, texture in biological friction surfaces have to be optimized and functional at inception (Aubret 2009), (Bauchot). This is because malfunction of body parts, or surfaces, can jeopardize the very survivability of the species. Survival of the species, is to be noted, represents the ultimate goal in biology (Bejan 2006a), (Bejan 2006b). It is no surprise, therefore, that biological systems contend with their environmental constraints through elegant solutions. It is no surprise, therefore, that biological systems offered many a lesson to engineers and designers in general. This is especially true when functional surfaces are considered (Baum 2009)-(Boback 2003). Here, the intimate link between surface macro-micro designs and efficiency of locomotion is visible.

Locomotion in biological species is very important. It aids the survivability of the species as it is the means of hunting, hiding, and escape. Friction, static or dynamic, is an essential component of the mechanics of locomotion. In mammals, for example, friction between footpads and substrate terrain is essential for propulsion and stability (Boback, Guyers 2003)-(Cowles 1956).

Consequently, surfaces of footpads in mammals include textural features that accommodate both load bearing and traction generation

(Cowles 1956). Many reptiles depend on adhesive forces generated, and regulated, through their surface textures for climbing, crawling, and mating (Cox 2011)-(Ashton 2003). Therefore, optimization of dynamic friction through surface texturing is advanced in legless reptiles (around 3000 species of snakes).

Legless reptiles have at least five well defined gaits that are basic modifications of sliding. Forces necessary for propulsion in snakes are generated by muscles. However, because of their sliding locomotion modes, their skin is the medium that transmits the muscular propulsion forces to the ground. In the meantime, the skin is also the medium that dissipates both reactions due to body weight, and frictional resistance forces arising because of contact with the ground. In addition, the skin also accommodates the friction tractions necessity for prevention of slip and moving forward. Locomotion in biology, in general, from an energetic point of view is a major expense. Thus, in general, surface designs for biological species contribute to the so-called economy of effort. To achieve economy of effort, a balance between frictional resistance and friction tractions have to exist while the reptile is in motion. This balance is accommodated through the microstructure of the skin in snakes. Moreover, snakes, in particular, are found almost everywhere on the planet. There habitat is diverse and in many cases is a style from the tribological point of view. Many of the snake species, therefore, show functional adaptations to accommodate the constraints of their local environment. These adaptations reflect within the micro-ornamentation of the ventral surface (which principally accommodate friction). Being that diverse implies that the ventral scale, across the species, manifests in natural database of designs which if harvested may potentially offer innovative fundamental solutions for many technical challenges currently faced in surface engineering. To that effect, within the past few years there have been an extensive effort to characterize the tribological behavior of many legless reptiles (Bogue 2014)-(M. Mühlberger 2015). Many of the results pointed at the potential of surface mimicry. More importantly, some of the independently man-developed surface texturing schemes converge to those already found on ventral scales of snakes both in shape and topography of the layout (Greiner 2015), (Anis).

REFERENCES

Abakan. (n.d.). https://www.marketwatch.com/Abakan-addressescrucialneedsfornextgenerationsurfaceengineering

Abdel-Aal, H. A. (2000). On the influence of thermal properties on wear resistance of rubbing metals at elevated temperatures. *Journal of Tribology*, *122*(3), 657–660. doi:10.1115/1.555417

Abdel-Aal, H. A. (2001). Thermal compatibility of dry sliding tribo-specimens. *Wear*, *250-251*(PART 2), 939–948. doi:10.1016/S0043-1648(01)00705-0

Abdel-Aal, H. A. (2001). On the influence of tribo-induced superheating on protective layer formation in dry sliding of metallic pairs. *International Journal of Thermal Sciences*, *40*(6), 571–580. doi:10.1016/S1290-0729(01)01245-5

Abdel-Aal, H. A. (2001). On the connection of thermal dilatation to protective layer formation in the fretting of metallic tribo-specimens. *Wear*, *247*(1), 76–87. doi:10.1016/S0043-1648(00)00518-4

Abdel-Aal, H. A. (2001). Correlating thermal dilatation to protective layer formation in fretting wear. *International Communications in Heat and Mass Transfer*, *28*(1), 97–106. doi:10.1016/S0735-1933(01)00217-2

Abdel-Aal, H. A. (2002). Thermal kinetics of protective oxide layer formation in the dry sliding of metallic tribo-specimens. *Tribology International*, *35*(11), 757–769. doi:10.1016/S0301-679X(02)00031-2

Abdel-Aal, H. A. (2003). Efficiency of thermal energy dissipation in dry rubbing. *Wear*, *255*(1-6), 348–364. doi:10.1016/S0043-1648(03)00355-7

Abdel-Aal, H. A. (2003). On the interdependence between kinetics of friction-released thermal energy and the transition in wear mechanisms during sliding of metallic pairs. *Wear*, *254*(9), 884–900. doi:10.1016/S0043-1648(03)00243-6

Abdel-Aal, H. A. (2005). On the role of intrinsic material response in failure of tribo systems. *Wear*, *259*(7-12), 1372–1381. doi:10.1016/j.wear.2004.11.008

Abdel-Aal, H. A. (2013). On Surface Structure and Friction Regulation in Reptilian Locomotion. *Journal of the Mechanical Behavior of Biomedical Materials*, *22*, 115–135. doi:10.1016/j.jmbbm.2012.09.014 PMID:23582565

Abdel-Aal, H. A. (2015). The Structure of Ventral Scale Textures in Snakes in Comparison to Texturing of Deterministic Tribological Surfaces. *Processing Techniques and Tribological Behavior of Composite Materials.*

Abdel-Aal, H. A., & El Mansori, M. (2013, September 1). Tribological analysis of the ventral scale structure in a Python regius in relation to laser textured surfaces. *Surface Topography: Metrology and Properties*, *1*(1), 015001. doi:10.1088/2051-672X/1/1/015001

Achtsnick, M., Drabbe, J., Hoogstrate, A. M., & Karpuchewski, B. (2004). Erosion Behaviour and Pattern Transfer Accuracy of Protecting Masks for Micro- Abrasive Blasting. *Journal of Materials Processing Technology*, *149*(1-3), 43–49. doi:10.1016/j.jmatprotec.2003.10.037

Anis Fatima, P. T. (n.d.). Mativenga, On the comparative cutting performance of nature-inspired structured cutting tool in dry cutting of AISI/SAE 4140. *Proceedings of the Institution of Mechanical Engineers, Part B: Journal of Engineering Manufacture*. doi: 10.1177/0954405415617930

Arnold, S., & Bennett, A. (1988). Behavioral variation in natural populations. V. Morphological correlations of locomotion in the garter snake. Thamnophis radix. *Biological Journal of the Linnean Society. Linnean Society of London*, *34*(2), 175–190. doi:10.1111/j.1095-8312.1988.tb01955.x

Ashton, K. G., & Feldman, C. R. (2003). Bergmann's rule in nonavian reptiles: Turtles follow it, lizards and snakes reverse it. *Evolution; International Journal of Organic Evolution*, *57*(5), 1151–1163. doi:10.1111/j.0014-3820.2003.tb00324.x PMID:12836831

Aspinwall, D. K., Wise, M. L. H., Stout, K. J., Goh, T. H. A., Zhao, F. L., & El-Menshawy, M. F. (1992). Electrical Discharge Texturing. *Journal of Machine Tools & Manufacture*, *32*(12), 183–193. doi:10.1016/0890-6955(92)90077-T

Aubret, F., & Shine, R. (2009). The origin of evolutionary innovations: Locomotor consequences of tail shape in aquatic snakes. *Functional Ecology*, *22*(2), 317–322. doi:10.1111/j.1365-2435.2007.01359.x

Bauchot, R. (n.d.). *Snakes: A Natural History*. New York: Sterling Publishing Co., Inc.

Baum, M. J., Heepe, L., Fadeeva, E., & Gorb, S. N. (2014). Dry friction of microstructured polymer surfaces inspired by snake skin. Beilstein Journal of Nanotechnology, 5, 1091-1103. doi:10.3762/bjnano.5.122

Baum, M. J., Kovalev, A. E., Michels, J., & Gorb, S. N. (2014). Anisotropic Friction of the Ventral Scales in the Snake Lampropeltis getula californiae. *Tribology Letters*, *54*(2), 139–150. doi:10.100711249-014-0319-y

Bejan, A., & Marden, J. H. (2006). a Unifying constructal theory for scale effects in running, swimming and flying. *The Journal of Experimental Biology*, *209*(2), 238–248. doi:10.1242/jeb.01974 PMID:16391346

Bejan, A., & Marden, J. H. (2006). b Constructing animal locomotion from new thermodynamics theory. *American Scientist*, *94*(4), 343–349. doi:10.1511/2006.60.1000

Benz, M. J., Kovalev, A. E., & Gorb, S. N. (2012). Anisotropic frictional properties in snakes. *Proc. of SPIE*, 8339. 10.1117/12.916972

Berthé, R. A., Westhoff, G., Bleckmann, H., & Gorb, S. N. (2009). Surface structure and frictional properties of the skin of the Amazon tree boa Corallus hortulanus (Squamata Boidae). *J. Comp. Phys. A*, *195*(3), 311–318. doi:10.100700359-008-0408-1 PMID:19137315

Boback, S. M. (2003). Body size evolution in snakes: Evidence from island population. *Copeia*, *2003*(1), 81–94. doi:10.1643/0045-8511(2003)003[0081:BSEISE]2.0.CO;2

Boback, S. M., & Guyer, C. (2003). Empirical evidence for an optimal body size in snakes. *Evolution; International Journal of Organic Evolution*, *3*, 195–211. PMID:12683530

Bogue, R. (2014). Snake robots: A review of research, products and applications. Industrial Robot. *International Journal (Toronto, Ont.)*, *41*(3), 253–258.

Brinksmeier, E., Glabe, R., Riemer, O., & Twardy, S. (2007). Precision Cutting Processes for the Manufacture of Micro Molds. *Proceedings of the 2nd International Conference on New Forming*.

Cecchetto, L., Serenelli, L., Agarwal, G., Izzi, M., Salza, E., & Tucci, M. (2013). Highly textured multi-crystalline silicon surface obtained by dry etching multi-step process. *Solar Energy Materials and Solar Cells*, *116*, 283–290. doi:10.1016/j.solmat.2013.04.015

Cecchetto, L., Serenelli, L., Agarwal, G., Izzi, M., Salza, E., & Tucci, M. (2013). Highly textured multi-crystalline silicon surface obtained by dry etching multi-step process. *Solar Energy Materials and Solar Cells, 116*, 283–290. doi:10.1016/j.solmat.2013.04.015

Choi. (2015). Fabrication of superhydrophobic and oleophobic Al surfaces by chemical etching and surface fluorination. *Thin Solid Films*, *585*(30), 76–80.

Choi, K. H., Meijer, J., Masuzawa, T., & Kim, D. (2004). Excimer Laser Machining for 3D Microstructure. *Journal of Materials Processing Technology*, *149*(1-3), 561–566. doi:10.1016/j.jmatprotec.2004.03.005

Cicconofri, G., & DeSimone, A. (2015). Limbless Locomotion on Solid Surfaces: A Case Study in Soft Bio-Inspired Robotics. *Mathematical Modelling*, *8*(1), 827–828.

Costa, H. L., & Hutchings, I. M. (2008). Inkjet Printing for Patterning Engineering Surfaces. *Proceedings of 24th Conference on Digital Fabrication*, 256–259.

Cowles, R. B. (1956a). Some of the activities of the side winder. *Copeia, 4*, 220-224.

Cowles, R. B. (1956b). Sidewinding locomotion in snakes. *Copeia*, *4*(4), 211–214. doi:10.2307/1440272

Cox, C. L., Boback, S. M., & Guyer, C. (2011). Spatial dynamics of body size frequency distributions for North American squamates. *Evolutionary Biology*, *38*(4), 453–464. doi:10.100711692-011-9132-9

Cuervo, P., López, D., Cano, J., Sanchez, J., Rudas, S., Estupiñan, H., . . . Abdel-Aal, H. A. (2015). *Topographic characterization of snake skin ventral scales for tribological purposes.* VIII Congress of Materials, Paipa-Boyacá, Colombia.

Dai, F. Z., Wen, D. P., Zhang, Y. K., Lu, J. Z., Ren, X. D., & Zhou, J. Z. (2015). Micro-dimple array fabricated on surface of Ti6Al4V with a masked laser ablation method in air and water. *Materials & Design*, *84*(5), 178–184. doi:10.1016/j.matdes.2015.06.137

Daniel, J. H., Moore, D. F., & Walker, J. F. (2000). Focused Ion Beams and Silicon-on- Insulator: A Novel Approach to MEMS. *Smart Materials and Structures*, *9*(3), 284–290. doi:10.1088/0964-1726/9/3/306

Datta, M. (1995). Fabrication of an Array of Precision Nozzles by Through-Mask Electrochemical Machining. *Journal of the Electrochemical Society*, *142*(11), 3801–3806. doi:10.1149/1.2048416

Defourny, J., & Jodogne, P. (1992). Lasertex, A Concept of Roughness—From the Rolling-Mill to the Coated Sheet. *Revue de Métallurgie*, *89*(5), 439–446. doi:10.1051/metal/199289050439

Dobrolyubov, A. I. (1986). The mechanism of locomotion of some terrestrial animals by travelling waves of deformation. In *Journal Theor* (pp. 457–466). Oxford: Biology. doi:10.1016/S0022-5193(86)80195-3

Feldman, A., & Meiri, S. (2013). Length-mass allometry in snakes. *Biological Journal of the Linnean Society. Linnean Society of London*, *108*(1), 161–172. doi:10.1111/j.1095-8312.2012.02001.x

Greiner, C., & Schäfer, M. (2015). Schäfer M., Bio-inspired scale-like surface textures and their tribological properties. *Bioinspiration & Biomimetics*, *10*(4), 044001. doi:10.1088/1748-3190/10/4/044001 PMID:26125522

Guan, X. S., Dong, Z. F., & Li, D. Y. (2005). Surface nanocrystallization by sandblasting and annealing for improved mechanical and tribological properties. *Nanotechnology*, *16*(12), 2963–2971. doi:10.1088/0957-4484/16/12/040

Humphrey, N. (1999). *A History of the Mind: Evolution and the Birth of Consciousness*. Springer Science & Business Media.

Lin, Li, Huang, & Houng. (2015). Using chemical wet-etching methods of textured AZO films on a-Si:H solar cells for efficient light trapping. *Materials Chemistry and Physics, 160*(15), 264-270.

Liu, W., Sun, L., Luo, Y., Wu, R., Jiang, H., Chen, Y., ... Liu, Y. (2013). Facile transition from hydrophilicity to super hydrophilicity and superhydrophobicity on aluminum alloy surface by simple acid etching and polymer coating. *Applied Surface Science*, *280*(1), 193–200. doi:10.1016/j.apsusc.2013.04.124

Lu, Y., Hua, M., & Liu, Z. (2014, July). The Biomimetic Shark Skin Optimization Design Method for Improving Lubrication Effect of Engineering Surface, ASME. *Journal of Tribology*, *136*(3), 0317031–3170313. doi:10.1115/1.4026972 PMID:25053867

Malshe, Rajurkar, Samant, Hansen, Bapat, & Jiang. (2013). Bio-inspired functional surfaces for advanced applications. *CIRP Annals - Manufacturing Technology, 62*(2), 607-628.

Mühlberger, Rohn, Danzberger, Sonntag, Rank, Schumm, … Bergmair. (n.d.). UV-NIL fabricated bio-inspired inlays for injection molding to influence the friction behavior of ceramic surfaces. *Microelectronic Engineering, 141*, 140-144. Doi:10.1016/j.mee.2015.02.051

Stout, K. J. (1997, January). A Review of nano-Technology. *The Korean Journal of Precision Engineering*, *14*(1), 29–51.

Stout, K. J., & Blunt, L. (2001). A contribution to the debate on surface classifications random, systematic, unstructured, structured and engineered. *International Journal of Machine Tools & Manufacture*, *41*(13-14), 2039–2044. doi:10.1016/S0890-6955(01)00069-4

Surface engineering market and technology, Howard Lang. (2013). *SulzerMetco*. www.sultzer.com

Time for strategic change: uk surface engineering and advanced coatings industry a report by the SEAC special interest group. (n.d.). https://www.uk-cpi.com/news/report-the-uk-surface-engineering-and-advanced-coatings-industry/October-8-2014

Wang, L., & Li, D. Y. (2003). Mechanical, electrochemical and tribological properties of nanocrystalline surface of brass produced by sandblasting and annealing. *Surface and Coatings Technology, 167*(2–3), 188-196. doi:10.1016/S0257-8972(02)00894-0

Yang, B., & Lee, M. (2013). Fabrication of honeycomb texture on poly-Si by laser interference and chemical etching Original Research Article. *Applied Surface Science*, *284*(1), 565–568. doi:10.1016/j.apsusc.2013.07.134

Chapter 2
Fundamentals of Sliding Interaction of Surfaces

ABSTRACT

This chapter reviews the laws governing the friction behavior of objects. The material starts with a historical view of the evolution of friction laws and how they shaped the science of tribology. The second part of the chapter provides a generalized overview of the mechanics of contact between complying solids. Major contact models are listed, and formulas for calculating area of contact, contact forces, contact stresses are also developed. Finally, the chapter applies the presented information to the contact of biological species. The information presents a summary of the major rules in biological attachments, their adhesion and friction, along with their contact mechanics.

INTRODUCTION

Friction is essential to performing numerous daily tasks. Interacting with frictional phenomena is natural. Rubbing hands seeking warmth on a cold morning takes advantage of frictional effects. Walking, ironing, and combing of hair are but some examples of friction driven daily activities. Frictional phenomena occur in and around all living beings. Our sense of texture is a friction-based function. Vibrations, resulting from sliding of human fingers on a real surface, cause the resonance of mechanical receptors, that match the frequency of the vibrations. The receptors in turn generate signals to the brain conveying the sense of roughness or smoothness.

DOI: 10.4018/978-1-7998-1647-8.ch002

Friction is also essential to many manufacturing operations. Surface generation by turning, milling and shaping for example are frictional processes. Some winter sports, such as skiing and bobsledding, depend on friction especially when high performance is required. Ice at low temperatures offers high resistance to motion due to a high coefficient of friction. When friction increases it leads to higher temperatures that cause melting of a thin layer into water. This melt layer lubricates the contacts and causes a drastic drop in friction whence less resistance to motion. Pedestrian slippage accidents constitute another example of the role friction assumes in daily life. In all, friction is a naturally occurring phenomenon that affects life activities to a greater extent than commonly thought. It is no accident that such a phenomenon and associated effects has been the subject of numerous investigations that date perhaps to existence of man on the planet.

Friction takes place between any two contacting surfaces. It can be external or internal. External friction is friction between solid bodies. It is the manifestation of the forces that resist the movement of a solid object upon embarking on motion. Thus friction may retard or completely halt movement of complying solid bodies. Internal friction, on the other hand, denotes energy losses inside of a body (solid, gas, or a liquid) due to the resistance of motion between the elements making up the material while it undergoes deformation (fluid friction is also used to describe the friction between layers within a fluid that are moving relative to each other).

Conventionally, friction is regarded as an undesirable effect since it is associated with wasted energy. However, in practice, whether friction is to be deemed undesirable or favored depends on the application. For example, playing the violin requires friction. In the case of brakes, high friction is desirable to dissipate the kinetic energy of a moving vehicle whence halting its' motion. Meanwhile, friction between pen and paper is not as desirable since high friction causes difficulty in writing.

Historical Prospective

Since surface interactions dictate or control the function of many devices developed by man, tribology has been of central importance for thousands of years, even though this has not always been generally recognized [4]. It is the engineering aspect of friction which has the longest history. The first practical aspects of friction, namely the use of frictional heating in the lighting of fires, has its roots in prehistory. More than 400 000 years ago, our hominid ancestors in Algeria, China, and Java were making use of friction when they chipped stone tools. By 200000 BC, Neanderthals had achieved a clear mastery of friction, generating fire by rubbing wood on wood and by striking of flint stones. The second practice, showing an understanding of frictional phenomena, was the use of liquid lubricants to minimize the work required to transport heavy objects which dates back more than 4000 years.

Figure 1. Handaxe extremely archaic from the Quaternary fluvial terraces of Duero river (Valladolid, Spain) and it is dated in Acheulean/Abbevillian period (Lower Paleolithic).

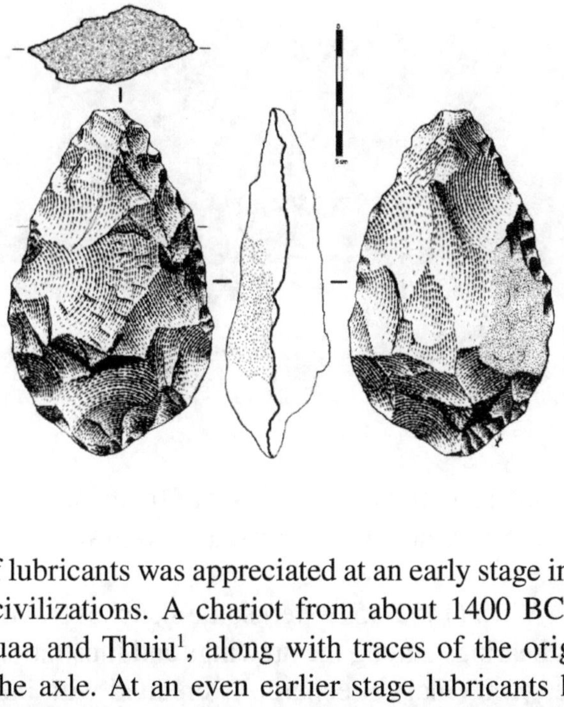

The value of lubricants was appreciated at an early stage in the Sumerian and Egyptian civilizations. A chariot from about 1400 BC was found in the tomb of Yuaa and Thuiu[1], along with traces of the original lubricant remaining on the axle. At an even earlier stage lubricants had been used to reduce friction between sledges used for the transportation of building blocks and the wooden logs or planks on which they moved. Thus, the use of grease, oil or mud as a lubricant was recorded as early as 2400 BC. In modern times, although friction was the subject of many studies, notably by DaVinci, Amonoton and Bernoulli a comprehensive theory for lubrication, which may be argued as the start of rigorous efforts in tribology, did not appear until late in the nineteenth century.

Many forms of bearings came into existence in response to attempts to reduce the friction resistance to motion in chariots. One of the oldest known forms was found in the Chinese village of Hsüh-Chia-Yai in Shansi (Needham 1991). The device contained annular objects divided into four or eight compartments by small transverse partitions. Several of these possible "ball-races" or rather "bearings" of different sizes were found. A leap into the middle ages (the golden age of Scholarship and craftsmanship in the Islamic World) reveals also the use of different kinds of bearings. In 1206, Al-Jazari used several kinds of bearings for his amazing designs.

Figure 2. The Tutankhamen chariots represent a late stage in an evolution that had begun around 1600 B.C.

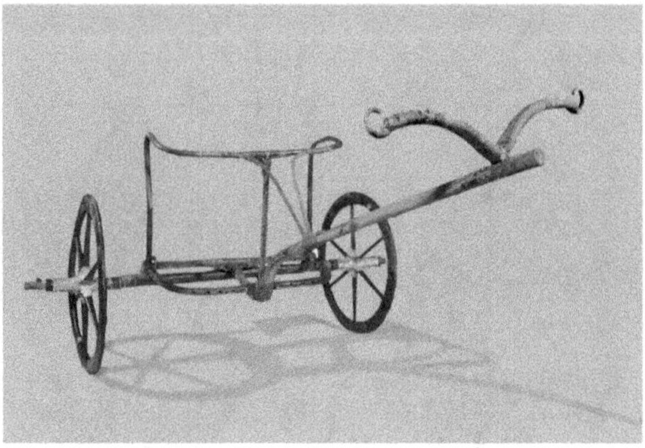

Among these, and mentioned in his book of Ingenious Machines was one bearing called "Al-Kharaza" (which translates from Arabic as "the spherical shaped pebble").[2]

Leonardo Da Vinci (1452-1519) left some sketches that detail designs of thrust and pivot bearings among the drawings and illustrations of the Codex Madrid volumes (Da Vinci).[3] Leonardo is credited with initiating the first study that investigates the nature of friction on scientific bases (hypotheses and deduction through experimentation). Da Vinci formulated the two basic laws of friction 200 years before Newton even defined what force is. Da Vinci rationalized that doubling the load of an object will double its friction; and the areas, of the rubbing solids, in contact have no effect on friction. The first statement implies that friction is proportional to the normal force, and the second states that friction is independent of cross-sectional area. The second statement is counterintuitive; since one is inclined to assume that friction does depend upon the cross-sectional area. Da Vinci also observed that different materials move with different "ease". He reasoned that such an occurrence is a result of the smoothness of the material in question. Accordingly, smoother materials will be subject to smaller resistance to motion. Da Vinci also made some quantitative statements about friction. He stated: "Resistance in the form of frictional force is generated whenever force is exerted to slide an object. The size of this frictional force is equal to one fourth of the objects weight (i.e., in contemporary terms, the coefficient of friction is 0.25)".

Typical of many incidents in the history of the sciences, the work of Da Vinci became obscure for almost two centuries to follow. This is when Guillaume Amontons (1663-1705) (1699), later Charles Augustine de Coulomb (1736-1806) (1785) independently rediscovered the same laws Da Vinci, discovered earlier. Amonotons contributed some original thoughts involving the roughness of surfaces. He proposed that friction was predominately a result of the work done to lift one surface over the roughness of the other. Amontons' work fundamentally guides most efforts to probe the multifaceted genesis of friction. According to his statements, two basic rules govern friction.

The first rule may be stated as:

The friction-induced force acting on a solid body is only proportional to the external forces acting on that body, i.e.

$$F_s = \mu F_N \tag{2.1}$$

where, FN is the external force (s) acting on the body, and μ[4] denotes the "coefficient of friction".

Whereas, the second rule states that

The friction force (and thereby the coefficient of friction) is independent of the apparent area of contact between the contacting bodies.

Consequently, two complying bodies, regardless of their physical size, will have the same coefficient of friction.

Coulomb learning about the work of Amonotons engaged in experimental verification. His interest extended to investigating the dependence of static friction on the time of rest.[5] He confirmed earlier findings and established that sliding friction, within a first order approximation, is independent of the sliding speed. That is, the kinetic friction force (and thereby the coefficient of kinematic friction μk) do not depend on the sliding velocity once motion starts. As such, one may distinguish between the static friction force, FS, and the kinematic (dynamic) friction force Fk. Zhurvalev (2013, 2010)credits the famous Mathematician Leonard Euler with discovering and formulating the independence of the kinematic friction from sliding speed and the law of dry friction in 1748 (i.e., when Coulomb was only 12 years old). Euler's interest appeared to be the mathematical formulation of friction laws. To him, is attributed the distinction between static and kinematic friction.[6,7] He was the first to introduce an analytical solution for a friction problem in 1750 when he solved the problem of rope friction (Euler 1748, Euler 1758). His solution considered interlocking between small triangular irregularities for which he considered that the friction coefficient is equal to their geometric gradient. Naturally, at the time of Euler surfaces were not defined in the precise manner dominant today. Nevertheless the notion of coupling between topography and friction was pioneering.

Refinement of the theory of dry friction continued beyond the era of Euler and Coloumb. The year 1895witnessed serious scrutiny of the laws of dry friction. That is, when Painleve (1954) attempted the solution of problems in mechanics that include dry friction. He encountered several difficulties that prompted him to propose that the laws of dry friction are logically incompatible with the fundamental laws of classical mechanics. The common convention of the time did not support the opinion of what came to be known as the "Painleve Paradox". It was settled that the difficulties encountered by him were not particular to the laws of dry friction, but are rather related to mathematical ill-posedness of mechanics problems. The particular problem posed by Painleve` included translation and spinning under which translational and rotational motions were non-linearly coupled. This coupling necessitated computations of complex nature.[8] The paradox triggered intense debate that attracted famous names in the history of mechanics. Among them was Ludwig Von Prandtl (who traditionally is associated with Boundary Layer Theory in fluid mechanics).

Prandtl designed an experimental setup (Pfeifer 1909) to investigate the so-called Painleve`-Klein system. The complete description of such a problem has to consider deformability of the contacting bodies (Dosaev 2015). Two opinions developed. The first (Klein [1910], Von Mises [1909] and Hamel [1909]) advocated the need to amend friction laws by considering the possibility of the discontinuity of sliding events at points of contact. Prandtl [1909], however, supported the view that the Euler-Coulomb laws are adequate. In 1928 (some 25 years past the introduction of the well-known Boundary Layer Theory) Prandtl [1909] proposed a simple model that considered the motion of a stage containing many elastically coupled "atoms". The stage moves along a periodic potential. In the same paper, he extended the model to an "ensemble" of atoms with different stiffnesses and to "systems", which form a macroscopic point of view simulating a multi-contact situation (Prandtl 1928). Traditionally, the work of Tomlinson (1929), which appeared in 1929, traditionally is considered the initiation of the atomic models of friction. Such a milestone, instead, should, be attributed to Prandtl.

Popov and Gray [2012] argue that the actual contribution of Tomlinson was to add an adhesive component to friction, whereas in actuality, Prandtl describes many fundamental properties of a dry friction system.[9] Almost a decade later, an improvement on the Tomlinson model, the Frankel-Kontorova model [1938, 1938], appeared in Russian literature. This work, which considers coupling between a one-dimensional chains of adsorbate atoms appears to have been developed in the absence of knowledge about the work of Prandtl.

Starting the second quarter of the twentieth century many researchers sought to find a theoretical justification for the rules of Amontons and Coulomb. One pioneering work that addressed the problem was that of Zhuravlev [1940] who attempted to generalize the work of Kontorova. Kontorova noted that Tomlinson's theory implied a coefficient of friction inversely proportional to the cube root of nominal (normal pressure). In practice, however, no load dependency was observed. She attributed the discrepancy to the assumption of dominant elastic contact, which implied that Tomlinson disregarded the possible change in the contact, from elastic to plastic and the effect of that change on the number of molecules coming into contact. Zhuravlev, noticed that such an explanation may be lacking since:

"if validity of the Amontons-Coloumb law at large normal pressures were caused by plastic deformations then it would be unclear why Amonotons' law is valid for brittle solids like glass?"

Recognizing the inadequacy of previous attempts, Zhuravlev embarked on driving Amontons' law through modified Tomlinson's assumptions. Namely, that the number of summits, within a contact surface, at a specific height, increases as the level of consideration goes deeper into the rough surface. Starting from Hertz contact theory, Zhuravlev deduced that the coefficient of friction scales with the inverse of the eleventh root of the nominal load, which he considered as weak dependence and can be neglected. Almost in the same period, Bowden [1939] experimentally verified the dependence of the friction force on the real area of contact and its independence from the size of the nominal area of contact. In effect, therefore, he verified the first law of Amontons.

Tribology

The concept of this science may be defined as: "the science and technology of interacting surfaces, in relative motion, and associated practices" (Fitch 2006). The word tribology was coined by Peter Jost in a British Government Report in March of 1966 (Jost 1966). The word originated from the Greek word "tribos" which means rubbing. Although of recent origin, the majority of the topics covered by this discipline are ancient and well known. It is often said that [31] "tribology was a name selected for a several-thousand year-old baby", and they include the study of lubricants, lubrication, friction, wear and bearings.

In 1885, a British railroad engineer, Beauchamp Tower, conducted experiments to determine friction of the journal of a railway-car wheel (Dowson). He observed the generation of hydrodynamic pressure in the first partial journal bearing. The pressure generated was sufficiently large to displace plugs placed in oil holes in the bearing casing. Osborne Reynolds became intrigued with Tower's results and showed that they could be well explained by applying the principles of fluid mechanics. In 1886 Reynolds published his classical theory of hydrodynamic lubrication (Tower 1886), which is widely used in the design of modem machinery. English biologist Hardy introduced the term boundary lubrication in 1922 (Hardy 1922a), (Hardy 1922b) after finding that very thin adsorbed layers, perhaps only 10 Ao thick, were sufficient to cause two glass surfaces to slide over each other with very low friction. He concluded that under such condition lubrication depended wholly upon the chemical composition of the fluid, not on its viscosity, and that a good lubricant was a fluid which was adsorbed by the solid surfaces. His experiments showed that, of paraffins, alcohols and fatty acids, the latter gave the lowest sliding friction in boundary lubrication experiments. Fatty acids have a polar head and a long hydrocarbon tail. The polar head tends to bind relatively strongly to polar surfaces such as metal oxides. Hardy pictured a sliding interface as consisting of solid surfaces covered by monolayers of fatty acids with the hydrocarbon tail pointing away from the surfaces; the sliding occurs at the plane where the hydrocarbon tails meet.

Tribology evolved into three major branches: mechano-tribology, thermo-tribology, and thermo-mechano tribology (Blok 1987). The first branch, mechano-tribology mainly investigates the mechanical aspects of rubbing and lubrication. This includes the mechanics of contact between solids, stress and strain development in sliding, mechanics of lubrication etc. The second branch thermo-tribology is concerned with the thermal aspects of rubbing solids. Thus, within this branch topics such as: temperature rise in dry or lubricated sliding, heat transfer between rubbing surfaces, and the effect of thermal energy on friction response are of major concern. The coupling between thermal and mechanical influences in rubbing, meanwhile, is a focus of thermo-mechano-tribology. Here topics such as: thermo-elastic stability of contacts and cases where a coupling between the thermal and the mechanical states of the rubbing solids (e.g., in machining and metal forming) affects the frictional response of materials, are a focus.

Mechano-tribology which is the origin of tribology as a science has its roots in the seminal works of Hertz (Hertz 1896), (Hertz 1882) and of Osborne Reynolds (1908). The first founded the theory of the concentrated and isothermal contact on counter formal surfaces. The second, meanwhile, established the classical theory of isothermal hydrodynamic lubrication of conformal rubbing surfaces. The theoretical aspects of investigating the mechanical aspects of tribology were further consolidated in the early twenties of the twentieth century. That is when Ragner Holm published his findings about the disperse nature of the contact between conformal surfaces (Holm 1908), (Holm 1946). The findings of Holm were, later on, considerably extended by Bowden and his team in Cambridge (Bowden 1950a), (Bowden 1950b). Although the analysis of thermal phenomena associated with frictional events is of recent origin, thermotribology is, by far, the oldest of the branches of tribology. It dates back to the moment in history when primitive man learned to "make fire" through rubbing two sticks.

Modern Perspective of Friction Laws

The technological leap witnessed starting the last few decades of the twentieth century resulted in advanced imaging devices that facilitated probing of surfaces and monitoring their interaction within unprecedented resolution. Application of modern technology gave rise to elaborate contact models backed by accurate experimental data and in-depth probing of the physical mechanisms involved. However, the universality of the laws and their ability to predict friction behavior in all possible scenarios is being scrutinized. Contemporary opinions affirm that the classical laws of friction apply to cases where roughness features are deemed "normal" (i.e, no hierarchical features are encountered). Once hierarchy is encountered, then the topographical arrangement of the roughness features plays a major role in determining the evolution of contact between surfaces and development of the friction force within the contact. Thus, in contrast to regimes considered by Amonotons, the true pressure within the area of contact becomes a vital parameter especially if the true area of contact is to be controlled (e.g., in case of ultra-precise modern surfaces).

Sliding of Complying Solids

Friction takes place between any two contacting surfaces in relative motion. It manifests as a force that resists contacting solids upon embarking on motion. Thus, friction may retard or completely halt movement of complying solid bodies. Friction forces act at the interface between any complying solids in sliding relative to each other, Figure 1. A "friction" force is different from a conventional applied force. Newtonian mechanics defines a force as an independent influence that acts on a body from the outside and causes it to accelerate. The friction force, however, is not an independent, nor is it an external influence. Rather, it is internal and opposes the externally applied force. Thus, it may be thought of as a reaction force rather than an action force.

Friction may be external or internal. External friction is friction between solid bodies. Internal friction, on the other hand, denotes energy wastage inside a body (solid, gas, or a liquid) due to the resistance of motion between the elements making up the material while it undergoes deformation (fluid friction is also used to describe the friction between layers within a fluid moving relative to each other).

In principal, there exists two types of external friction: dry friction and lubricated friction. Dry friction takes place between solid surfaces without lubrication. Whereas, lubricated friction takes place between solid surfaces in the presence of an interstitial fluid layer that separates the two contacting surfaces.

Dry friction may be static or kinematic (dynamic). Static friction takes place at the brink of relative sliding of solid surfaces. It represents the resistance to motion at the very beginning of sliding contact. The frictional force within such a regime is that developed before sliding occurs. This force represents the greatest resistance to motion that has to be passed in order to initiate sliding. Kinematic friction, on the other hand, represents the magnitude of the frictional force that acts during the relative sliding motion. That is, if a tangential force (F) is applied to a contacting complying solid pair, such as that shown in figure 3, then the value that is required to initiate motion of one of solid relative to the other is at least the static friction force Fstatic. Whereas, once motion starts, the force required to maintain motion is the kinetic (or dynamic) force Fkinetic. Normally the time duration of static friction is in the order of a few milliseconds, figure 3.

Figure 3. Schematic illustration of the forces acting on a solid body sliding over a horizontal surface (left) and a free body diagram (right). FN is the normal force, FT is the tangential frictional force, W is the weight of the base body (sliding object), and μ is the coefficient of friction.

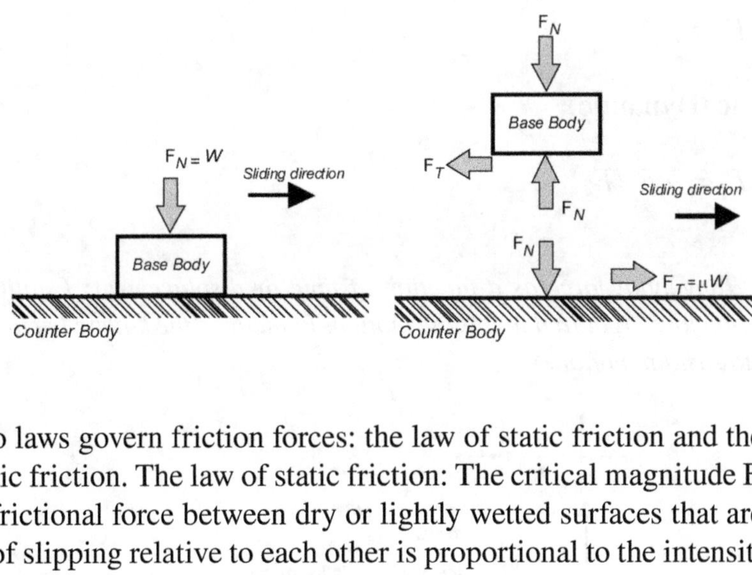

Two laws govern friction forces: the law of static friction and the law of dynamic friction. The law of static friction: The critical magnitude Fs of the static frictional force between dry or lightly wetted surfaces that are on the verge of slipping relative to each other is proportional to the intensity FN of the mutual resultant force between them:

$$F_S = \mu_S F_N \tag{2.1}$$

The constant μS is called the coefficient of static friction, is independent of the interfacial contact area, it depends only on the nature of the contacting surfaces. The law of dynamic friction: The magnitude Fk of the frictional force between two dry or lightly wetted surfaces sliding relative to one another is proportional to the intensity FN of the mutual resultant force between them:

$$F_k = \mu_k F_N \tag{2.2}$$

The constant μk is the called the coefficient of static friction, is independent of the interfacial area of contact, and of the relative sliding speed of the contacting surfaces; it depends only on the nature of the contacting surfaces.

The first law, thus, determines the "greatest" friction force that can develop between contacting surfaces "before" sliding occurs, whereas, the second law determines the magnitude of the frictional force that acts "during" the relative sliding motion. If sliding motion between two bodies has not occurred, and is not imminent, then the magnitude FT of the frictional force is always less than the critical force Fs. Mathematically we can express these remarks in the form:

Static:

$$0 \leq F \leq F_S = \mu_S F_N \tag{2.3}$$

Kinematic (Dynamic):

$$0 \leq F \leq F_k = \mu_k F_N \langle F_S \tag{2.4}$$

Figure 4. Tangential force as a function of time or displacement; Fstatic is the static friction force required to initiate motion, Fkinetic is the kinetic friction force required to sustain motion.

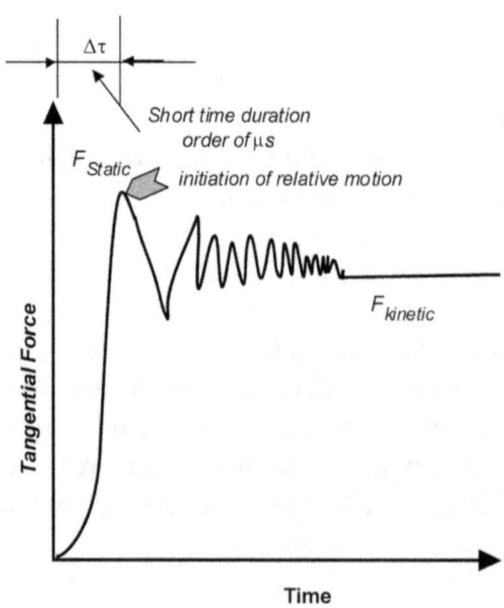

Figure 5 provides a schematic summary of these rules. Note that F= Fs holds only when relative slip is imminent; and F=Fk holds in equation 2.4 only when sliding occurs. Further, F=0 holds only for "ideal" perfectly smooth surfaces for which μs=μk= 0. In all other cases, since μs › μk, once slip is achieved, a smaller force acts to oppose motion. That is once motion starts then the force needed to maintain motion, Fk, would always be less than the static frictional force Fs

Figure 5. Graphical interpretation of Coulomb's laws of static and dynamic friction.

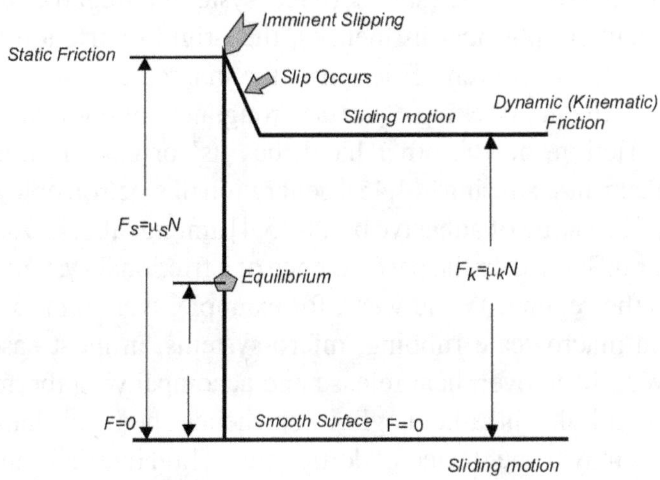

Dry friction rarely takes place in reality since adsorption usually forms a liquid film on the surface of the solid. Dry systems, in essence, are models that simplify analysis. There are physical justifications for considering a sliding system as a dry frictional system. Among these are cases where the normal force is so large, the so-called heavily loaded contacts e.g., in machining and material removal processes in general, that it presses any applied lubricant out of the contact. In this case, the surfaces come directly into contact. Dry sliding can be generated in a vacuum chamber or by hydrophobization of the surface of sliding samples.

Friction regimes are distinguished according to the scale of occurrence of events. Thus, one may identify a "micro-scale regime" and a "macro-scale" regime. Table 1 presents a summary of the main differences between the two regimes.

Table 1. Summary of the differences between "macro" and "micro" scale friction.

Parameter	Macro-scale Regime	Micro-scale Regime
Contact normal force (N)	> 1	< 1
Ratio of real to apparent area of contact	Ar << A app	Ar £ A app
Influence of Wear	Dominant	May be neglected
Influence of surface properties	May be neglected	Dominant
Influence of thermal events	Dominant	May be neglected

Although all frictional systems (tribo-systems), regardless of scale, experience identical physical influences, the origin of friction within any one regime is rather different. Friction at the macro-scale for example is a consequence of the interlocking of surface roughness of both rubbing solids. Micro-scale friction, on the other hand, has its' origins in atomic lattice vibrations, electron scattering [44,45], generation of electron hole pairs or the formation and breaking of adhesive bonds [3,4] among others. Dominance of one particular influence on the performance of a frictional system varies also according to the regime. While wear, for example, is a common dominant occurrence in macro-scale rubbing, micro-systems, in most cases, exhibit negligible wear. Moreover, heat release and accompanying thermal effects, in micro-scale rubbing is rather negligible, whence chemical changes within the adsorbates may be neglected. Additionally, while inertia is an important influence in macro-systems, surface tension and adhesion exert considerable influence on surfaces sliding in the micro-regime.

Equations 2.2 and 2.4 describe the behaviour of the majority of rubbing surfaces involving both dry and lubricated contacts, ductile and brittle, both rough and smooth surfaces (as long as they are not adhesive), and both macroscopic and microscopic contacts. These general rules imply that the static and dynamic coefficients of friction are constants, which, in turn, imply that the coefficient of friction is a material property. Numerous tables that list static and dynamic coefficients of friction for various rubbing material pairs are available in open literature. Table 2 provides an example of such listings.

Modern studies, however, indicate that the coefficient of friction, static and kinematic, will depend on the nature of the contacting material surfaces and their interaction with their environment. That is, on the materials of which the bodies are made, their surface roughness quality, their degree of lubrication, their temperature, and on the cleanliness of the surface (i.e., whether the surface contains oxides, adsorbed layers etc.,). It will also

Table 2. Static and kinematic friction coefficient data for different materials tested in air at room temperatures.

Material		Test Geometry	Friction	Coefficient
Fixed Specimen	Moving Specimen	Static	Kinetic	
Ag	Ag	IS	0.5	...
Al	Al	IS	0.57	...
	Ti	IS	0.54	...
Al, alloy 6061-T6	Al, alloy 6061-T62	FOF	0.42	0.34
	Cu	FOF	0.28	0.23
Au	Ag	IS	0.53	...
	Au	IS	0.49	...
Brass, 60Cu-40Zn	Steel, tool	POR	...	0.24
Ni	Cr	IS	0.59	...
Steel, 52100	Ni3Al, alloy IC-396M	RSOF	...	1.08
	Steel, 1015 annealed	BOR	...	0.74
	Steel, O2 tool	BOR	...	0.49
Steel, mild	Steel, mild	BOR	...	0.62
Ti	Al	IS	0.54	...
	Ti	FOF	0.55	...
	Ti-6Al-4V	POF	0.43	0.36
Ti-6Al-4V	Al, alloy 6061-T6	FOF	0.41	0.38
	Cu-Al (bronze)	POF	0.36	0.27
Ag	Alumina	RPOF	...	0.37
Al	Zirconia	RPOF	...	0.63
Alumina	Alumina	SPOD	...	0.5
Boron carbide	Boron carbide	POD	...	0.53
Cr	Alumina	RPOF	...	0.5
Silicon nitride	Silicon carbide	SPOD	...	0.54
Steel, M50 tool	Boron carbide	POD	...	0.29

BOR, flat block pressed against the cylindrical surface of a rotating ring;
FOF, flat surface sliding on another flat surface;
IS, sliding down an inclined surface;
POR, pin sliding against the cylindrical surface of a rotating ring;
RSOF, reciprocating, spherically ended pin on a flat surface;
SPOF, sliding spherically ended pin on a flat coupon.

depend on the environment of sliding (ambient air, inert gas, or in vacuum), and whether sliding takes place within a magnetic or electrical field. Thus, the coefficient of friction in reality is not a constant, and whence it is not a material property. Moreover, many researchers observed the deviations of some frictional systems from behavior predicted by the Da Vinci-Amontons rules. The observed deviations are pronounced at the extremities of the force-velocity domain. That is when the ratio of the normal force, FN to a particular system parameter is relatively large and a quasi-linear relationship is no longer valid. Mathematically, the limit that determines the validity of a linear relationship between the normal force and a corresponding system parameter stems from a Taylor series expansion of the expression of the friction force around the point of interest. Thus, if Ff = f (μ, FN) then the linearity of the relationship between the friction force and normal loading may be determined from the following:

$$F_t = f\left(\mu, F_N\right) \approx f\left(0\right) + f'\left(0\right) F_N + \frac{1}{2} f''\left(0\right) F_N^2 + ... = \mu F_N + \frac{1}{2} f''\left(0\right) F_N^2 \tag{2.5}$$

A linear (or quasi linear) relationship is admissible (i.e, Ft = μ FN) if:

$$F_N \ll \frac{2\mu}{f''\left(0\right)} \tag{2.6}$$

If the ratio of the normal force to the particular system parameter is large then a nonlinear relationship is more accurate. Such is the case in dry friction where the so called-Coulomb regime dominates.

The Coulomb law of friction, while apparently linear, in essence reflects non-linearity. In fact, it is one of the most complicated constitutive equations in rigid body dynamics (Wiercigroch 2000). To illustrate this point we express the Coulomb law in vector form, viz:

$$\vec{F}_f = \mu \vec{F}_N \frac{\vec{V}}{\left|\vec{V}\right|}, \tag{2.7}$$

where V is the relative velocity. The friction force Ff, in equation (2.8) is defined as the force to lift off the contacts of the upper surface over the contacts of the lower surface. The positive direction of the friction force is positive for positive velocity V. Equation (2.8) entails a so-called switching function formulation of the velocity vector. The inclusion of the switching function is necessary to represent the transition from a static to a kinematic friction regime. A switching function may assume the form of a "signum function", where:

$$Sgn(V)\begin{cases} +1 & \forall \quad V>0 \\ 0 & \forall \quad V=0 \\ -1 & \forall \quad V<0 \end{cases} \tag{2.8}$$

A switching function may also be expressed as a Heaviside function H (V), where:

$$\begin{gathered} \{H(V)-H(-V)\}V \neq 0 \\ H(V)\begin{cases} -1 & \forall \quad V>0 \\ 0 & \forall \quad V<0 \end{cases} \end{gathered} \tag{2.9}$$

Regardless of the form used to express the Coulomb friction regime a fundamental question immediately arises upon studying equation (2.8). This concerns the transition from a static to a kinematic friction regime. Precisely, when the sliding velocity does exceed the zero value and which path exactly does the velocity follow during initiation of motion.

Note that since at zero velocity Sgn (V) is zero, then equation (2.9) collapses indicating that the friction tractions disappear. This, mathematically, amounts to a discontinuity that prevents mathematical formulation of the dynamics of a dry rubbing frictional system as a well-posed problem. Furthermore, due to the ambiguity at zero value, a unique solution of the problem, in the mathematical sense, is virtually unattainable. This situation has lead to several paradoxes in rigid body dynamics (perhaps the most famous is the Painlèvè Paradox [47-48]).

To this end, currently, it is widely accepted that these rules are not valid over large ranges of loads and sliding velocities and that they completely break down for atomically smooth surfaces in strongly adhesive contacts.

Nature of Surfaces

Friction is essentially an interfacial phenomenon. The forces developing at the interface of the contact with the mating counterpart will thus influence the frictional behaviour of a sliding material. Consequently, the behaviour of rubbing solids experiencing will depend on the physical, chemical properties and on the geometrical characteristics of the interfacial real area of contact. The properties of the interface need not be the same as those of the bulk of the mating materials. Moreover, the mechanical response of the interface (e.g., strain hardening or strain rate response) may also differ from that of the material bulk. In many cases, especially in sliding of metals, the state of cleanliness of the interface (i.e., the presence or absence of contaminants or oxides), plays a significant role in determining the path of the friction process and the level of damage that may be incurred upon either of the sliding materials.

The friction force between a sliding counter face and a soft material consists of a shear component, S, and a ploughing component, P, such that:

$$F_f = S + P = \tau A_{real} + HA', \tag{2.10}$$

Where τ is the shear strength of the soft material, Areal is the real area of contact, H is the hardness of the soft material, and A' is the cross-sectional area of the indentation caused by the counter face onto the soft material. For hard materials under sufficiently low contact pressures, the macroscopic indentation of the counter face (i.e., the indentation-induced deformation) onto the surface is insignificant. In such a case, deformation occurs only at the microscopic scale, i.e., it is limited to the asperity tips. Under these conditions, both the real area of contact and the indentation cross sectional area may be assumed equal. It follows that in the case of a hard-solid sliding on another hard solid the frictional force approximates to,

$$F_f = [\tau + H]\mathrm{A}r_{eal} \tag{2.11}$$

Equation (2.12) implies that when the material properties of the sliding solids remain unchanged, the frictional force only depends on the real area of contact. Consequently,

$$\frac{F_f^{Text}}{F_f^{flat}} = \frac{A_{real}^{Text}}{A_{real}^{flat}} \tag{2.12}$$

The superscripts text and flat refer to a textured (rough) surface and a completely flat surface. Under constant normal load, the Amonton-Coulomb laws imply that the friction forces will be proportional to the real area of contact, thus we may write,

$$\frac{F_f^{Text}}{F_f^{flat}} = \frac{\mu_{Text}}{\mu_{flat}} \tag{2.13}$$

The terms μtext and μflat denote the coefficient of friction for the textured and the flat surfaces respectively. Comparing equations (2.13) and (2.14) we may write,

$$\frac{\mu_{Text}}{\mu_{flat}} = \frac{A_{real}^{Text}}{A_{real}^{flat}} \tag{2.14}$$

The real area of contact is a function of the roughness of the surface. Therefore, the coefficient of friction between the rubbing surfaces will also depend, among other factors, on the roughness of the contacting solids.

Contact Between Sliding Solids

Solid surfaces, irrespective of their method of formation, contain irregularities or deviations from a prescribed geometrical form. The high points on the surfaces are referred to as asperities, peaks, summits, or hills and the low points are referred to as valleys. When describing the topography of a surface, three contributions are considered: small scale deviations (individual asperities), large surface deviations (waviness), and directional pattern (the so called "lay" of the surface). The smoothest of metallic surfaces have irregularities of the order (100-1000Ao) in height. General workshop finish, meanwhile, milled or ground surfaces, have peak-to-valley heights in the order (10-20 μm). These may be superimposed on surface waves in the order (1000-10,000 μm) wavelength. Figure 8 shows the dimensions of some features of engineering surfaces compared (using a scale of 108) with some features of the earth's surface referred to as valleys (Figure 9).

When two rough surfaces acted upon by a normal force come into contact, the opposite surface peaks to make contact first, are those for which the sum of the heights is the longest. As the load is increased, new pairs of opposite peaks having an even smaller sum of heights will be coming in contact. Once in contact, the surface peaks become deformed. This deformation at first is elastic. When the load exceeds a critical threshold, however, the deformation will change into plastic or rather, elastic-plastic, because the material under a permanently deformed peak will deflect elastically. The increasing load gives rise to elastic deformation of surface waves on which surface asperities are located. This deformation leads to an increase in the contour area of contact, and as a result, to an increase in the number of peaks sustaining the load. Since the peaks differ in height, the deformation of various peaks on the same surface will be different at any instant of time. The pairs of the highest peaks in contact will be deformed the most; whereas peaks whose height is smaller than average usually do not come into contact even at high loads.

The irregularities of the mating surfaces are out of contact over a considerable portion of the apparent contact area because of surface waviness and form errors. For this reason, the contact area hardly depends on the geometric area of the mating parts. It is exactly this fact that lies behind the conclusion by Coulomb that the frictional force is independent of the apparent contact area of the rubbing surfaces. Waviness results in the contact spots grouping in separate areas at wave front crests (fig 9-b); the total of these areas forms the contour area of contact Ac. This area may be defined as the area over which the surface waves make contact. The contact is obviously intermittent because of surface roughness. The contour area is to be taken to mean the area within which the contact of the irregularities occurs, the distance between the contact spots not exceeding the sampling length specified by the standard for measuring a given magnitude of surface roughness. Where one of the mating surfaces has waviness and the other is relatively flat, the contour area of contact will characterize the frictional area over which the surfaces wear as a result of interaction of the irregularities. Approximate measurements of the area of contact evaluate the contour area. The real area of contact, Ar, implies the area within which the irregularities forming surface roughness are in contact. The real area of contact approaches in magnitude the area where the atoms and molecules interact, although these two areas are not identical in some cases. In what follows, this fact would be considered and the real area of contact between two surfaces.

Figure 6. Some dimension of importance in rubbing contacts compared with terrestrial objects (scale 1/108)

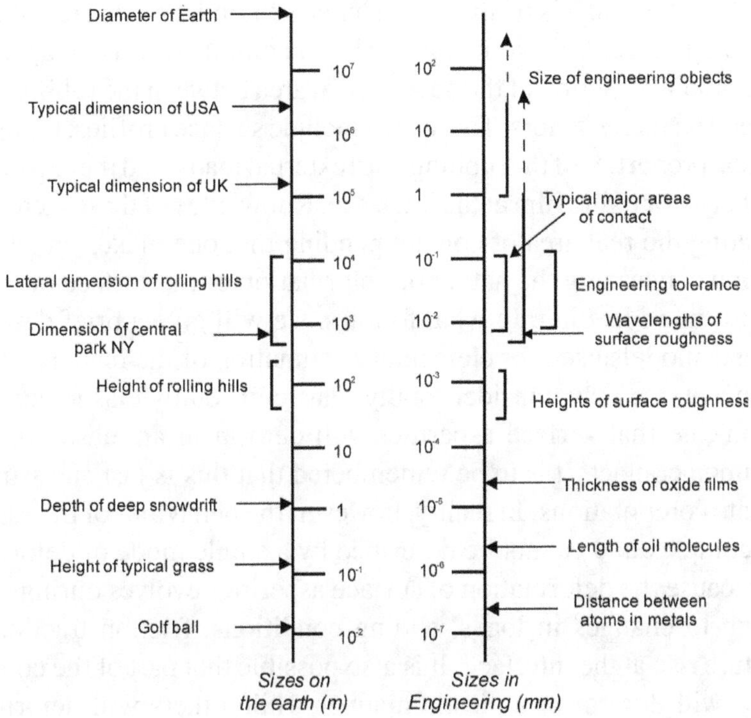

Figure 7. Contact between rough surfaces, (a) General geometry of surfaces, summits, asperieties, and peaks; o.l outer line, m.l mean line, i.l inner line. (b) discrete contact resulting from contact of the asperities (summits) of each of the contacting surfaces.

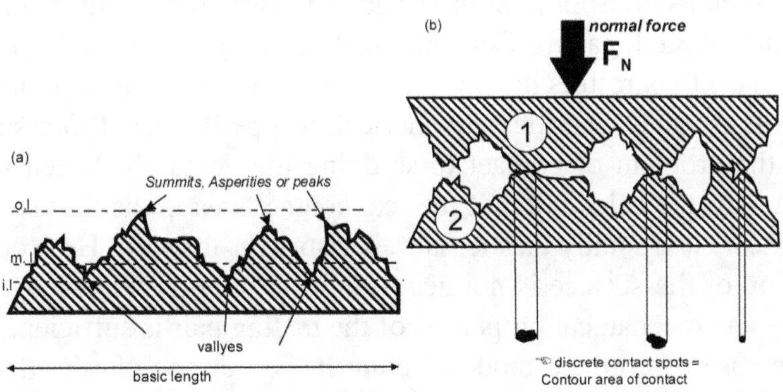

The Area of Contact Between Solids

Of interest in the analysis of sliding systems is to find the real area of contact between the two rubbing solids, Ar,. The magnitude of this area, which in general is a small fraction of the geometrical area between the rubbing solids, will depend on many factors. These may include surface profiles, topography, mechanical properties of the rubbing pair, external loads, and the existence of a lubricant layer or oxide film at the interface. Knowledge of these factors helps in estimating the real area of contact pending that one makes an additional assumption concerning the nature of deformation of the surface protrusions (asperities) upon loading. In what follows, we will give a brief description of the basic models used for elementary estimation of the real area.

In concept, we may consider totally plastic or totally elastic conditions. These indicate that surface asperities will deform in an elastic or plastic manner upon contact. It is to be remembered that this is just an assumption to facilitate computations. In reality, however, the behaviour of the asperities is more complex and cannot be described by a single mode of deformation. This is because the deformation of surface asperities evolves during rubbing in relation to changes in loads, sliding conditions, friction tractions, and temperature rise at the interface. It is also possible that part of the contacting asperities will deform in a plastic manner while others will deform in an elastic manner. In such a case, we may speak of an elastic-plastic contact. Another possibility that due to severe thermal effects, rheological behaviour may ensue and the surface can deform in a visco-elastic manner.

Contact between rough surfaces entails the engagement of a large number of surface asperities from both mating surfaces. These may be of different shapes, lengths, heights and spacing. Considering the exact geometry and topography of the surface thus results in considerable difficulties. Conventionally, it is assumed that surface asperities are spherically capped bodies. This assumption reduces the problem of contact to studying the contact between spheres. Consequently, the calculation of the real contact area in the first two cases, totally plastic and totally elastic, is rather straight forward. Here complete description of the surface is not necessary and only the knowledge of the geometry and mechanical properties of the mating pair is sufficient.

In the case of a mixed mode of contact, i.e., elastic-plastic, the actual topography of the mating surfaces is influential. Accordingly, sufficient knowledge of surface topography, including distribution of heights of the surface protrusions, in addition to mechanical properties is essential. This complicates the analysis and in general, statistical treatment is essential.

Plastic Contact

Consider two solid bodies, body 1 and body 2 in Figure 10, that are brought into contact under the influence of a normal external force P. Initially, as the surfaces approach each other, few asperities of one surface, that of body 1 say, will establish contact with some asperities on the surface of body 2. At this stage, the forces resisting the deformation of these asperities are usually not of a sufficient magnitude to support the normal load P. Assuming that body 2 is softer than body 1, then the force P will push body 1 downwards for closer contact. As body 1 further approaches body 2, the touching asperities will deform and more asperities will come into contact. Throughout this process, the area of contact between the surfaces, δA say, will grow as more asperities come into contact and begin to deform. This growth continues until the forces resisting further deformation of the asperities exactly counterbalances the external force P. That is the growth of the contact area continues until static equilibrium takes effect. Upon establishment of static equilibrium, growth of the area of contact will seize and there will exist a definitive area of body 1 that is in real contact with body 2. This will constitute the real area of contact Ar between the two surfaces. This area need not be a continuous batch; rather it may be formed of several islands distributed over the interface between the two bodies.

Considering static equilibrium between the external force P and the forces resisting deformation of the asperities we may write:

$$P = \int_A \sigma \delta A \tag{2.15}$$

where, σ denotes the local pressure acting on individual junctions.

If we assume that each of the contacting junctions is in a state of incipient plasticity flow, then we may replace σ in equation (2.16) by the yield pressure of the softer material in compression. This is equivalent to the penetration hardness of the material. Performing the integration in (2.16), substituting for σ by H, and re-arranging the terms the area of contact is expressed as,

$$A_r = \frac{P}{H_{Sof}} \tag{2.16}$$

The subscript "sof", in equation 2.17, indicates that the hardness of the softer material is used in calculations. The radius of the contact patch, Figure 10-b, in such a case will thus be given by, a= (P/π H) 1/2.

Equation (2.17), originally developed by Bowden and Tabor (1939), implies that the real area of contact is linearly dependent on the normal force P.

The force needed to shear the contacting junctions is

$$F_s = \tau_s A_r \tag{2.17}$$

where τs is the yield stress during shear. Substituting for Ar from equation (2.17) the shearing force is written as,

$$F_s = \frac{\tau_s}{H} P \tag{2.18}$$

Figure 8. Area of contact between two complying solids in case of plastic deformation of junctions, (a) real contact geometry (b) idealized contact where the true shape of the asperity is modeled as a spherically capped body and the roughness of the mating body, body 2, is imposed on the counter body, body 1. The resulting area of contact is a circle of radius a. H1 and H2 are the penetration hardness of solids 1 and 2 respectively.

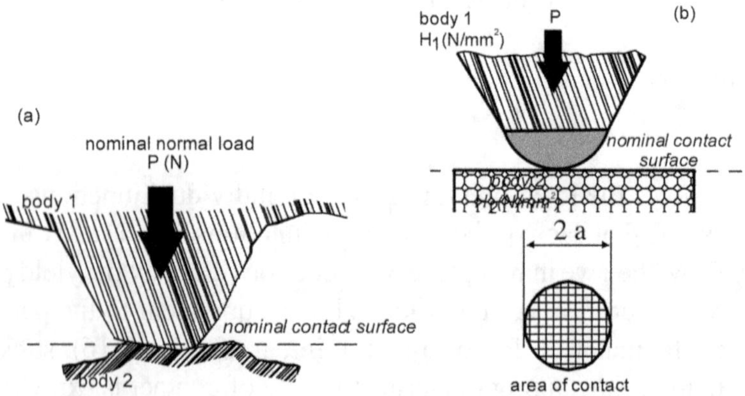

Comparing equation (2.18) to equation (2.19), Fs=μsFN, then the coefficient of friction μ is expressed as,

$$\mu = \frac{\tau_s}{H} \tag{2.19}$$

One explanation of the friction laws of Amonton is to postulate that the friction force is directly proportional to the area of real contact, Ar, between the contacting solids Such a postulate reduces the problem to explaining why the real area of contact should be proportional to the load. The explanation based on equations (2.17)-(2.20), is straight forward: because of surface roughness, the initial area of contact increases until the contact pressure falls to the "flow pressure" of the asperities on the softer of the contacting solids. The friction coefficient, equation (2.9), thus becomes the ratio of two strength properties, the shear strength of the junctions formed (τs) and the flow pressure of the solid (H). This yields an explanation of why coefficients of friction are much the same for hard or soft metals. Further, the equations explain why typically μ ~ 1 for the sliding of clean surfaces, which follows from the fact that τs and H are of similar magnitude.

Elastic Contact

The theory developed by Hertz (1882), continues to provide the fundamental theoretical aspects necessary to analyse many contact situations. The theory relates the force acting on a single asperity to its deformation and contact area. It considers two rough solid bodies brought into physical contact through the action of applied forces (Figure 10). The dimensions of the contact area between the bodies are small compared to the dimensions of each body and to the radii of curvature of the surfaces. The strains are sufficiently small so that the provisions of linear elasticity apply. Finally, the problem considers frictionless contact so that only normal pressure is transmitted. The analysis simplifies considerably when contact between symmetric shapes, such as spheres, cylinders or flat planes, is considered.

Assume that two bodies of revolution (spheres) are brought into contact by the action of an external force F, which results in a deformation field. During compression by the normal load, distant points T1 and T2 displace distances δ1 and δ2 respectively parallel to the Z–axis towards O. The quantity δ= δ1+ δ2, is the normal approach or the interference. The expression for the deformation field in both solids results from minimization of the elastic deformation energy. For two contacting spheres, the contact area is circular and the radius of the contact region is given by:

$$a = \left(\frac{3FR}{4E_c}\right)^{1/3} \tag{2.20}$$

where,

$$\frac{1}{E_c} = \frac{1-\nu_1^2}{E_1} + \frac{1-\nu_2^2}{E_2} \text{ and } \frac{1}{R} = \frac{1}{R_1} + \frac{1}{R_2}$$

Figure 9. Hertz model for contact between two non-conforming elastic bodies

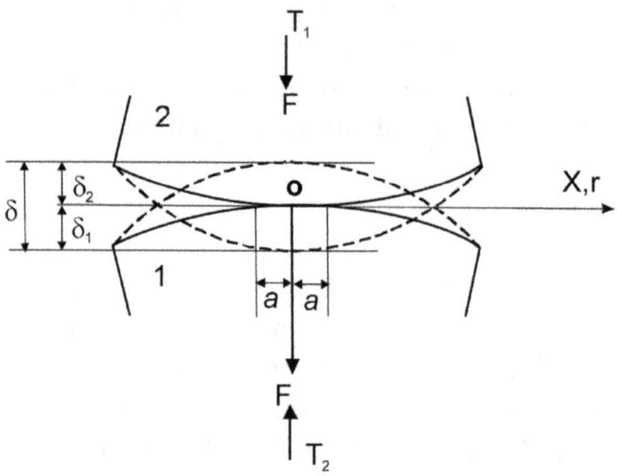

The distance the two spheres approach each other (i.e, the penetration) is given by,

$$\delta = \left(\frac{3F}{4E_c R^{1/2}}\right)^{2/3} \tag{2.21}$$

whereas, the pressure distribution in the contact area and the maximum pressure are given by,

$$p_{\max} = \left(\frac{6FE_c^2}{\pi^3 R^2}\right)^{\frac{1}{3}}$$

$$P(r) = P_o\left[1 - \left(\frac{r}{a}\right)^2\right]^{\frac{1}{2}} \quad r \le a \tag{2.22}$$

$$P_o = \frac{F}{\pi a^2}$$

Note that P_0 is the average pressure on the contact spot.

Equations 2.23, are special cases of the more general results for nonconformal contact of bodies of general ellipsoidal profiles. The contact area is elliptical and the contact pressure distribution is semi-ellipsoidal.

Detailed results for contact area and interference vs. normal force are complicated and are given by Johnson (1985) and Cooper (1969). For moderately elliptical contacts, Greenwood (1985) showed that the contact pressure and approach can be approximated by using the circular contact formulas with an equivalent radius of curvature equal to (AB)-1/2, where A and B are the principal relative curvatures. Solutions for the Hertz contact problem remain valid until the applied load is large enough to initiate plastic deformation.

In general, the theory of Hertz applies to the static contact of two stationary elastic solids. Consequently, cases where a tangential force, such as a friction force, static or dynamic, are not directly obtained from the theory. Table 3 provides a summary of the equations used to evaluate the area of contact in case of the contact between symmetric three-dimensional solids.

The theory of contact between two conforming, nominally rough surfaces is based on modelling the asperities as spherically capped summits. Additional simplifications are obtained by ascribing the roughness to one surface only, and taking the mating surface to be smooth. These assumptions reduce the problem to the study of contact between a sphere and a flat, see Figure 11. Further, if one assumes that the contact between the two surfaces takes place without transition to plasticity then Hertz contact theory may be applied to calculate the area of the contact spot between the two solids.

Table 3. Summary of equations for analysis of contact between three-dimensional elastic solids

Contact Geometry	Dimensions of Contact Spot	Maximum Contact Pressure
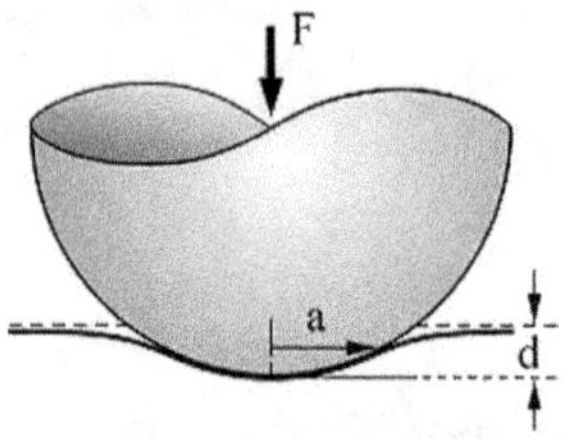	$a = \sqrt{Rd}$	$F = \frac{4}{3}E^{*}\sqrt{Rd^{3}}$
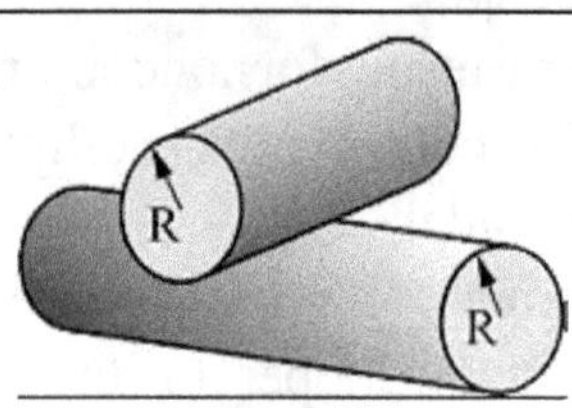	$a = \sqrt{Rd}$	$F = \frac{4}{3}E^{*}\sqrt{Rd^{3}}$
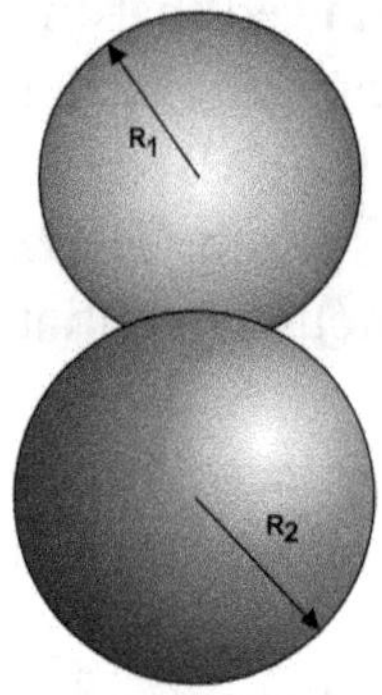	$c = \left[\frac{3PR_{c}}{4E_{c}}\right]^{1/3}$	$P_{\max} = \frac{1}{\pi}\left[\frac{6PE_{c}^{2}}{R_{c}^{2}}\right]^{1/2}$
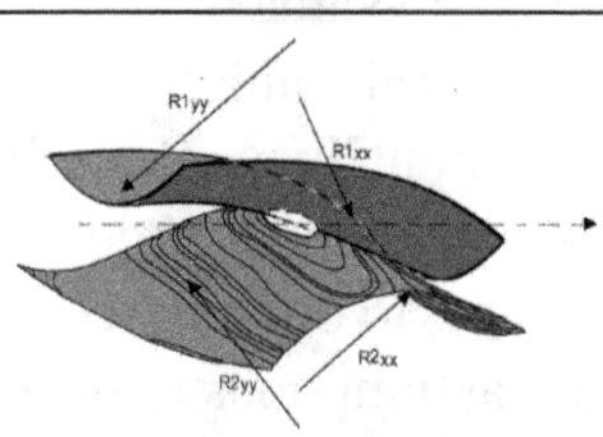		

continues on following page

Table 3. Continued

Contact Geometry	Dimensions of Contact Spot	Maximum Contact Pressure
$R_a = \frac{1}{(A+B)-(B-A)}$ $R_b = \frac{1}{(A+B)+(B-A)}$ $A+B = \frac{1}{2}\left(\frac{1}{R_{1_{xx}}} + \frac{1}{R_{1_{yy}}} + \frac{1}{R_{2_{xx}}} + \frac{1}{R_{2_{yy}}}\right)$ $B-A = \frac{1}{2}\left\{\left(\frac{1}{R_{1_{xx}}} - \frac{1}{R_{1_{yy}}}\right)^2 + \left(\frac{1}{R_{2_{xx}}} - \frac{1}{R_{2_{yy}}}\right)^2 \ldots\right.$ $\left.\ldots 2\left(\frac{1}{R_{1_{xx}}} - \frac{1}{R_{1_{yy}}}\right) + \left(\frac{1}{R_{2_{xx}}} - \frac{1}{R_{2_{yy}}}\right)\cos(2\alpha)\right\}^{1/2}$ $e^2 = 1 - \left(\frac{R_b}{R_a}\right)^{4/3}$ major and minor axis $a = C\left[1-e^2\right]^{-1/4} \quad b = C\left[1-e^2\right]^{1/4}$	$P_{\max} = \frac{3P}{2\pi C^2}$ $C = \left(\frac{3PR_c}{4E_c}\right)^{1/3} F_1$ $F_1 = \left[1 - \left(\frac{R_a}{R_b}\right)^{0.0602} - 1\right]^{1.456}$	
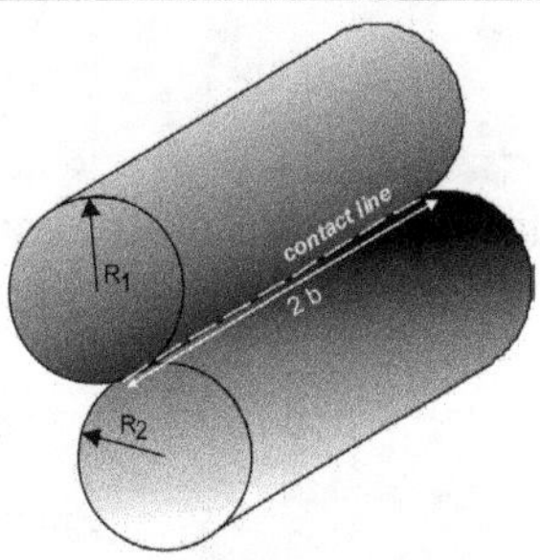	$b = \left[\frac{4PR_c}{\pi E_c}\right]^{1/2}$ $\frac{1}{R_c} = \frac{1}{R_1} + \frac{1}{R_2}$	$P_{\max} = \frac{2P}{\pi b}$
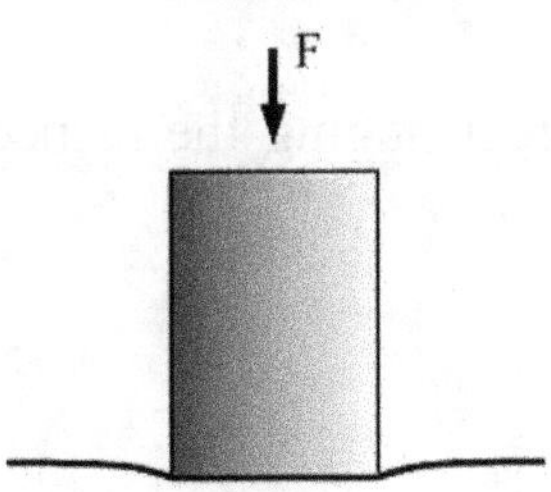		$P_o = \frac{1}{\pi} E_c \frac{d}{a}$ $P(r) = P_o\left(1 - \frac{r^2}{a^2}\right)^{-\frac{1}{2}}$ $F = 2aE_c d$

continues on following page

Table 3. Continued

Contact Geometry	Dimensions of Contact Spot	Maximum Contact Pressure
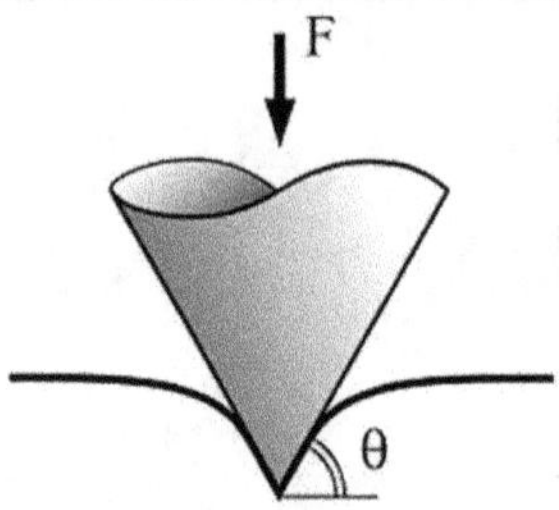	$a = \frac{2}{\pi} d \tan(\theta)$	$P(r) = -\frac{Ed}{\pi a(1-\nu^2)} \times \ln\left(\frac{a}{r} + \sqrt{\left(\frac{a}{r}\right)^2 - 1}\right)$ $F_N = \frac{2}{\pi} E_c \frac{d^2}{\tan(\theta)}$

$$\frac{1}{E_c} = \frac{1-\nu_1^2}{E_1} + \frac{1-\nu_2^2}{E_2}$$

Ec combined elastic modulus
E1, E2 Young's modulus of elasticity for bodies 1 and 2 respectively.
ν1, ν2 Poisson's ratio for bodies 1 and 2 respectively.
P contact pressure
R1, R2 radii of curvature for bodies 1 and 2 respectively.

Referring to Table 3, the case of a circular contact spot, the radius of the contact area is given by:

$$a = \left(\frac{3}{4}\frac{R_c p}{E_c}\right)^{1/3} \quad (2.23)$$

For an elastic contact the maximum shear cannot exceed the yield stress in shear for the material, as such the maximum shearing force τs, may be written as before,

$$F_{s(\max)} = \tau_s A_r \quad (2.24)$$

Upon substitution from equation (2.24) and rearranging the respective terms, the maximum shear is written as,

$$F_{s(\max)} = 0.3538 \tau_s E_c^{3/2} R_c^{2/3} P_N^{2/3} \quad (2.25)$$

Here we have replaced P with FN for consistency. It is seen that the friction force in such a case is proportional to $F_N^{2/3}$.

Figure 10. Schematic of the contact between a single asperity and a flat substrate. The asperity is modelled as a hemispherical caped structure and the substrate is assumed flat.

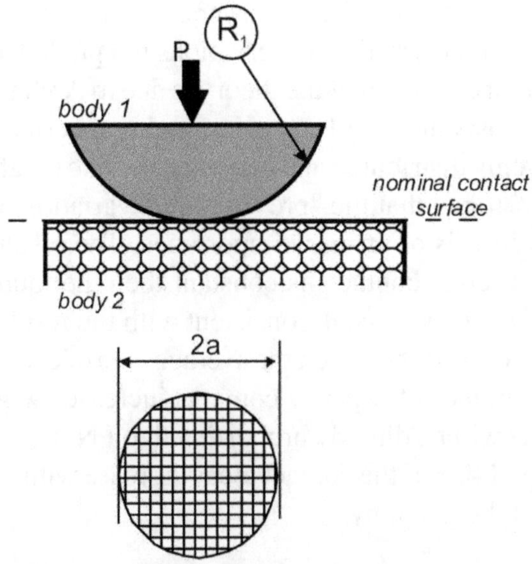

Equation (2.26) implies that the real area of contact in case of elastic loading depends on the simultaneous effect of the normal load, the initial geometry (the radii of curvature of the contacting solids, and the elastic moduli (stiffness of the contact). That is the contact area will constitute the response of the material to the interaction between the structural properties of the contacting solids and the external applied force. This represents a more complex dependency than in the case of a plastic contact where the area of contact depends on the ratio of the external force to the hardness of the softer material.

In its' simplest form, the elastic contact assumption stems from studying the contact of a single sphere. This raises the question of what happens if multiple contacts (spheres) from each solid engage upon contact. Further, in practice, pure elastic contact or pure plastic deformation upon contact do not strictly materialize. Rather, both occurrences manifest as extremes. This poses the question of when would a surface establish contact in a pure elastic mode and under what conditions the transition from elasticity to plasticity takes place. Finally, the derivation of elastic and plastic contact formulas considers an idealized asperity. That is, the analysis considered a smooth asperity, which, in effect, represents a surface where no distribution of heights or asperity sizes is present. In other words, the analysis did not consider real surface effects.

The previous considerations lead to several extensions of contact analysis in order to accommodate practical observations not predicted by elementary theory. An important extension of the theory is due to Archard [54] who used a model surface consisting of a flat covered with spherical asperities. The asperities were evenly distributed and had intermediate heights. The analysis of Archard demonstrated that the "pre" assumed variation of the number of contact area with load is of primary importance. The results indicated that if the number of discrete contacts is constant then the total area will be in proportion to FN2/3. This value is consistent with the results obtained from Hertzian contact theory. If, however, the average area of each contact remains constant and the number of asperity contacts increases with load, then the real area of contact will be directly proportional to FN. If, on the other hand, both the number and size of the contact areas increase with load then the real area of contact will be given by,

$$A_r = K\left(\frac{F_N}{E}\right)^n \tag{2.26}$$

where K is a constant of proportionality and n is an index the value of which depends on the particular model chosen. Table 4 presents a summary of the values of the index n in relation to the chosen model.

A consequence of considering multiple areas of contact is the realization that the number of contacts increases in proportion to the normal load FN. Additionally we note that the mean size of the contacts, aav say, remains constant. That is, the enlargement of existing contacts always exactly balances the increase in the load through the creation of new small areas (Archard 1986). It follows that,

$$A = \sum \pi a^2 = N\pi a_{av}^2 \propto F_N \tag{2.27}$$

aav and Pm, the mean contact pressure throughout the total contact area remain constant. The explanation of Amonton's laws of friction follows also from considering multiple areas of contact. This is because the real area of contact, in such a case, is proportional to the load FN. This is also consistent

Table 4. Variation in the proportionality index n with the level of detail considered in a surface contact model

Model	n	N (no of Contacts)
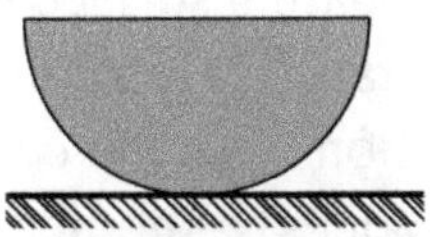	2/3 Hertzian contact	$A_r \propto F_N^{2/3}$ N=1
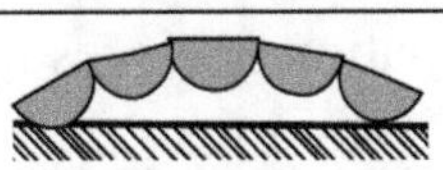	4/5	$A_r \propto F_N^{4/5}$ $N \propto F_N^{2/3}$
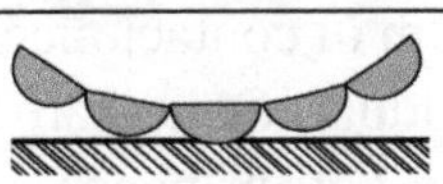	8/9	$A_r \propto F_N^{8/9}$ $N \propto F_N^{4/5}$
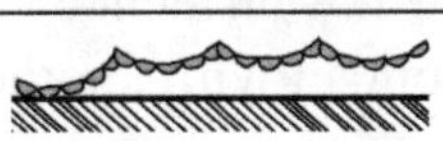	14/15	$A_r \propto F_N^{14/15}$ $N \propto F^{8/9}$
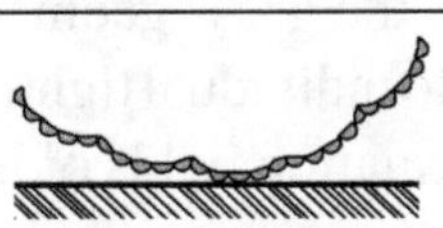	26/27	$A_r \propto F_N^{26/27}$ $N \propto F_N^{14/15}$
	44/45	$A_r \propto F_N^{44/45}$ $N \propto F_N^{26/27}$

with the case of plastic contact conditions where upon the inception of plasticity, the area of an individual contact depends on the applied load. These findings prompted Archard to conclude that the explanation of Amonton's laws lies in the random roughness of the surfaces rather than in their mode of deformation. Greenwood (1966) echoed such an opinion, and indicated that randomness of surfaces is synonymous to their fractal nature.

The fundamental argument in studying contact of real surfaces is not proper selection of a contact model. Rather, the main query is to determine the effect of any load increase on the real area of contact. That is to determine whether an increase in the nominal load creates new contact areas or, to the contrary, load increase induces a corresponding increase in the size of existing contact spots. For the majority of practical surfaces, an elastic contact model that considers the number of contact spots to be constant, throughout the duration of contact, yields an area of contact that is proportional to $F_N^{\frac{1}{3}}$. On the other hand, keeping the average (not the actual) size of the surface roughness constant within a given model while varying the number of contacting spots, results in a real area of contact that is proportional to the nominal load. In all, Archards' argument highlights the importance of considering the state of the roughness for accurate estimation of the real area of contact between the two solids.

Analysis of a contact situation involving two rough surfaces is not a straightforward problem. This is because accommodating surface roughness requires special mathematical treatment (mainly statistical), and unlike flat surfaces, one may encounter mixed contact conditions (i.e. the contact may include a mix of elastic and plastic deformation zones). Another difficulty arises from structural interference between deformation fields resulting from loading.

One definition of a flat surface, within the realm of contact mechanics, is that its' area of contact is large enough so that individual contacts are dispersed such that forces acting on neighbouring spots do not influence each other (Buckley 1981). A curved surface establishing contact with another curved surface does not display such a trait. In this case, the gross geometry of the touching surfaces limits the apparent contact area to individual tightly packed. Tight packing leads to interference between deformation fields of individual clusters and thereby the interaction between the forces cannot be neglected. In all, it is essential to incorporate roughness features upon analysing the contact of real surfaces.

Elastic-Plastic Contact

Contact of elastic plastic bodies at small loads involves surface deformation. Initially, this deformation is elastic with the maximum shear stress τ_{max} occurring some distance beneath the surface, below the centre of the contact region. Increasing the load increases the maximum shear stress until a critical load is reached. At this critical load, τ_{max} exceeds the critical shear stress of the solid. This is the point of plastic deformation onset. Increasing the load further, leads to the growth of the plastic zone (in response to the increase in the contact pressure). This process continues with further increase in the nominal contact load until the plastic zone reaches the surface and the entire region around the indentation becomes fully plastic (Figure 11-a). In all, the deformation in a contact zone evolves in relation to the nominal load (and thereby the shear stress) from: elastic to elastic-plastic, and finally to fully plastic (Figure 11-b). Plastic deformation initiates within one of the contacting solids. It further develops within the solid until the mean contact pressure reaches a critical limit that surpasses the yielding limit (approximately 1.1 σy) for the mating solid. At this critical limit, the mating solid starts to deform plastically as well. Plastic deformation continues to evolve until both solids are permanently deformed. The elastic plastic transition occupies a sizeable region on a map of deformation as that of Figure 11 c.

Figure 11. Regimes of deformation for a sphere and a flat (the example illustrated is a hard steel ball, 6.3 mm diameter, and a mild steel flat.) recalculated in SI units after Archard [27].

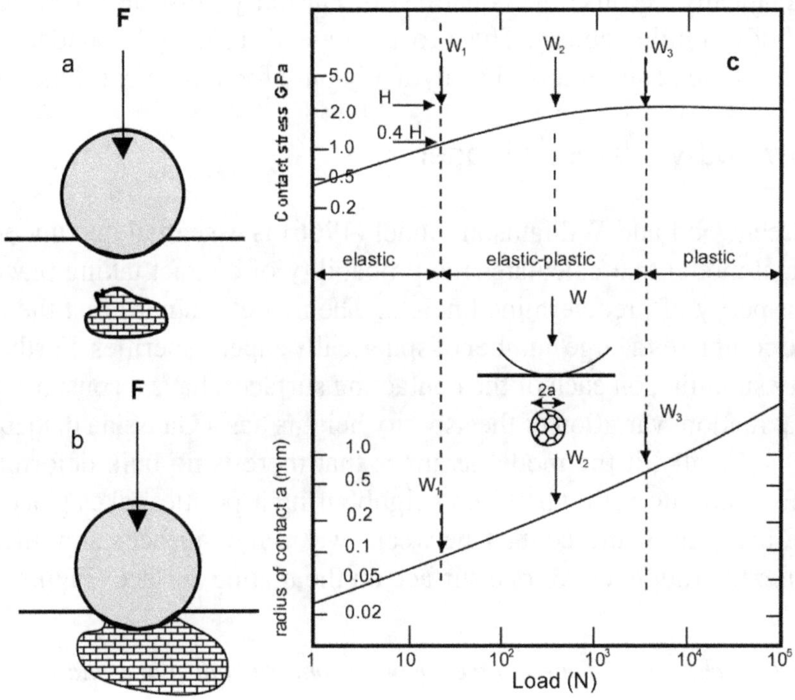

Multi-Asperity Contact Models

Multi-asperity contact models are either predominately uncoupled or completely coupled. Uncoupled contact models represent surface roughness as a set of asperities, often with statistically distributed parameters such as height or summit curvature (Figure 12). The effect of each individual asperity is local and considered separately from other asperities; the cumulative effect is the summation of the actions of individual asperities.

Figure 12. A rough half-space and a flat body about to establish contact

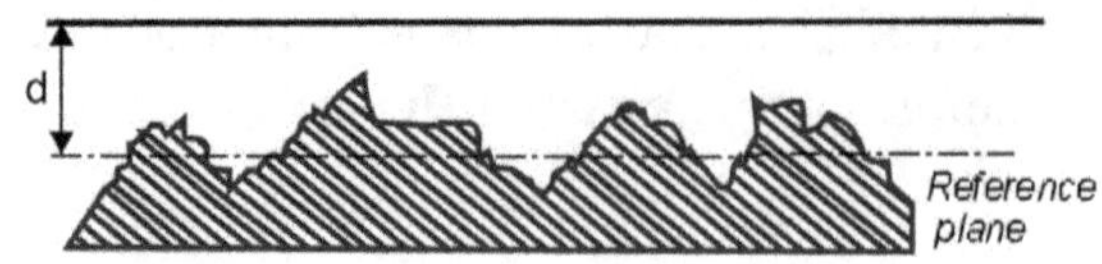

Coupled contact problems with rough surfaces are more complicated mathematically because the equations of elasticity must be solved for the entire body simultaneously. This procedure leads to mixed boundary value problems which can be solved analytically only for simple configurations.

Greenwood-Williamson Model

The Greenwood and Williamson model (1966) is essentially an uncoupled statistical model that calculates the probability of contact taking place at a given asperity of predetermined height. The model assumes that the rough surface comprises a large number of spherically caped asperities. Further, the asperity summits, on each of the contacting surfaces, have a constant radius and that random variation of the asperity heights has a Gaussian distribution of peaks. Finally, it the model assumes that there is no bulk deformation. Therefore, no interaction between neighbouring asperities takes place.

The analysis of the contact between two rough surfaces simplifies by imposing the roughness of one surface on the mating surface (Figure 13).

Figure 13. Schematic of the contact between one smooth plane and one rough surface with equal asperities

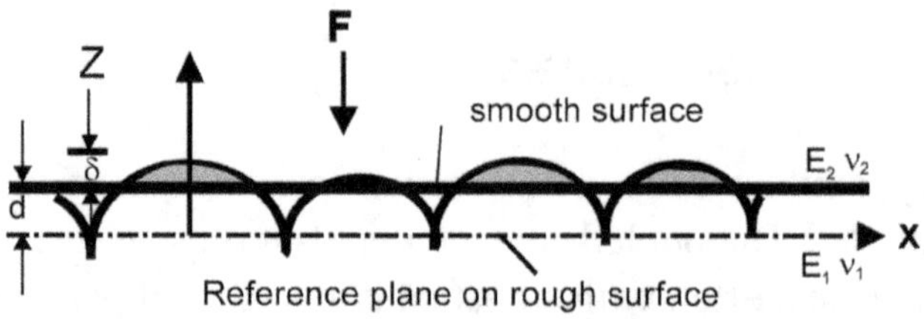

The problem, therefore, reduces to studying the contact between one smooth surface and another of which roughness is the equivalent roughness of the two surfaces. This equivalent rough surface has the asperity curvature, Rp, equal of the sum of the curvatures of the two rough surfaces, i.e,

$$\frac{1}{R_p} = \frac{1}{R_{p1}} + \frac{1}{R_{p2}} \tag{2.28}$$

and the standard deviation of the equivalent rough surface is given by:

$$\sigma_p = \sqrt{\left(\sigma_{p1}^2 + \sigma_{p2}^2\right)} \tag{2.29}$$

The building block of the analysis is the behaviour of an individual asperity, which is known from Hertzian theory. Thus, in terms of the compliance, δ, the load on individual asperity Fi, the radius of a contact spot, ai, and the area of a contact spot for an individual asperity, i.e., are expressed as,

$$F_i = \frac{4}{3} E_c R^{\frac{1}{2}} \delta^{\frac{3}{2}} \tag{2.30}$$

$$a_i = \sqrt{R_p \delta} \tag{2.31}$$

$$A_i = \pi R \delta \tag{2.32}$$

For N asperities the total load Ft will be equal to NFi, and the total real area At=NA therefore,

$$F_t = \frac{4E_c A_t^{3/2}}{3R\sqrt{\pi N}} \tag{2.33}$$

When the complying force compresses the two contacting surfaces until their reference planes are separated a distance d, any asperity whose height originally exceeds the separation distance will establish contact. In such a case, the probability of any asperity of height z, establishing contact is calculated by,

$$P(z > d) = \int_d^{\infty} \phi(z) dZ \tag{2.34}$$

It follows that, for N asperities in the surface, the expected number of contacts, n, will take the form,

$$n = N\int_{d}^{\infty} \phi(z)dZ \tag{2.35}$$

Now, since δ=z-d and Ai=πδRp the mean area of contact will be given by:

$$\pi R_P \int_{d}^{\infty} (z-d)\phi(z)dz \tag{2.36}$$

and the expected total area of contact will be given by,

$$A_t = \pi N R_P \int_{d}^{\infty} (z-d)p(z)dz \tag{2.37}$$

Similarly, the expected total load in this case is given by,

$$P_t = \frac{4}{3} N E_c \sqrt{R_P} \int_{d}^{\infty} (z-d)^{3/2} \phi(z)dz \tag{2.38}$$

Describing heights in terms of the standard deviation of the height distribution, σ, and introducing the surface of asperity summits per unit area: N=ηAn, where η is the density of asperities, allows for expressing the number of contact spots, total area of contact and the load in the standardized form,

$$\begin{aligned} n &= \eta A_n F_o(D) \\ A_t &= \pi \eta A_n R_p \sigma_p K_1(D) \\ F &= \frac{4}{3} \eta A_a E_c R_p^{\frac{1}{2}} \sigma_p^{3/2} K_{3/2}(D) \end{aligned} \tag{2.39}$$

D in equation (2.999) is the standardized separation: D=d/σp, η is the density of the asperities, and Km(D) is a parabolic cylinder function given by:

$$K_m = \int_D^{\infty} \left(s - D\right)^m \phi^*(s)ds \tag{2.40}$$

The standardized height distribution, ϕ*, in equation 2.55 denotes the height distribution scaled so that its' standard deviation is unity[10].

The Greenwood Williamson model defines the contact through calculating three fundamental parameters (σp, Rp and η). These parameters are essentially similar to those resulting from the Hertz model, except that the surface parameters σp and η result from statistical treatment of the roughness.

Figure 14 (a and b) examines the relationship between some of the contact parameters and load. Figure 14-a, depicts the relation of the separation distance to the load whereas Figure 14 b plots the relation between the area of contact and the load. The separation in Figure 14-a, is expressed in non-dimensional form obtained through dividing the separation by the standard deviation of the roughness.

Figure 14. Effect of load on contact parameters (a) relation between separation and load, curve shows expected value of the load for a given separation assuming Gaussian distribution of asperity heights. (b) Relation between real area of contact and load. Values recalculated for SI units from Greenwood and Williamson (1966).

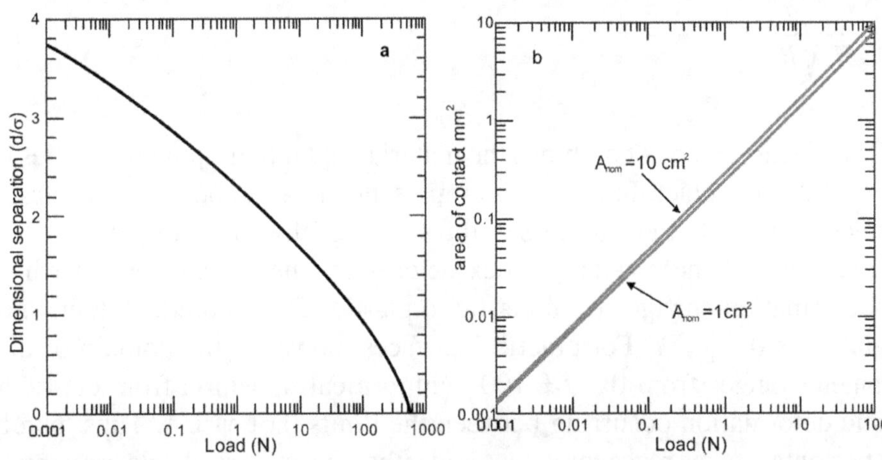

The plot depicts the expected value of the load for a given separation assuming Gaussian distribution of asperity heights. Data imply weak dependence of the separation on nominal load. For an area of 100 mm2, an increase in load from 10 KPa to a heavy pressure 1MPa reduces the separation from 2.6 σ to around 0.9 σ, which indeed is a modest change relative to the change of the load.

Figure 14-b shows that the nominal area of contact does not affect the real area of contact. Both plots in the figure (for 100 mm2 and 1000 mm2) yield almost identical real area of contact at the same loading. Noting that the calculations were performed under the assumption of ideal plastic flow, Greenwood and Williamson suggested that an "elastic contact hardness" controls the real area of contact in elastic contact situations. This parameter, they proposed, is equivalent to $0.25E_c\sqrt{\sigma / R}$. The elastic hardness plays an equivalent role to that of the hardness in describing elastic contact conditions. It allows precise calculation of the area of contact directly from knowledge of the limiting load necessary to exceed elastic contact conditions.

As mentioned earlier, for two curved bodies in contact the non-interference assumption does not apply. In this case, Hertzian contact theory represents the upper limit. The real pressures acting on individual contacts, however, will still be proportional to the elastic contact hardness. This implies that elastic contact theory still applies in situations that where partial plastic deformation takes place. To this end Greenwood and Williamson proposed so-called plasticity index (ψ) given by:

$$\psi = \frac{E_c}{H}\sqrt{\frac{\sigma_p}{R_p}} \tag{2.41}$$

The plasticity index combines the material and topographic properties of both solids in contact. In essence, it represents the ratio between the elastic contact hardness to the hardness of the softest of the contacting pair.

The value of the plasticity index determines the critical load at which the deformation changes from elastic to plastic. The theoretical domain of the index is 0£ ψ £ ¥. For practical surfaces, however, the domain of that parameter ranges from 0£ ψ £ 100, with critical transition from elastic to plastic deformation occurring between the limits 0.6£ ψ £ 1. Thus, purely elastic contact takes place for values lower than 0.6 and purely plastic contact takes place when the index exceeds unity.

Adhesive Contact Between Elastic Bodies

So far, the discussion was concerned with the contact of complying static engineering surfaces. If the touching surfaces are to slide relative to each other, they must first overcome a resistive tangential or frictional force, Ft. As such, it is clear that this frictional force must arise from the force interactions between the two surfaces that act only at the contact spots. Contacting surfaces in relative dry sliding are influenced by two factors: the interfacial bonds at the contact interface, and the deformation of the material in and around the contact regions.

The first factor has been found to have a strong influence on the friction force. This influence depends on both surface cleanliness and the compatibility of the sliding materials (Buckley 1981, (Rabinowicz 1971). It also contributes directly to the frictional energy through the irreversibility associated with making and breaking of adhesive bonds within the real area of contact.

The second factor includes two sources of plastic deformation, that due to the ploughing of hard surface asperities through the surface of a softer material (Bowden 1964) and that due to near surface plastic deformation owing to adhesive surface tractions (Rigney 1979). Although the first source, ploughing, has long been considered an important contributor to sliding friction, the second has come into prominence especially with the advent of micro and nano devices. In fact, currently, there is strong emphasis on the characterization of the exact relations that govern adhesive and small-scale tractions between surfaces (see for example Bhusan (1995) and the references therein). In all, both sources contribute to frictional energy dissipation through the work done in plastic deformation of the two sliding materials and of the material making up any third body present at the sliding interface (i.e., transfer films, wear debris, etc.,). The deformation contribution is therefore a volumetric one, in contrast with that of adhesion, which occurs only on the surface.

Adhesion takes place at the interface between two contacting solids and in contact situations comprising two solids interposed with liquids. The strength of adhesion depends on the state of the surface of the two solids. If both surfaces are clean of chemical films and adsorbed layers, strong adhesion (i.e. bonding between the two solids) may take place. The presence of surface contaminants or films reduces the strength of adhesion in general.

In general, adhesion may originate from physical or chemical interactions. Chemical interactions entail covalent, ionic (electrostatic) bonds, and metallic bonds. Physical interaction, on the other hand, entails intermolecular forces than manifest as hydrogen bonds and van der Waals bonds. In all, adhesion is a function of a multitude of factors, which include material pair, interfacial conditions, surface cleanliness, crystallographic structure, temperature, and duration of contact, among other factors.

The real area of contact between the solids influences the strength of adhesion as well. As explained previously, the real area of contact between two solids depends on the normal load, the roughness of the two surfaces, and the mechanical properties. In essence, the force of adhesion increases linearly with an increased normal load and an increased area of contact. The increase in adhesion due to an increase in area of contact is common when materials with low modulus of elasticity and less roughness establish contact. Soft materials (such as most of the skins or biological attachments) creep under load and that increases their area of contact and thereby adhesion. When an elastomer contacts a solid surface, and as a result of interatomic attraction (van der Waals, vdW, forces the real area of contact increases.

Adhesion plays an important role in the Biological world. Mechanical interaction of animals with their environment influences locomotion. The principle challenge for most insects and reptiles is to regulate and optimize their locomotry forces through complex irregular environments. Optimization of forces often involves the application of capillary and regular forces through dry or wet attachment systems of various shapes. One famous example is the Geko, which depends on a hierarchical arrangement of setae to regulate Van der Waals forces and thereby changes strength of adherence according to the contacting surface (Gorb 2008). Many engineering problems, such as bio-fouling, arise from the detrimental effects of biological adhesion.[11] Analysis of the micro-mechanisms of contact and the response of surfaces to bio-adhesion requires inclusion of adhesive effects in contact analysis (i.e. application of the models used to investigate contact between solids under the combined effects of a normal and tangential loading).

Van der Waals Interactions

The van der Waals force, generally attractive in nature, is a short-range force that decays rapidly to zero away from a surface. This force is a sum of non-covalent forces (intermolecular) forces. These forces, which act between stable molecules, are weak compared to those appearing in chemical bonding (Stone.1996). All atoms and molecules, even in an inert gas such as helium and argon, exhibit weak, short-range attractions due to vdW forces. Friction, surface tension, viscosity, adhesion and cohesion, are also related to vdW forces (Bhushan 1995), (Kim 2008). These phenomena arise from fluctuations in the electric dipole moments of molecules, which becomes correlated, as the molecules come closer together, giving rise to an attractive force (Leckband 2001).

In 1893, Johannes D. van der Waals (1837–1923) (Van der Waals 1893) developed a thermodynamic theory of capillarity to explain the behavior of liquids, after having introduced unspecific forces for gas molecules. He established the minimization of free energy as the criterion for equilibrium in a liquid-gas system and applied this to surface tensions, introducing the long-range vdW forces as resulting from dipole and quadrupolar interactions between molecules that make up gases, liquids or solids (French 2000). Van der Waals forces are the general name given to a set of forces characterized by the same power dependence on distance, having the dipole moment and the atomic polarizability as the important parameters (Garbassi 1994). They include three forces of different origins (Tabor 1969): Keesom, Debye, and London forces.

The first contribution, Keesom force (after Willem Hendrik Keesom (1915), is due to electrostatic interactions between charges (in molecular ions), dipoles (for polar molecules), quadrupoles (all molecules with symmetry lower than cubic), and permanent multipoles. Often, these forces occur between two molecules with permanent dipoles, and are temperature dependent.

The second contribution is the induction or Debye force (also known as polarization (Gerschel), arises from interactions between rotating permanent dipoles and from the polarizability of atoms and molecules (induced dipoles). These induced dipoles occur when one molecule with a permanent dipole repels another molecule's electrons. The forces between induced and permanent dipoles are not as temperature dependent as Keesom interactions because the induced dipole is free to shift and rotate around the non-polar molecule. The Debye induction effects and Keesom orientation effects are referred to as polar interactions.

The third and dominant contribution is the dispersion or London force (fluctuating dipole-induced dipole) (London,1930), due to the non-zero instantaneous dipole moments of all atoms and molecules. Such polarization can be induced either by a polar molecule or by the repulsion of negatively charged electron clouds in non-polar molecules. London interactions are caused by random fluctuations in electron density in an electron cloud. An atom with a large number of electrons will have a greater associated London force than a smaller atom. The dispersion (London) force is the most important component because all materials are polarizable, whereas Keesom and Debye forces require permanent dipoles. The London interaction is universal and is present in atom-atom interactions as well. Hamaker (1937) developed the theory of vdW between macroscopic bodies in 1937 and showed that the additivity of these interactions renders them considerably more long-range. For two parallel surfaces separated by a distance d, the force per unit area is

$$Fvdw = \begin{cases} \dfrac{A}{6\pi d^3} & d \leq 30nm \\ \dfrac{B}{d^4} & d > 30nm \end{cases} \tag{2.42}$$

Table 5. Values of Hamaker Constant (A/10-20J) for some materials

Materials	Vacuum	Water
Polystyrene	7.9	1.3
Hexadecane	5.4	-
Gold	40	30
Silver	50	40
Al2O3	16.75	4.44
Copper	40	30
Water	4	-
Pentane	3.8	0.34
Decane	4.8	0.46
Hexadecane	5.2	0.54
Water	3.7	-
Quartz (Fused)	6.5	0.83
Quartz (Crystalline)	8.8	1.7
Fused Silica	6.6	0.85
Calcite	10.1	2.23
Calcium Fluoride	7.2	1.04
Sapphire	15.6	5.32
Poly (Methyl Methacrylate)	7.1	1.05
Poly (Vinyl Chloride)	7.8	1.3
Polyisoprene	6	0.74
Poly (tetrafluoroethylene)	3.8	0.33

(Hamaker 1937).

where A and B are the Hamaker constants. Table 5 provides representative values of the Hamaker constant for some materials (Hamaker 1937), whereas Table 6 gives the variation of the Van der Waal force versus particle diameter.

Table 6. Variation of Forces (N) versus particle diameter, d (μm)

Diameter	Air	Water
d (μm)	Fv	Fv
0.2	3 x 10-8	2x10-9
2	3 x 10-7	2x10-8
20	3 x 10-6	2x10-7

Free Surface Energy

A detailed calculation of van der Waals forces is difficult. A simpler approach is to use the concept of free surface energy. If one cleaves a crystalline solid along its cleavage plane, two highly chemically active surfaces are generated. The cleavage process causes the fracture of cohesive bonds across the cleavage interface, and these fractured bonds leave the surface in a highly energetic state. Thus, the energy that normally would be associated with bonding to other atoms like other atoms in the bulk solid is now available at the atoms on the surface. This energy required to create a new surface, expressed over an area consisting of many atoms in the surface lattice, is referred to as free surface energy. It is a function of the material as well as the surface orientation. Because the atoms at the surface have this unused energy, they can interact with each other, with other atoms from the bulk, and with species from the environment. Free surface energy influences adhesive bonds for solids in contact and, hence, friction and wear. In addition, it determines the nature of the interaction of lubricants with solids. When a bond is formed between two materials having free surface energies per unit area in air $(\gamma_{SA})1$ and $(\gamma_{SA})2$ or simply γ1 and γ2] in contact, the surface energy of the interface per unit area changes to γ_{12} . Based on early work by Bradley [75,76] work of adhesion or the energy of adhesion per unit area is defined as $\omega_{ad}=\Delta\gamma=\gamma_1+\gamma_2-\gamma_{12}$ where Δγ is equal to a reduction in the surface energy of the system per unit area always negative, in mJ/m2.

Thus, Δγ represents the energy that must be applied to separate a unit area of the interface or to create new surfaces. For two similar materials, Δγ becomes the work of cohesion, and is equal to 2γ (γ_{12}=0). The higher the surface energy of a solid surface, the stronger the bonds it will form with a mating material.

Analysis of Contact with Adhesion

In typical elastic contact problems (Figure 15), three macroscopic contact variables (load, penetration, and contact radius) are the object of the analysis. The expressions obtained by Hertz (see Table 2) dominated the field of analysis perhaps until the early years of the twentieth century. The advent of thermoplastics in the 1930's triggered considerable interest in the essence of bonding between polymeric particles. This is because successful applications of the resulting polymers required a suitable level of adhesion to enhance bonding, printing, and coating. Consequently, initial efforts to model adhesive contacts focused on calculation of adhesive forces between small particles, which were modelled as smooth spheres that do not undergo any deformation upon contact.

Bradley considered the full interaction potential between two surfaces, including attractive and repulsive contributions to the potential. Summation of the interaction stresses over the curved bodies, assumed rigid, provides the interaction force. When surface deformation is taken into account, the Bradley method evolves into the so-called full self-consistent method (Bradley1936). Bradley showed that if two rigid spheres of radii R1 and R2 are placed in contact, they will adhere with a force 2πΔRγ, where R is the equivalent radius (R=R1R2/(R1+R2)) and Δγ is the surface energy or 'work of adhesion' (Δγ =γ1+γ2-γ12).

Derajaguin (1979) was the first to consider the problem of adhesive contact with deformation. He considered the attractive part of the potential but treated the repulsive part differently. Further, he assumed that this repulsive contribution is so steep that the surfaces cannot interpenetrate (i.e. contact deformations do not increase the adhesion force needed to interrupt the contact). As a result, the boundary conditions specified the surface normal displacement instead of the surface normal stress inside the contact zone, which permitted the use of the Hertz model for the adhesion-less elastic contact. One drawback of this earlier attempt was neglecting the energy of the noncontact adhesive forces acting within the ring-shaped zone located immediately outside the real area of contact (see Figure 16 and 17). To date, the models of Bradely and Derjanguin along with their contrasting foundations, continue to define theoretical efforts in that topic. This early work of Derajaguin went un-noticed for quite some time. However, in the late sixties of the twentieth century some experimental contradictions to Hertz theory were reported in literature during experiments with smooth rubber spheres, glass spheres (Derajaguin 1934). In particular, at low loads the observed contact areas were considerably higher than the predictions of Hertz theory. Furthermore, the contact areas tended towards a finite value as the load reduced to zero. At high loads, however, the discrepancy with the predictions of Hertz disappeared. These observations suggested that attractive surface forces are of significant influence at light contact loads, whereas their influence diminishes considerably as the load was increased. Further, the influence of the operative surface forces peaked when the nominal loads were reduced towards zero. Perhaps unaware of earlier contribution by Derajaguin, Johnson, Kendall, and Roberts, JKR, (Kendall 1971) introduced a theory that accounts for surface energy effects in the contact of elastic bodies. The JKR theory proved the existence of a finite area of contact between surfaces even under zero load. The theory also predicted the external force required to separate two bodies of given surface energy and geometry.

Figure 15. Contact between elastic sphere and hard flat surface with no applied force: a) in the absence of attractive forces between the two bodies, b) in the presence of attractive forces, surfaces are drawn together to make contact over a circle of radius a, and c) pressure distribution in the presence of attractive forces.

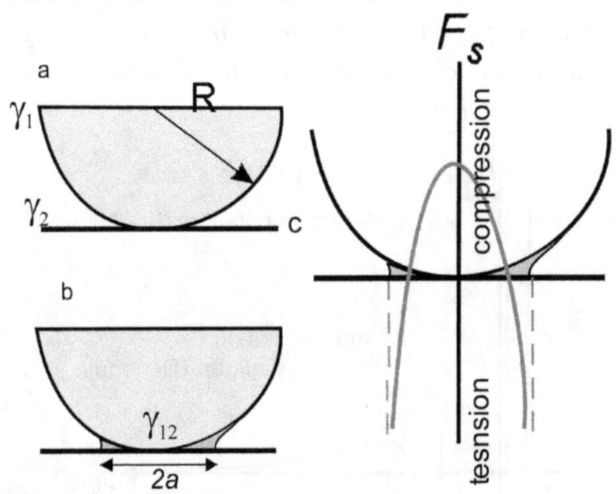

Figure 16. Derjaguin approximation, a: The geometry of the Hertzian contact. b: interaction between curved surfaces before contact.

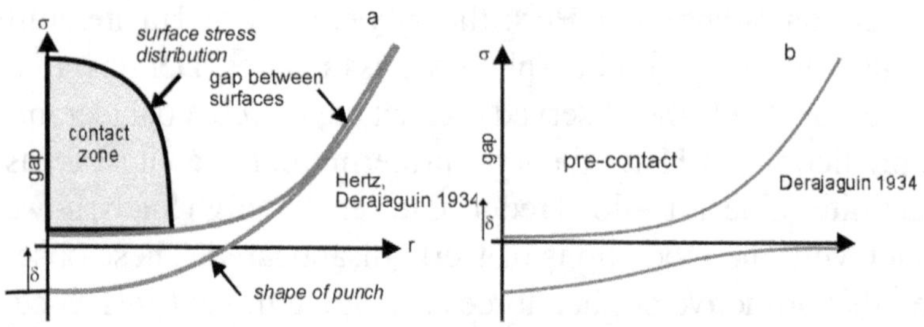

Derjaguin, Muller and Toporov followed the work of Johnson and coworkers in 1975 (Derjaguin 1975) and introduced another model (DMT) that considered that molecular force interaction outside the contact zone. The JKR model and the DMT model appeared at first to contradict each other. Through a Griffith energy argument (assuming that contact over a circle of radius a, introduces a surface energy-πa2Δγ) the JKR theory shows that the force needed to separate two spheres is equal to (3/2) πΔγR. This value being independent of the elastic modulus appears to be universally applicable. However, this value contradicted Bradley's estimation (2πΔγR). In 1977, Tabor (1997) explained this discrepancy through identifing a parameter that governs the transition from the Bradley pull-off force to the JKR value.

Figure 17. Typical interaction potential between °at surfaces (Bradley, full self-consistent model) and approximation (dashed) to the repulsive part (Derjaguin, restricted self-consistent model) V(z) the interaction potential between flat surfaces of unit area separated by a distance z

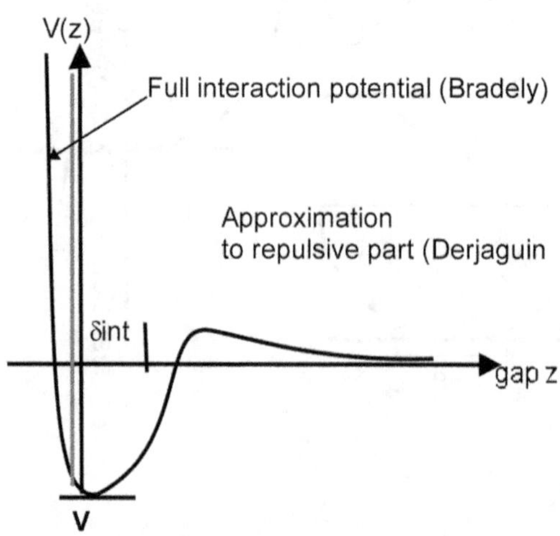

Subsequently Muller et al. (1980) performed a comprehensive numerical study of the problem in terms of surface forces rather than surface energy, (combining the Lennard–Jones law of force between surfaces with the elastic equations for a half-space). The results confirmed the validity of Tabor's parameter and his argument about transition. Maugis (1992) provided a semi-analytical solution to the contact problem with interactions inside and outside the contact zone using the Dugdale approximation and showed the transition from DMT to JKR behavior as two opposite ends of a continuous spectrum based on a parameter λ (the Maugis parameter), which is equivalent to the Tabor parameter ($\lambda = 1.16\mu T$). Since then, Greenwood and Johnson (2006) developed simple algebraic equations, which conveniently extend the analysis to viscoelastic contacts. Table 7 provides a comparison between the major assumptions used in developing the well-established models in the field, whereas Table 8 provide a summary of the formulas resulting from each model.

Figure 18. Interactive forces for the Hertz, JKR, DMT, MD and MYD models. The Hertz model does not consider the adhesion in contact. The JKR model includes only a short-range adhesion in the contact area, actually it is a function with the work of adhesion. The DMT model shows a long range surface force which acts outside the radius of the circle of contact. The MD model considers the Dugdale (square well) potential to describe attractive forces. The MYD model, however, includes both the short-range and long-range forces acting both inside and outside the circle of contact, just like the Lennard-Jones potential.

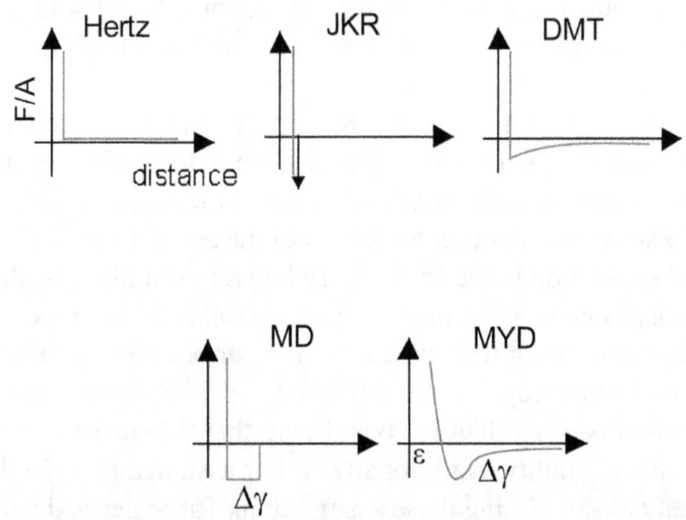

Table 7. Comparison of the various contact theories

Theory	Assumptions	Limitations
Hertz (a)	No surface forces	Not valid for low loads if surface forces are present
JKR (b)	Short-ranged surface forces act only inside contact area Contact geometry allowed to deform Applies to high λ_s systems only	May underestimate loading due to surface forces
DMT (c)	Long-ranged surface forces act only outside contact area	May underestimate contact area due to restricted geometry Applies to low λ_s systems only
Maugis–Dugdale (d)	Periphery of tip–sample interface modeled as a crack that fails at its theoretical strength	Analytical solution, but parametric equations Applies to all values of λ_s

Transition Between Adhesive Regimes

Tabor Coefficient

The pull-off force in case of the JKR model, does not depend on the elastic modulus of the materials involved. Rather, it depends only on the equivalent radius of curvature and the work of adhesion so that it appears to be universally applicable. As such, it should also apply to the adhesion contact of rigid spheres. Moreover, it should yield similar predictions to those of the DMT model. Since the expressions differ in the constants, 3/2 and 2 respectively, then for the same work of adhesion the separation load will be different[12]. This apparent controversy was resolved by Tabor (1997)] who established that expressions resulting from the JKR theory and the DMT theory express the extremes of two contact regimes.

Upon comparing both theories, JKR and DMT, Tabor concluded that the main defect of the later was neglecting the deformation resulting from adhesive forces outside of the contact[13], whereas the defect of the former was to neglect these very forces (i.e. forces out of the contact zone). Considering the height of separation in the JKR theory he proposed that one should pass from one theory to another continuously without singularities. Tabor simplified the analysis by assuming that the contacting surfaces are so smooth to the extent that molecular contact is established over the entire contact region. This assumption is equivalent to considering that the surfaces involved are initially of optical quality. Additionally, Tabor assumed that the deforming solid is ideally elastic. Using these assumptions Tabor derived a parameter, μ, given by

Table 8. Summary of contact parameters for established adhesive contact theories

Theories	Contact Radius a	Relative Approach δ	Pull-off Force Pc	Closure Condition
	$K=\frac{4}{3}\left(\frac{\left(1-\nu_1^2\right)}{E_1}+\frac{\left(1-\nu_2^2\right)}{E_2}\right)^{-1}$, $\omega_{ad}=\Delta\gamma=\gamma_1+\gamma_2-\gamma_{12}$, P (external load)			
Hertz	$\left(\frac{RP}{K}\right)^{\frac{1}{3}}$	$\frac{a^2}{R}$		$\sigma(a)=0$
JKR	$\left[\frac{R}{K}\left(P+3\omega_{ad}\pi R+\sqrt{\left(6\omega_{ad}\pi RP+(3\omega_{ad}\pi R)^2\right)}\right)\right]^{\frac{1}{3}}$	$\frac{a^2}{R}-\sqrt{\frac{8\omega_{ad}\pi a}{3K}}$	$-\frac{3}{2}\omega_{ad}\pi R$	$\delta U=0$ U (P.E)
DMT	$\left[\frac{R}{K}\left(P+2\omega_{ad}\pi R\right)\right]^{\frac{1}{3}}$	$\frac{a^2}{R}$	$-2\omega_{ad}\pi R$	
Maugis-Dugdale	$\bar{a}^3-\lambda\bar{a}^2\left(\sqrt{m^2-1}+m^2\tan^{-1}\sqrt{m^2-1}\right)=\bar{P}$ $\bar{a}^2-\frac{4}{3}\lambda\bar{a}\sqrt{m^2-1}=\delta$ $\frac{1}{2}\lambda\bar{a}^2\left(\sqrt{m^2-1}+(m^2-2)\tan^{-1}\sqrt{m^2-1}\right)+\frac{4}{3}\lambda\bar{a}^2\left(\sqrt{m^2-1}\tan^{-1}\sqrt{m^2-1}-m-1\right)=1$ $\bar{a}=a\left(\frac{K}{\pi\omega_{adh}R^2}\right)^{\frac{1}{3}}$, $\bar{c}=c\left(\frac{K}{\pi\omega_{adh}R^2}\right)^{\frac{1}{3}}$, $\bar{P}=\frac{P}{\pi\omega R}$, $\bar{\delta}=\delta\left(\frac{K^2}{\pi^2\omega_{adh}^2R}\right)^{\frac{1}{3}}$, $\lambda=2\sigma_o\left(\frac{R}{\pi\omega_{adh}K^2}\right)^{\frac{1}{3}}$, $m=\frac{c}{a}$		$-n\omega_{ad}\pi R$, $\frac{3}{2}<n<2$	$K_I+K_m=0$

$$\mu=\left(\frac{R\Delta\nu^2}{E_c^2 z_0^3}\right)^{\frac{1}{3}} \tag{2.43}$$

where zo is the equilibrium spacing in the Lenard-Jones potential and Ec is the composite modulus. The Tabor parameter defines the bounds of two regimes. The first represents the adhesive contact of stiff solids, solids with small radii of curvature and contact under weak energy of adhesion. Whereas the second manifests cases of compliant solids, contact under the influence of large adhesion energy, and contact of solids with large radii of curvature. For the first regime, the parameter μ is considerably less than unity (i.e. $\mu \ll 1$) and the DMT theory applies. For the second regime, however, the tabor parameter μ is greater than unity (i.e $\mu \gg 5$) and the JKR theory applies.

Barquins and Maugis [87] used the Sneddon solution for the contact of axi-symmetric punches to analyze Transition between the JKR and DMT theories. They showed that if the punch undergoes a rigid body displacement then the singularities of stresses and the discontinuities of the displacements, which appear are those of fracture mechanics. The stress intensity factor, under this condition KI, is proportional to the rigid body displacement.

Maugis (1999) improved Tabors' idea of smooth transition. To this end, he represented the surface force in terms of a Dugdale cohesive zone approximation such that the work of adhesion is given by

$$\Delta\gamma = \sigma_o h_o \tag{2.44}$$

where σo is the maximum force predicted by the Lennard-Jones potential and ho is the maximum separation obtained by matching the areas under the Dugdale and Lennard-Jones curves. This implies that the attractive force is constant for the interval $zo \leq z \leq zo + ho$, and there is no further penetration in compression (i.e. extreme rigidity in compression). Perfect contact, therefore, occurs in an area of radius a, and adhesive forces of magnitude σo extend to an area of radius c >a. In the region a<r<c, the two surfaces are separated by a distance h(r) with h(a)=0 and h(c) =ho. By so doing, Maugis recovered the JKR-DMT transition analytically. Maugis studied the transition in terms of a parameter, λ, given by,

$$\lambda = \frac{2\sigma_o}{\left(\pi\Delta\gamma K^2 / R\right)^{1/3}} \tag{2.45}$$

The parameter λ, similar to Tabors parameter μ, represents the ratio of elastic to deformation to the range of surface forces, and is related to μ by (l=1.16 m). According to Maugis criterion, the JKR and the DMT theories represent two extremes for $\lambda \to \infty$ and $\lambda \to 0$ respectively (see Figure 20).

Figure 19. JKR–DMT transition. Surface stress distribution for P = 3πwR and for various values of the parameter λ. JKR–DMT transition. Surface stress distribution for P = 3πΔγR and for various values of the parameter λ.JKR–DMT transition. Surface stress distribution for P = 3πΔγR and for various values of the parameter λ.

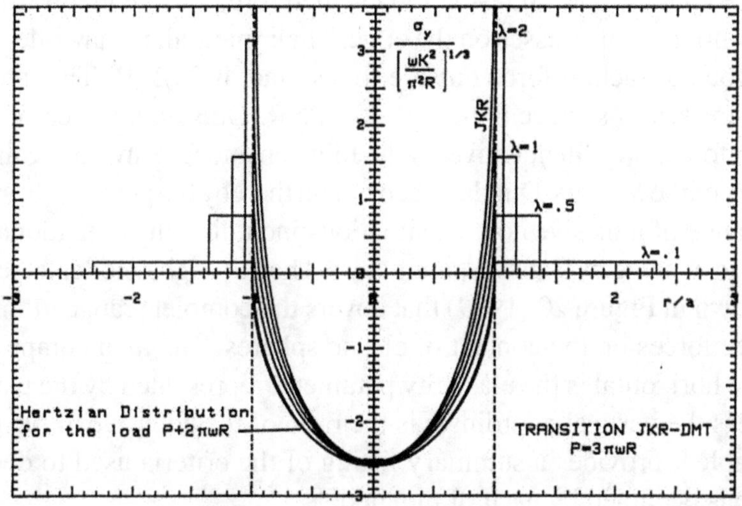

Figure 20. Adhesion map depicting the range of application of the ewll established adhesive contact models in terms of the Tabor parameter.

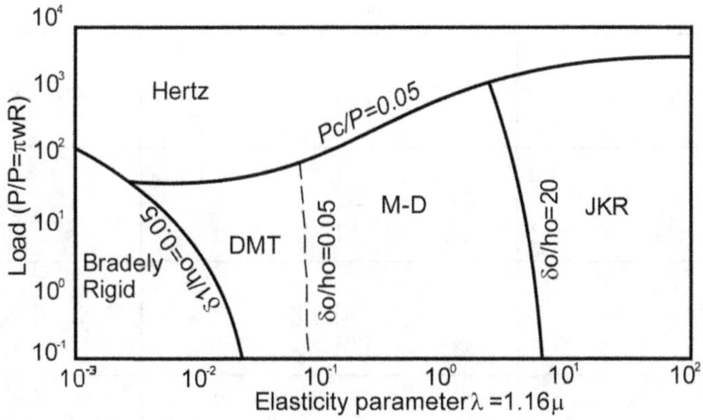

In the JKR theory, adhesion stresses outside the area of contact are reduced to a peak infinitely narrow of zero measure. This is equivalent to reporting the joining point of stresses on the edge at infinity. The total adhesion force outside the contact being zero, the integral of stresses in the area of contact is equal to the applied load P, and negative loads are sustained by tensile elastic stresses. The theory of DMT assumes Hertzian stress distribution within the contact. The adhesion forces, therefore, must tend toward zero to ensure continuity of stresses on the edge. Their integral, meanwhile, is finite and the total attraction force outside the contact is $2\pi\Delta\gamma R$. The integral of normal stresses in surface is then $P + 2\pi\Delta\gamma R$. Outside the area of contact adhesion forces sustain negative loads. Johnson and Greenwood realized the generality of the Muagis-Dugdale theory, and thereby the potential to studying a wide range of adhesive contact situations including the transitional zones difficult to analyze by Tabors' parameter). They used the model to develop a map (shown in Figure 20 (1999) that covers the complete range of influence of surface forces on the contact of elastic spheres. The map comprises two axes. The horizontal is the elasticity parameter represented by the parameter λ, whereas the vertical coordinate is the ratio of the adhesive to the pull-off force. Table 9 provides a summary listing of the criteria used to determine the regions occupied by each of the models.

Table 9. Criteria for well-established adhesive contact models

Zone	Criterion	
Hertz	$\left\lvert \frac{P_a}{P} \right\rvert < \xi$	$\xi \ll 1$
JKR	$\left\lvert \frac{\delta_a}{h_o} \right\rvert > \eta$	$\eta \gg 1$
DMT	$\left\lvert \frac{\delta_a}{h_o} \right\rvert < \eta$	$\eta \ll 1$
Bradely (Rigid)	$\left\lvert \frac{\delta_1}{h_o} \right\rvert < \zeta$	$\zeta \ll 1$
Separated Surface	$\frac{\delta_a}{h_o} = \frac{\bar{P}}{2} - 1$	

Contact Analysis of Biological Attachments

Adhesion plays a fundamental role in the biological world. It ensures locomotion on many tribologicaly challenging surfaces, it aids in survival functions such as prey capturing and defense. In the past decade or so, the topic of adhesion within biological systems received considerable attention. To date, however, the exact physical mechanisms underlying the precise function of many biological attachments are still unsettled. This is because many species, insects, gekos, spiders, and snakes, have the ability to engage a complex array of structural, material, and surface topology features in synchronized function.

Traditionally, the analysis of the mechanics of contact in biological attachments borrowed from the mechanics of solids. In particular, the statistical provisions made to simplify the analysis of manufactured rough surfaces have affected the analysis of biological attachments. One example is the assumption of hemispherical capped attachments, which mimics the classical assumption of asperities in a solid surface in the Greenwood Williamson model. Adopting a generalized assumption about the geometry of the attachment tips helped mask the mechanistic variations among different species. Several microscopic studies reveal the existence of a vast array of contact shapes (attachment tips) (see Figure 22). Most of the attachment terminal ends differ considerably from spherical. Therefore, to pinpoint the operative micro-mechanisms during function of the species, there is a need to adjust the applied models to reflect the diversity observed in nature.

In the realm of biological contact mechanics (similar to elastic contact problems), the analysis relies on the JKR and the DMT theories. For biological attachments, the ranges of accessible parameters fall within the transition zone between the two theories (Figure 23). Howerver, despite such a fact, generally the JKR theory is favored. This is because the JKR theory offers the advantage of its framework being consistent with linear fracture mechanics (Arzt 2002).

Two approaches apply to the analysis of biological contacts. The first is thermodynamic in essence. It entails minimization of the balance between the elastic and the surface energy at the contact. The second entails modelling the detachment of a contact as an unstable crack. This allows application of the theory of linear elastic fracture mechanics and thereby calculating the stress intensity factor KI assuming plane strain conditions.

The JKR theory assumes that short-range contact forces act only inside the contact area. The contact geometry differs from the original Hertzian shape without adhesive forces. At the point of pull-off the incremental decrease in elastic energy Uelastic is counterbalanced by the incremental increase in surface energy Usurface, such that

$$\frac{dU_{elas}}{dA} + \frac{dU_{surf}}{dA} = 0 \tag{2.46}$$

where A is the true contact area of a single contact. When the elastic energy gained through detachment by dA equals the work of adhesion, the contact will become unstable and will detach.

Figure 21. Examples of fibrillar arrays fabricated by different methods (SEM micrographs). Direct nanofabrication without the use of a template by e-beam lithography (a)[5] or chemical vapor deposition (b) Replication of templates by means of molding, using either porous membranes (c)micromachined (d,e),etched (f),or photolithographic templates (h–l). Copyright © 2010 WILEY-VCH Verlag GmbH & Co. KGaA, Weinheim. Reproduced with permissions from, Kamperman, M, Kroner, E, del Campo, A, McMeeking, R. M., Arzt, E, Functional Adhesive Surfaces with "Gecko" Effect: The Concept of Contact Splitting, Adv. Eng. Mater. 12 (5), 335-348, 2010.

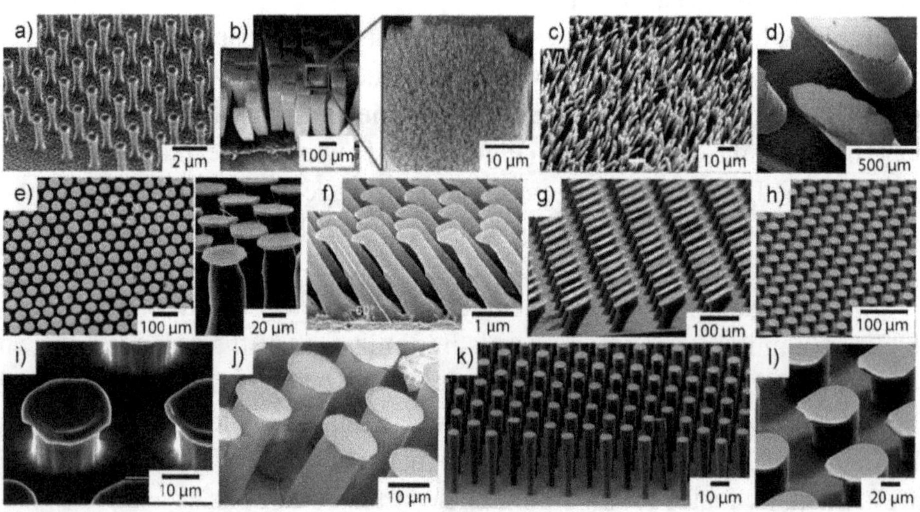

Figure 22. Schematic depicting the transition between the DMT model and the JKR model: Tabor's parameter μ as a function of contact radius R and effective modulus. The solid line indicates the transition between the DMT and the JKR theory. The two ellipses delineate roughly the range of parameters in natural attachment devices and the range accessible by artificial attachment systems.

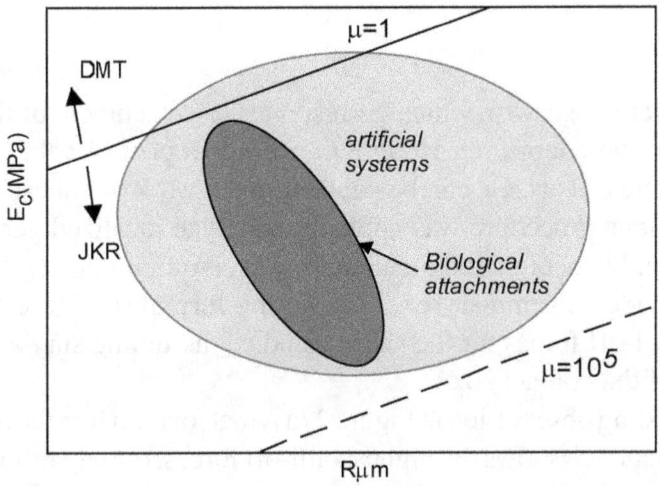

Figure 23. Side and plan view of contact shapes studied: (a) sphere, (b) flat punch/ vertical cylinder, (c) torus/horizontal cylinder, (d) suction cup, (e) elastic band (in each case, the shade area denotes contact).

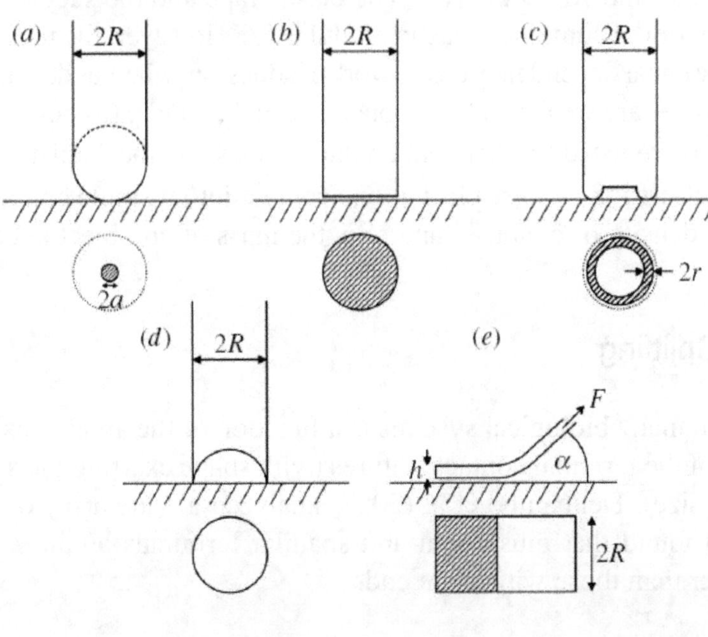

Alternatively, one can apply the principles of linear-elastic fracture mechanics. For plane strain, the stress-intensity factor KI relates to the energy release rate G in the following way:

$$G = \frac{1}{2E_c} K_I^2 \tag{2.47}$$

Unstable crack growth, which corresponds to detachment of the contact, occurs when the energy release rate G exceeds a critical value Gc. In the absence of rate effects, Gc can be equated to the work of adhesion γ.

The previous procedure was applied to analyze idealized geometries of some frequently encountered attachment pads (Autumn 2002) (Figure 23). Table 9 provides a summary of the resulting formulas. Figure 5 plots the predicted pull-off forces for the shapes studied, assuming single contact, as a function of the contact size).

An interesting observation in Figure 27a is that for radii R in the micrometre range the flat punch shows the highest pull-off forces, closely followed by the torus. A spherical contact would have a greatly diminished adhesion force, roughly by a factor of three with regard to the torus. Under the numerical assumptions made, these differences would decrease for smaller contact sizes and become negligible for R < 100 nm.

Table 11, presents a summary of the functional dependencies and scaling for each of the studied geometries. The elastic tape and the suction cup are independent of the combined elastic modulus Ec. However, the tape and the sphere show linear dependency on the work of adhesion, whereas dependencies for other shapes are weaker. The exponents s, m, k, r and p (as defined in the first column) are listed for different contact shapes and mechanisms. P_c^{tot} is the total pull-off force after splitting one contact into n single contacts. NA is the areal density of contacts; and m is the mass of an object adhering to the ceiling.)

Contact Splitting

Adhesion in many biological systems is a function of the areal density, size and shape of the terminal contact features (with shape exerting the strongest effect than size). Del campo et al (2007) analyzed a wide array of contact shapes and found that mushroom and spatular terminals attain values 30 times higher than those with a flat end.

Table 10. Predicted pull-off forces for the shapes in Figure 27.

	Load P1=	Stress Intensity Factor KI=	Pull Off Force Pc=
Sphere	$\frac{4}{3}\frac{\pi E_c a^3}{R}$	$\frac{P_a}{2a\sqrt{\pi a}}$	$-\frac{3}{2}\pi\gamma R$
Horizontal Cylinder and Torus	$\frac{\pi E_c a^2}{4R}$	$\frac{P_a}{\sqrt{\pi a}}$	$-\frac{3}{2}\left(\pi\gamma^2 E_c R/2\right)^{\frac{1}{3}}, r << R$ $-3\pi R\left(\pi\gamma^2 E_c r/2\right)^{\frac{1}{3}}, a = R$
Vertical Cylinder Flat Punch		$\frac{P_a}{2\sqrt{\pi a^3}}$	$-\left(8\pi\gamma^2 E_c R^3\right)^{\frac{1}{2}}$
Elastic Punch		$\frac{P_a}{2\sqrt{\pi a^3}}\beta, \beta \propto (R-a)$	$-\left(8\pi\gamma^2 E_c R^3\right)^{\frac{1}{2}}$
General Axisymmetric Profiles $h(r) = \frac{r^q}{R^{q-1}}$			$\propto R^{\frac{3(q-1)}{(2q-1)}} E_c^{\frac{(q-2)}{(2q-1)}} \gamma^{\frac{(q+1)}{(2q-1)}} + veq > 1,$ $-ve\ 1 \leq q \leq 2 \quad \propto R^{\frac{3}{2}} E_c^{\frac{1}{2}} \gamma^{\frac{1}{2}}, q \to \infty$
Film peeling			$-2\gamma R g(\alpha,\lambda)$ $g(\alpha,\lambda) = \frac{2\sin(\alpha)}{\sqrt{(1-\cos\alpha)^2 + 2\lambda} + (1-\cos\alpha)}$
Suction cup			$\pi \Delta PR2$

a: peel angel,
λ:γ/Eh
h: film thickness over distance δx)

Arzt et al. (2003) pointed out that the pull-off force in case of a spherical terminal, which is given by,

$$Po = \frac{3}{2}\pi\gamma R \tag{2.48}$$

scales with a geometrical length, which gives rise to a size effect. To that effect replacing one large spherical contact of radius Ro say by N smaller ones of radius $R=R_0N^{-1/2}$ with the same total (Hertzian) contact area the pull-off force for simultaneous detachment of these N spheres will then be enhanced by a factor of $N^{1/2}$:

$$P_o = \frac{3}{2}\pi\sqrt{N}R_o\gamma \tag{2.49}$$

This concept is based on two assumptions. The first is the validity of the JKR model even though the ratio a/R grows rapidly as the spheres become smaller. An estimate shows that this equation over predicts somewhat the benefits of contact splitting when the hemispherical radii become small. The other critical assumption is that all spherical elements detach simultaneously. Instead, it is possible that the elements will peel (i.e., begin to detach at one side of the large array). This resembles steady state peeling of a fibrillar surface consisting of N hemispheres of radius R=RoN-1/2 backed by stiff platens. Peeling will occur when,

$$P_d = \frac{4}{3}\pi\sqrt{N}R_o\gamma \tag{2.50}$$

Comparing equations 2.50 and 2.51 indicates that peeling under the specified condition slightly affects the peel off force, and the benefit of contact splitting is retained.

Figure 24. Pull-off forces for single contacts for various shapes as a function of feature size. Parameters assumed: Ec = 1 MPa, γ = 50 mJ m−2, r = R/10, Δp = 50 kPa. (b) Same as part (a) except Ec= 1 GPa (all other parameters unchanged).

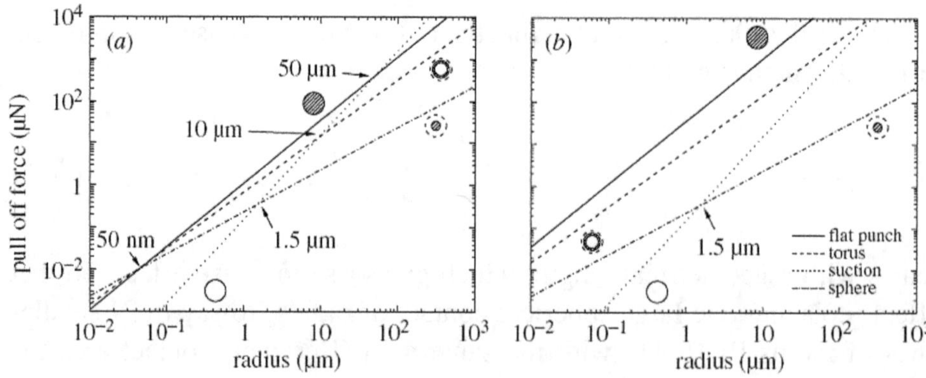

Table 11. Functional dependencies for various contact shapes

	Hemisphere	Torus	Elastic Punch	Flat Punch	Suction Cup	General
$P_c \propto R^s$	1	4/3	1	3/2	2	$\frac{3(q-1)}{2q-1}$
$P_c \propto E^m$	0	1/3	0	1/2	0	$\frac{q-2}{2q-1}$
$P_c \propto \gamma k$	1	2/3	1	1/2	0	$\frac{q+1}{2q-1}$
$P_c^{tot} \propto n^r P_c$	1/2, 1a	1/3, (1/2)a	1/2	1/4	0	$\frac{(q+1)}{2(2q-1)}$
$N_A \propto m^P$	2/3, (1/3)a	1, (2/3)a	2/3	4/3	¥	$\frac{2(2q-1)}{2(q+1)}$

Figure 25. (a) Comparison of pull-off forces for different shapes in terms of contact-splitting efficiency. The total apparent contact area is 100 cm2. Thin dashed lines: contact-invariant scaling (Arzt et al . 2003). Note that contact-invariant scaling must never cross the flat-punch curve. The horizontal dashed black line indicates the force necessary for human attachment. Parameters: Ec = 1 MPa, γ = 0.05 J m−2. (b) Same as part (a) except Ec = 100 MPa (allother parameters unchanged).

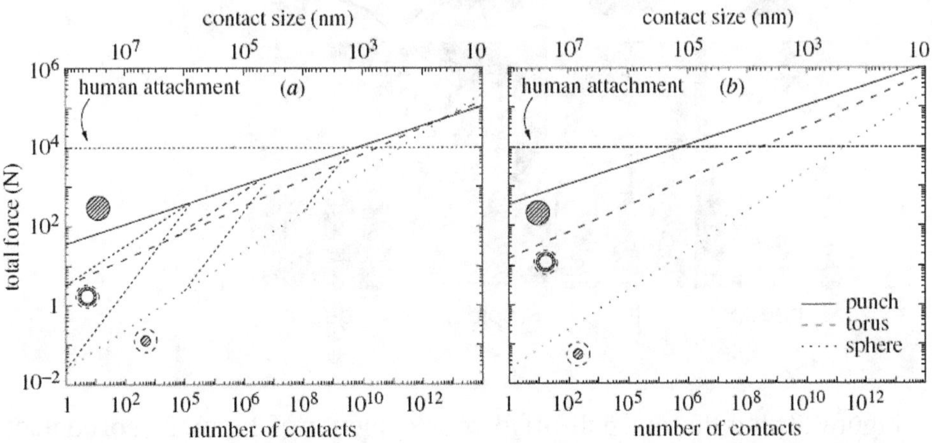

Shape of the terminal ends affect adhesion as well. Spolenak at al (2005) analyzed contact splitting effect for a number of shapes (refer to Figure 27). They assumed negligible friction at the interface and a modest degree of adhesion in order to apply the analysis of Kendall (1975). The pull-off load under the conditions of the analysis for a flat punch is given by,

$$P_o = \sqrt{8\pi E_c \gamma R^3} \tag{2.51}$$

The pull off-force, therefore, varies in proportion to N1/4 for N smaller fibrils with the same area of the fibrillar surface. It is seen that for the case of the flat punch, the contact splitting power is less than the case of a spherical terminal shape. Similar considerations for hemispheres suggest that little penalty is sustained as far as the pull-off force is concerned when peeling occurs through the forest of fibrils, rather than simultaneous detachment.

Figure 26. Terminal elements (circles) in animals with hairy design of attachment pads. Note that heavier animals exhibit finer adhesion structures
Arzt, E.; Gorb, S.; Spolenak, R., From micro to nano contacts in biological attachment devices", PNAS, 100, 19, 10603-10609, "Copyright (2003) National Academy of Sciences, U.S.A.", reprinted with permission.

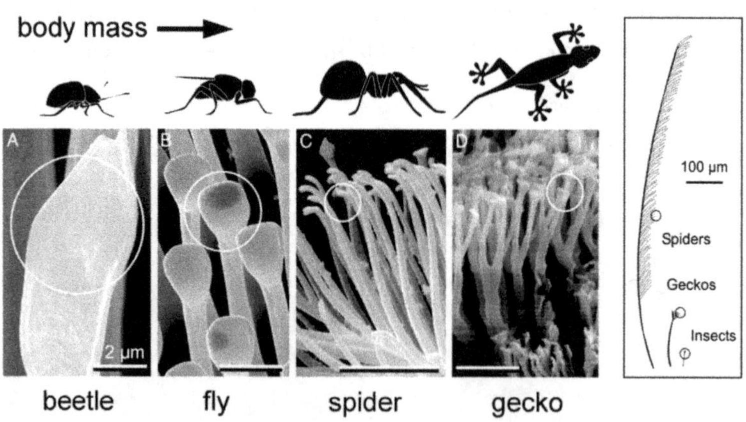

Figure 23 plots the total pull-off force as a function of the number of contact elements in an apparent contact area of 100 cm2. In Figure 28-a, the values for Ec and γ are the same as in Figure 28a. Shapes with lower pull-off forces as single contacts show higher contact-splitting efficiency so that the forces converge at contact sizes of ca. 100 nm. If one wanted to provide the force necessary for attachment of a weight comparable with a human being (e.g. 10 000 N, for safety), the flat punch or the torus would still be the preferred shapes. Note that over a small variation in contact size the splitting efficiency can be increased by using the criterion of curvature invariance. This changes the slope in Figure 4 from ½ to 1 for the spherical contact and from 1/3 to 1/2 for the toroidal case. However, the limit of this strategy is reached when the contacts have been pressed flat, i.e. at the line for the flat punch. Figure 28-b shows the same plot for Ec=100 MPa. Note that, as the flat punch and the torus depend on the modulus their pull-off forces increase significantly.

One interesting finding of Spolenaks' study is devising a single "master" curve that relates mass and density of the terminal pads given by,

$$\log N_A(m^{-2}) = 13.8 + 0.699 \log m(kg) \qquad (2.52)$$

A consequence of this equation is that increased mass requires finer adhesion structures (see Figure 27). The master equation (2.53) is plotted in Figure 19 for several groups.

Figure 27. Dependence of the terminal element density (NA) of the attachment pads on the body mass (m) in hairy-pad systems of diverse animal groups (logNA(m-2)=13.8 +0.699 logm(kg), R= 0.919).

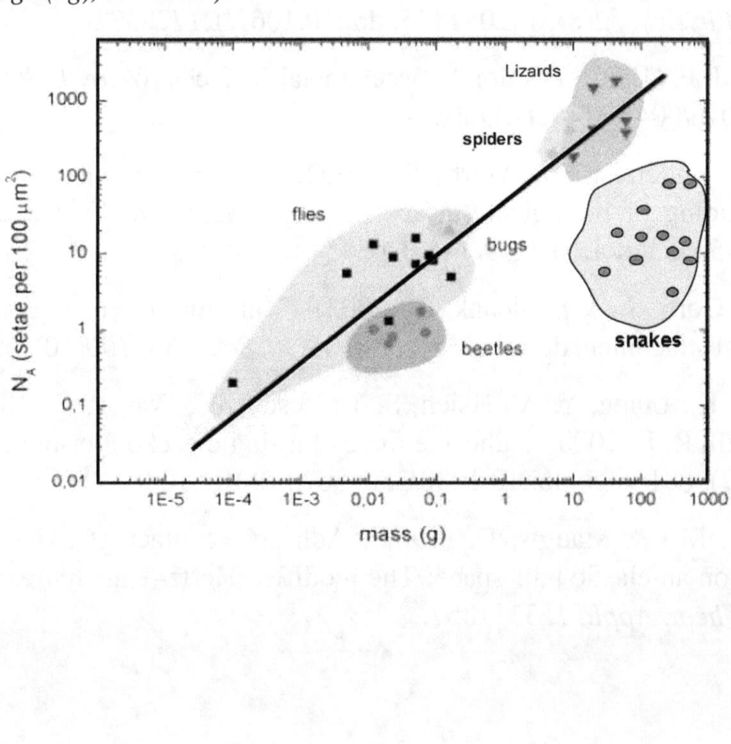

Research data for fibrillar arrays collected over the last years reveal some general trends summarized as follows. The adhesion force and strength of fibrillar surfaces increase significantly with decreasing fibril radii. Fibrils of varying lengths at constant fibril radius show an increase in adhesion force with increasing aspect ratio. This is attributed to the higher elastic energy dissipated at pull-off. Most of the stored elastic energy dissipated upon detachment, for low aspect ratio fibrils, comes from the deformation of the backing layer. For high aspect ratio fibrils, however, it comes from bending of the fibrils. Mushroom shaped tips have the highest adhesion more so than all examined tip shapes. They also show the highest potential for adhesion enhancement by contact splitting.

REFERENCES

Amontons. (1699). De la Résistance CauséeDans les Machines. *Mem. Acad. Roy*, 206-222.

Anitescu, M., & Potra, F. A. (1997). Formulating dynamic multi-rigid-body contact problems with friction as solvable linear complementarity problems. *Nonlinear Dynamics*, *14*(3), 231–247. doi:10.1023/A:1008292328909

Archard, J. F. (1961). Single contacts and multiple encounters. *Journal of Applied Physics*, *32*(8), 1420–1425. doi:10.1063/1.1728372

Archard, J. F. (1986). Friction between metal surfaces. *Wear*, *113*(1), 3–16. doi:10.1016/0043-1648(86)90052-9

Arzt, E., Enders, S., & Gorb, S. (2002). Towards a micromechanical understanding of biological surface devices. *Zeitschrift fur Metallkunde*, *93*(5), 345–353. doi:10.3139/146.020345

Arzt, E., Gorb, S., & Spolenak, R. (2003). From micro to nano contacts in biological attachment devices. *Proc. Natl Acad. Sci. USA*, *100*(10), 603–606.

Autumn, K., Liang, Y. A., Hsieh, S. T., Zesch, W., Wai, P. C., Kenny, T. W., ... Full, R. J. (2000). Adhesive force of a single gecko foot-hair. *Nature*, *405*(6787), 681–685. doi:10.1038/35015073 PMID:10864324

Barquins, M., & Maugis, D. (1982). Adhesive contact of axisymmetric punches on an elastic half-space: The modified Hertz–Huber stress tensor. *J. Méc. Théor. Appl.*, *1*, 331–357.

Bhushan, B. (Ed.). (1995). *Handbook of Micro/Nano tribology*. CRC Press.

Bhushan, B., Israelachvili, J. N., & Landman, U. (1995). Nanotribology—Friction, wear and lubrication at the atomic-scale. *Nature*, *374*(6523), 607–616. doi:10.1038/374607a0

Blok, H. (1987). Thermo-Tribology Fifty Years On. *Int. Conf. on Tribology, Ins. Mech. Eng*, 1-8.

Bowden, F. P., & Tabor, D. (1939). The area of contact between stationary and between moving surfaces. *Proceedings of the Royal Society of London. Series A, Mathematical and Physical Sciences*, *169*(938), 391–413. doi:10.1098/rspa.1939.0005

Bowden, F. P., & Tabor, D. (1964). *The friction and lubrication of solids*. Oxford: Clarendon.

Bradley. (1932). The cohesive force between solid surfaces and the surface energy of solids. *Phil. Mag., 13*, 853-862.

Bradley, R. S. (1936). The cohesion between smoke particles. *Transactions of the Faraday Society*, *32*, 1088–1090. doi:10.1039/tf9363201088

Buckley, D. H. (1981). *Surface effects in adhesion, Friction, Wear, Lubrication*. Elsevier.

Cooper, D. H. (1969). Hertzian contact-stress deformation coefficients. *ASME J Appl Mech*, *36*(2), 296–303. doi:10.1115/1.3564624

Coulomb, C. A. (1785). Theorie des Machines Simples. *Mem. Math. Phys. Acad. Sci., 10*, 161–331.

Da Vinci, L. (1975). Codex Madrid, reprint (tratado de estiticaymechanicaenitaliano). *Iwanami Shoten, 20*, 101.

Del Campo, A., Greiner, C., & Arzt, E. (2007). Contact shape controls adhesion of bioinspired fi brillar surfaces. *Langmuir*, *23*(20), 10235–10243. doi:10.1021/la7010502 PMID:17722937

Derjaguin. (1934). Untersuchungen Äuber die reibung und adhÄasion. *Kolloid Z., 69*, 155-164..

Derjaguin, Muller, & Toporov. (1975). Article. *J. Colloid Interface Sci., 53*, 314-326.

Dosaev, M. (2015). *On instability of a brake shoe*. The 14th IFToMM World Congress, Taipei, Taiwan. 10.6567/IFToMM.14TH.WC.OS4.015

Euler, J. A. (1758). RecherchesPlusExactessurl'Effect des Moulinà Vent. *Mem. Acad. Roy. Sci. Berlin*, *12*, 165–234.

Euler, L. (1748). Sur la Diminution de la Résistance du Frottement. *Histoire de l'Académie Royale des Sciences et Belles-Lettres de Berlin*, *4*, 133–148.

Fitch, J. (2006). Interview with luminary professor H. Peter Jost, the man who gave birth to the word tribology. *Machinery Lubrication, 1*. www.machinerylubrication.com

French, R. H. (2000). Origins and applications of london dispersion forces and hamaker constants in ceramics. *Journal of the American Ceramic Society*, *83*(9), 2117–2146. doi:10.1111/j.1151-2916.2000.tb01527.x

Garbassi, F., Morra, M., & Occhiello, E. (1994). *Polymer Surfaces from Physics to Technology*. New York, NY: John Wiley & Sons.

Gerschel, A. (1987). Dipole-induced static and dynamic liquid structures. J. Chem. Soc., Faraday Trans. 2 Mol. *Chemical Physics*, *83*, 1765–1776.

Gorb, S. N. (2008). Biological attachment devices: Exploring nature's diversity for biomimetics. *Phil. Trans. R. Soc. A*, *336*(1870), 1557–1574. doi:10.1098/rsta.2007.2172 PMID:18192171

Greenwood, J. A. (1985). Formulas for moderately elliptical Hertzian contacts. *ASME J Tribol*, *107*(4), 501–504. doi:10.1115/1.3261116

Greenwood & Johnson. (2006). Oscillatory loading of a viscoelastic adhesive contact. *J. Colloid Interface Sci, 54*, 296-284.

Greenwood, J. A., & Williamson, J. P. B. (1966). Contact of Nominally Flat Surfaces. *Proceedings of the Royal Society of London. Series A*, *295*, 1442.

Grigoryan, S. S. (2001). The Solution to the Painleve Paradox for Dry Friction. *Doklady Physics, 46*(7), 499-503.

Hamaker, H. C. (1937). The London – van der Waals attraction between spherical particles. *Physica*, *4*(10), 1058–1072. doi:10.1016/S0031-8914(37)80203-7

Hamel, G. (1909). Bemerkungenzu den vorstehendenAufs¨atzen der HerrenF. Klein und R. v. Mises, Ztschr. f. *Mathematics and Physics*, *58*(1-2), 195–196.

Hardy, W.B., & Doubleday, L. (1922a). Boundary Lubrication-The Paraffin Series. *Proc. R. soc. London A, 100*, 25-39.

Hardy, W.B., & Doubleday, L. (1922b). Boundary Lubrication-The Temperature Coefficient. *Proc. R. soc. London. A, 101*, 487-492.

Hertz, H. (1896). Über die berührung fester elastischer Körper [On the contact of rigid elastic solids]. In Miscellaneous Papers. Macmillan.

Hertz, H. R. (1882). Ueber die Beruehrung elastischer Koerper [On Contact Between Elastic Bodies]. In Gesammelte Werke (vol. 1). Academic Press.

Holm, R. (1908). *Experimentelle Untersuchungen über die geschichtete positive Glimmlichtsäule: insbesondere über das Schichtenpotential in H2, N2, He*. Diss. Uppsala Univ.

Holm, R. (1946). *Electric contacts*. Stockholm: Gebers.

Hughes & White. (1979). 'Soft' contact problems in linear elasticity. *Q. J Mech. appl. Math., 32*(445), 71.

*al-Jazari. (1974). The Book of Knowledge of Ingenious Mechanical Devices [Kitab fimaa'rifat al-hiyal al-handasiyya]*Hill, D. R., Trans.). Springer Netherlands. doi:10.1007/978-94-010-2573-7

Johnson, K. L. (1985). *Contact mechanics*. Cambridge: Cambridge University Press. doi:10.1017/CBO9781139171731

Johnson & Greenwood. (1997). An adhesion map for the contact of elastic spheres. *J. Colloid Interface Sci., 192*, 326-333.

Johnson, K. L., Kendall, K., & Roberts, A. D. (1971). Surface energy and contact of elastic solids. *Proceedings of the Royal Society of London. Series A, Mathematical and Physical Sciences, 324*(1558), 301–320. doi:10.1098/rspa.1971.0141

Casulleras. (2007). BanuMussa. In T. Hockey, V. Trimble, T. R. Williams, K. Bracher, R. A. Jarrell, & J. D. Marché II, (Eds.), *The Biographical Encyclopedia of Astronomers* (pp. 92–94). doi:0.1007/978-0-387-30400-7

Jost, P. (1966). *Lubrication (Tribology): A report on the present position and industry needs*. Presented to UK Dep. Of Education and Science.

Keesom, W. H. (1915). The second virial coefficient for rigid spherical molecules whose mutual attraction is equivalent to that of a quadruplet placed at its center. *Proc. R. Acad. Sci, 18*, 636–646.

Kendall, K. (1971). The adhesion and surface energy of elastic solids. *Journal de Physique (Paris), D4*, 1186–1195.

Kendall, K. (1975). Thin film peeling—The elastic term. *Journal de Physique (Paris), D4*, 1450–1452.

Kim, D. I., Grobelny, J., Pradeep, N., & Cook, R. F. (2008). Origin of adhesion in humid air. *Langmuir, 24*(5), 1873–1877. doi:10.1021/la702412y PMID:18193903

Klein, F. (1910). ZuPainlev'esKritik der CoulombschenReibgesetze, Ztschr. f. *Mathematics and Physics, 58*(1-2), 186–191.

Kontorova, T. A., & Frenkel, Y. I. (1938). On the theory of plastic deformation and crystal twinning [in Russian]. *Zh. Eksp. Teor. Fiz., 8*(1), 89–95.

Kontorova, T. A., & Frenkel, Y. I. (1938). Article. *Zh. Eksp. Teor. Fiz., 8*(12), 1340–1348.

Kragelskii, V., & Shchedrov, V. S. (2010). *Progress in Science of Friction. Dry Friction.* Moscow: Izd-vo AN SSSR.

Leckband, D., & Israelachvili, J. (2001). Intermolecular forces in biology. *Quarterly Reviews of Biophysics, 34*(2), 105–267. doi:10.1017/S0033583501003687 PMID:11771120

London, F. (1930). Zur theorie und systematik der molekularkräfte. *Zeitschrift für Physik A Hadrons Nuclei, 63*, 245–279.

Maugis, D. (1992). Adhesion of spheres: The JKR–DMT transition using a Dugdale model. *Journal of Colloid and Interface Science, 150*(1), 243–269. doi:10.1016/0021-9797(92)90285-T

Maugis, D. (1999). Contact, adhesion and rupture of elastic solids. In *Solid state science* (p. 234). Springer.

Mises, R. (1909). ZurKritik der Reibungsgesetze, Ztschr. f. *Mathematics and Physics, 58*(1-2), 191–194.

Muller, Yushchenko, & Derjaguin. (1980). On the influence of molecular forces on the deformation of an elastic sphere. *J. Colloid Interface Sci., 77*, 91-101.

Needham, J., Wang, L., Robinson, K. G., & Needham, J. (1991). Science and Civilization in China: Vol. 4. Physics and physical technology. Cambridge Univ. Press.

Painlevé, P. (1895). Leconssur le Frottement. Hermann.

Pfeifer, F. (1909). ZurFrage der sogenanntenCoulombschenReibungsgesetze, Ztschr. f. *Mathematics and Physics*, *58*(3), 273–311.

Popov, V. L., & Gray, J. A. T. (2012). Prandtl-Tomlinson model: History and applications in friction, plasticity, and nanotechnologies. *Zeitschrift für Angewandte Mathematik und Mechanik*, *92*(9), 683–708. doi:10.1002/zamm.201200097

Prandtl, L. (1909). Bemerkungenzu den Aufs¨atzen der HerrenF.Klein, R. v.Mises und G. Hamel. Ztschr. f. *Mathematics and Physics*, *58*(1-2), 196–197.

Prandtl, L. (1928). EinGedanken model zurkinetischenTheorie der festenKörper. *Zeitschrift für Angewandte Mathematik und Mechanik*, *8*(2), 85–106. doi:10.1002/zamm.19280080202

Rabinowicz, E. (1971). The determination of compatibility of metals through static friction tests. *ASLE Transactions*, *14*(3), 198–205. doi:10.1080/05698197108983243

Reynolds, O. (1886). On the theory of lubrication and its application to Mr. Beauchamp Tower's experiments, including an experimental determination of the viscosity of olive oil. *Phil. Trans. R. Soc.*, *177*, 157–234. doi:10.1098/rstl.1886.0005

Rigney, D. A., & Hirth, J. P. (1979). Plastic deformation and sliding friction of metals. *Wear*, *53*(2), 345–370. doi:10.1016/0043-1648(79)90087-5

Semen, K. K. (1774). книга, содержащаяучение о равновесии и движениител [A book containing the doctrine of equilibrium and motion of bodies]. St. Petersburg: Academic Press.

Spolenak, R., Gorb, S., & Arzt, E. (2005). Adhesion design maps for bio - inspired attachment systems. *Acta Biomaterialia*, *1*(1), 5–13. doi:10.1016/j.actbio.2004.08.004 PMID:16701776

Stone, A. J. (1996). *The Theory of Intermolecular Forces* (Vol. 32). Oxford, UK: Clarendon Pres.

Tabor, D. (1977). Surface forces and surface interactions. *Journal of Colloid and Interface Science*, *58*(1), 2–13. doi:10.1016/0021-9797(77)90366-6

Tabor, D., & Winterto, R. H. (1969). Direct measurement of normal and retarded van der waals forces. *Proc. R. Soc. Lond. Ser. A: Math. Phys. Sci*, *312*, 435–450.

Tomlinson, G. A. (1929). A molecular theory of friction. *The London, Edinburgh and Dublin Philosophical Magazine and Journal of Science*, *7*(46Supplement), 905–939. doi:10.1080/14786440608564819

Tower, B. (1883). First report on friction experiments (friction of lubricated bearings). *Proceedings - Institution of Mechanical Engineers*, *34*(1), 632–659. doi:10.1243/PIME_PROC_1883_034_028_02

Van der Waals, J. D. (1893). *Thermodynamische theorie der capillariteit in de onderstelling van continue dichtheidsverandering; Verh.* Amsterdam, The Netherlands: K. Akad. Wet.

Visser. (1972). On Hamaker constants: A comparison between Hamaker constants and Lifshitz-van der Waals constants. *Advances in Colloid and Interface Science, 3*(4), 331-363.

Wiercigroch, M., & Zhilin, P. A. (2000). On the Painleve Paradoxes. *Proceedings of the XXVII Summer School "Nonlinear Oscillations in Mechanical Systems",* 1–22.

Zhukovskii, N. E. (1948). *Equilibrium Condition for a Rigid Body Contacting with an Immovable Plane by a Site and Able to Move along This Plane with Friction. In Complete Papers* (Vol. 1, pp. 399–354). Moscow: Gostekhizdat. (in Russian)

Zhukovskii, N. E. (1948). *Friction of Railway Wheel Treads on Railways. In Complete Papers* (Vol. 7, pp. 426–478). Moscow: Gostekhizdat. (in Russian)

Zhuravlev, V. A. (1940). On question of theoretical justification of the Amontons–Coulomb law for friction of unlubricated surfaces. *Zh. Tekh. Fiz.*, *10*(17), 1447–1452.

Zhuravlev, V. F. (2010). History of the law of dry friction. *Doklady Akademii Nauk, 433*(1), 46–47.

Zhuravlev. (2013). On the history of the dry friction law. *Mechanics of Solids, 48*(4), 364.

ENDNOTES

1 The tomb of Yuya and Tuyu was one of the most important burials to be found in the Valley of the Kings. It was discovered on February 5'th, 1905, by James Quibell and Theodore M. Davis, the tomb contained one of the most complete and beautifully made sets of funerary equipment then known. Dating from the 18th Dynasty, it contained the burials of Yuya and Tuyu, the parents of Queen Tiye, and its many intact objects give an idea of how a high-status burial from the time of Amenhotep III was provisioned. Although not of royal lineage, Yuya and his wife Tuyu were probably individuals of immense political influence during the reigns of Amenhotep III and his father, Tuthmosis IV. The scarabs issued to commemorate the marriage of Amenhotep III to Tiye name her parents prominently in an unprecedented fashion, as though intentionally stressing the royal alliance with this Akhmim-based family. C. N. The many impressive titles of Yuya and Tuyu (including "God's Father," "His Majesty's Lieutenant Commander of Chariotry," "Master of the Horse," "Priest of Min," "Overseer of the cattle of Min, Lord of Akhmim" for Yuya and "King's Mother of the Great Royal Wife, "Chief of the Harem of Min," "Priestess of Amen," "Chief of the Harem of Amen," and "Chantress of Hathor," for Tuyu) as well as their sumptuous burial in the Royal Valley.

2 The complex nature of the designs illustrated and presented by Al-Jazari and his predecessors the Banu-Musa Brothers reflect in-depth knowledge of friction and its effect on motion. This is so because their precise design of clocks and their work on Automata, in general, depended on precise knowledge of frictional effects on motion for proper along with some means to calculate this effect, and how it affects time keeping. In addition they were capable mathematicians, which imply analytical aptitude and familiarity with formulation of equations (Casulleras 2007).

3 The Codex Madrid I includes on page 20 sketches of thrust and ball bearings that use wooden balls in place of rollers to move heavy objects, and spools with shafts for controlling the balls so that they roll smoothly and not deviate. In addition, on page 101, of the first volume, there are sketches of a pivot ball bearing with instructions favoring the placement of three balls below the shaft and arguments to why three balls are optimal in this case.

4 The introduction of this Greek letter as a symbol for the coefficient of friction carried from the work of the Russian mathematician-Physicist Semen Kirilovich Kotelinkov (1723-1806) who was a former student of the famous mathematician Euler. Kotelnikov conducted some of the earliest Russian studies of friction in the late 1700s (Semen) where the first usage of the letter "mu" appeared, he wrote "… If we denote the friction content F and the applied force P as unknowns, in the ratio μ:1, then friction F=μP."

5 Coloumbs findings, published in Essaisur la theorie du frottement (Coulomb), won him the Grand Prix of the Academie des Sciences in 1781.

6 Zhuravlev compared two paragraphs that introduce the statement of the friction laws as expressed by Euler and by Coulomb. He showed that the formulation of Coulomb is a word-for-word replication of the statement of Euler.

7 Euler paid attention to the unusual form of the dry friction law, which had a discontinuity of the first kind in the dependency of the force on velocity. This discontinuity was not typical of any force known at the time.

8 These computations were performed in 1894 by Zhukovski.

9 Prandtl finishes the discussion of his paper by "to conclude, it may now be suggested that our conceptual model is also suitable for the treatment of kinetic friction between solid bodies. This model is even better adapted for application to the case of two straight edges sliding against one another than for the case of plastic deformation"

10 Many practical surfaces, including reptilian surfaces, the height distribution is Gaussian to a very good approximation. For Gaussian distribution the standardized height distribution is given by $\phi^*(s) = \frac{1}{\sqrt{2\pi}} \exp\left(-\frac{s^2}{2}\right)$ which yields the cylindrical function k_m as, $K_m = \frac{m!}{\sqrt{2\pi}} \exp\left(-\frac{D^2}{4}\right) U\left(m + D/2\right), m \geq 0$

11 It is estimated that over 1700 species comprising over 4000 organisms are responsible for biofouling.

12 The solution given by the JKR model for the adherence between a rigid sphere and a rigid plane was incompatible with the calculations introduced by Bradley in 1932 at a point contact.

13 Recall that in the DMT theory the adhesive forces act in an annular zone around the contact but do not deform the profile. Thus, this theory does not entertain singular stresses. Separation in such a case takes place when the contact is reduced to a point.

Chapter 3
Classification of Surface Constructs

ABSTRACT

This chapter discusses routes for the classification of surfaces and the evolution of surface metrology. There are two basic ingredients to the surface of a manufactured workpiece: the manufacturing process and the production technique. Each of the ingredients has an effect on the functionality of a rubbing part. The material provides an answering scheme for a question that often arises about how to characterize the influence of roughness features and how to utilize characterization metrics to predict, then monitor, performance.

INTRODUCTION

The layer encompassing the roughness layout of the surface (i.e. the asperities) undergoes intense changes during friction. The action starts at the inception of contact between the two mating bodies. When two surfaces approach each other (figure 1-a), the asperities are the first parts of the surface layer to experience the influence of the operation environment (load, speed, etc.). Further, the evolution of the contact process, and the behavior of materials thereafter, will depend to a great extent on the geometry, distribution and properties of the experience.

DOI: 10.4018/978-1-7998-1647-8.ch003

Upon compliance, the asperities within the surface layers will deform to support the acting normal nominal load (figure 1-b). Because the asperities of a solid are not generally of the same length, or size, only the highest asperities will initially establish contact with like asperities within the counter body, to support the nominal load. However, because the initial true contact stress would be higher than the yield strength of the solid severe deformation will take place. The initial group of contacting asperities will be compressed relative to the stress acting on them. Compression will bring more asperities, those of lesser height, into contact to support the load. The new wave of asperities, in turn, will continue to undergo compression which will bring more asperities into contact. The compression-deformation process continues until the nominal load is fully supported. The initial engagement of asperities also implies that the only parts of the surface that establish contact are those parts supporting the load on the micro-level. This renders the true area of contact, that which is influential to the evolution of the friction process, only a small fraction of the joint apparent area between the two complying bodies.

Figure 1. Scenario of contact of real surfaces. Initially, the separation gap between the surfaces will diminish in proportion to the nominal normal load FN. The highest asperities will establish contact. Depending on the value of the true contact stress more asperities may come into contact until the nominal load is supported.

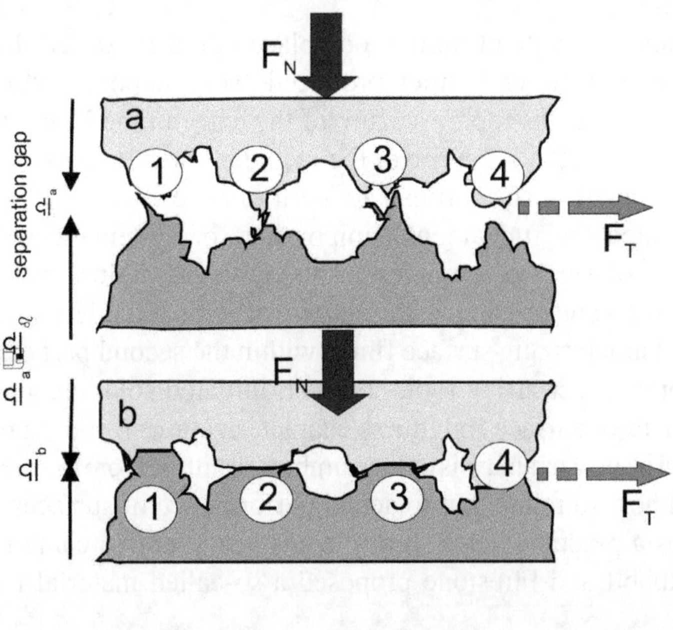

The scenario of contact described above, points at the crucial role the roughness of the body assumes in establishing the true area of contact and thereby in guiding the evolution of the rubbing process. It also points at the crucial role of roughness in initiating and controlling any friction -induced damage befalling the rubbing bodies. It is no surprise, therefore, that considerable attention has been directed toward the study of roughness and how it evolves in rubbing. Moreover, that has been extended effort, perhaps since the Industrial Revolution, to classify and characterize roughness and texture.

Historical Note

The realization that rubbing performance of a surface depends on the roughness have traditionally led to more thorough investigations into the connection between size of the textural features and damage accumulation in sliding. Initially, the common view was that the smaller (i.e. minute) the roughness features are, the less friction and moving surface will encounter. Thereby that surface will exhibit less wear damage. This view prompted the production of industrial surfaces based on a subjective scale of "smooth" for" ideal" performance and avoidance of "rough" to prevent a "poor" performing surface. This trend, which may follow intuition, in fact lacks a tangible measure that quantifies the extent of rough and smooth. As a sequence it proved disastrous when applied.

A notorious example of failure took place in the 1930s. At the time, the British car manufacturer Bentley produced "very smooth" cylinder bores. This resulted in a catastrophic seizure of the entry in the Le-Mans 24-hour race (*Jiang et. Al. 2007*). Failure of the cylinder bores in this event signaled the importance of surface finish to performance of sliding surfaces. It also highlighted the fundamental importance of systematic and accurate measurement of surface roughness. This notorious failure also prompted many companies and research institutions to develop reliable instrumentation capable of characterizing surface finish within the second part of the 1930s.

Developing a measuring tool, however, initiated solutions only for part of the problem of surface roughness characterization. A more pressing and fundamental issue was analysis of the numbers resulting from the measurement device, and how to relate that to actual performance of surfaces. An initial step towards a practical solution of the analysis problem came in 1933. In this year Abbott and Firestone proposed a so-called material issue curve.

This simple curve was suggested as a representation of the surface and as a source of realistic numbers, that if determined, would indicate the quality of the produced surface. Once the quality-indicating numbers are found from the curve, the suitability of the surface for a given application could be determined. Such a development was a major step toward linking function, performance, and topography. It further formed the base for the statistical analysis of roughness which was subsequently developed. Statistical analysis advanced the characterization of surfaces. It further, through developing the well-known ISO-family of parameters, advanced the knowledge concerning the relationship between topography of roughness and tribological performance. However, advancement of knowledge concerning the roughness and function led to modification of the view of surface functionality and in the role of roughness within surfaces. Eventually these dynamic views of surfaces triggered an interesting debate about developing criteria for classification of surfaces.

The need for revising, or introducing a new, classification of surfaces is, in essence, a consequence of advances in instrumentation; evolution of application demands; and the maturity of advanced enabling technologies. The combination of the previous factors has motivated a considerable shift in surface characterization paradigms to cope with the ever-evolving demands in surface functionality.

Traditionally, the importance of surface metrology has its roots in the need to optimize surface functionality. The advancement in control of roughness, and development of meaningful measures, have always led to advances in performance of tribological surfaces. Control of surface texture led to short the running-in periods for locomotive engines and prompted fuel efficiency. These advances, however, were mainly product of refinement of surface measuring instruments and considerable advances in the theoretical mathematical framework which formed the basis for analyzing measurements. On the other hand, the leap in measurement and analysis have also led to deeper understanding of the interplay between roughness and function. To that end, many researchers have developed routes to enhance function of surfaces for rubbing applications. Many of the envisioned surfaces have come to existence through a leap in technologies that enable manipulation of matter at minute scales. Introduction of many surface engineering techniques, combined to the plethora of designs that emanated within the past two decades have led to a paradigm shift in characterization, classification, and analysis of surfaces.

GENERIC CLASSIFICATIONS OF SURFACES

The question of surface classification has been entertained perhaps since the dawn of age. One can envision an early function-driven classification of satisfactory-unsatisfactory based on touch, or visual inspection or even aesthetics. Current definitions, classifications, and associated terminology, however, are a product of the early twentieth century. That is, when more thorough understanding of the makeup of surfaces was reached. Stout and Blunt (2000) estimate that the word "texture" may have started to appear in the context of surface classification in the 1930s. Stout et al, (Stout and Blunt. 1995a, b, 2001, Stout et al., 1993, Stout 1997) developed a comprehensive classification of surfaces (figure 2). They distinguished between classical surfaces (i.e. surfaces that result from a traditional manufacturing process) and engineered surfaces. The main difference lies in the complexity of the functional goal of the surface. In case of classical surfaces, the main function of the surface is to provide a predetermined geometry of a workpiece (e.g. to define boundaries of a box or a casting). Engineered surfaces, meanwhile, according to Stout and Blunt (1997), are customized to perform a predetermined task (e.g. hydrophobic, hydrophilic, etc.). Engineered surfaces may further be classified into several subcategories. They can be random or structured and their topography can result from a manufacturing process that yields random roughness (e.g. turning or milling).

Structured surfaces, meanwhile, are those surfaces for which topography is prearranged deliberately to meet the specified function requirement (i.e. the texture is arranged in a deterministic pattern that meets the functional requirements). Advances in instrumentation, demand by applications, and advances in production technologies always drove revisions to existing surface classifications. One major motivation for seeking new classification is the maturity of enabling technologies. Introduction of designer surfaces, for example, rendered old classifications less meaningful or impractical (Stout and Blunt 1997).

Introduction of designer surfaces triggered an interesting debate concerning what was meaningful and practical within surface classifications of the day. Stout and Blunt (1995-a) revised the classification of some manufactured surfaces, as follows:

"Structured surfaces" are those where the surface structure is a designer feature intended to give specific functional performance. Whereas "Engineered Surfaces" are those produced in a specific way that alter the surface and subsurface layers to give required performance. Thus, according to this definition, the distinction factor of whether a surface is "engineered' or "structured" is the inclusion of the sub-surface layer during the customization process. The ambiguity of such a definition motivated using different language for establishing the category of a surface. An example is the alternative definition was proposed by Evans and Bryan (1999) who considered "structured surfaces" as those with a deterministic pattern of usually high aspect ratio geometric feature designed to give a specific function and "Engineered surfaces" as surfaces where the manufacturing process is optimized to generate variation in geometry and/or near surface material properties to give specific function. Stout and Blunt argued that the Evan-Bryan definition excludes products resulting from "non-traditional" methods (such as MEMS where the fabrication route is masking-etching which is not deemed conventionally as a "surface production" process). This motivated the authors to propose the more comprehensive classification shown in figure 2. Table 1 summarizes the definitions of the surfaces associated with the classification of figure2.

Figure 2. Classification of surfaces according to texture. The figure presents a time line of the evolution of classification. In the 1920s of the twentieth century surfaces were considered a consequence of whatever manufacturing process used. Starting the 1930s engineers and designers started to recognize the potential impact a surface has on performance of an engineering system. Efforts to identify that relationship and enhance performance continued. The 1990s marked the introduction of the surface-by-design concept. Consequently, surfaces were classified as structural and non-structural. The target within the 21st century is a complete customization of surfaces.

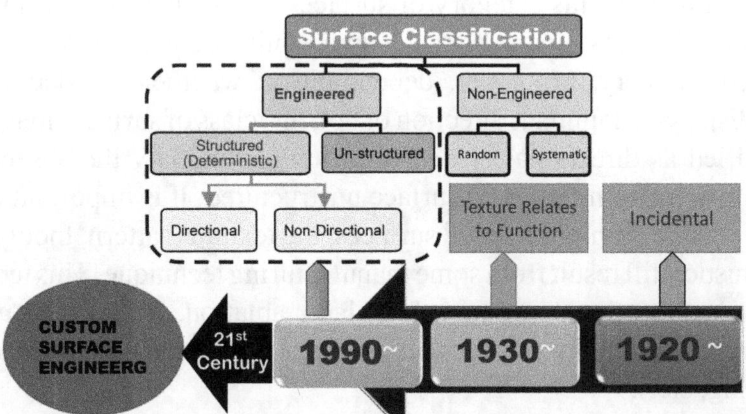

Table 1. Definitions of the surfaces associated with the classification of Figure 2.

Surface Classification	Definition
Engineered Surfaces	surfaces produced in specific ways that deliberately alter surface and sub-surface layers to give a specific functional performance.
Non-Engineered Surfaces	surfaces produced as a direct consequence of the manufacturing process where little or no attempt is made to influence surface character.
Structured Surfaces	surfaces with a deterministic pattern of usually high aspect ratio geometric features designed to give a specific function.
Directional Surfaces	surfaces with a deterministic pattern which exhibits specific directionality.
Non-Directional Surfaces	surfaces with a deterministic pattern but without specific directionality.
Unstructured Surfaces	surfaces where a deliberate attempt has been made to impart texture through semi-control of the manufacturing process without achieving a deterministic pattern.
Random Surfaces	surfaces produced by random and pseudo-random processes often with the specific intention of removing systematic features.
Systematic Surfaces	surfaces exhibiting some repetitive features which are a consequence of the natural constraints of the process by which they have been produced.

The principle distinction between surfaces in the figure is whether the resulting surfaces are altered, customized, to meet a given pre-conceived functional performance. In this sense, any surface that is a direct consequence of a manufacturing process, where no functional-driven specialization took place falls under the non-engineered surface category.

The main criterion for considering a surface as “engineered”, according to table 2, is deliberate adaptation to meet functional goals (i.e. custom tailoring surface and subsurface layers). In this sense, one can consider a functional surface (that which occurs in nature or man–made) also as an “engineered surface”. Hither to, this category of surfaces will be the focus. Engineered (functional) surfaces, upon acquiring a deterministic pattern, fall under the structured category. Moreover, depending on whether the deterministic pattern displays a dominant direction or not, this class of surfaces may further be classified as directional or non-directional. Naturally the absence of a deterministic pattern renders a surface unstructured. It is important to note, however, that in man-engineered surfaces, the textural pattern, though being deterministic, still result from some manufacturing technique. This technique can be a “conventional” one (such as laser ablation, or laser texturing in general) or alternatively may result from a non-traditional method (chemical, ion etching, etc.).

Tracing the origins of the production method over surface provides a facility for an alternative classification of surfaces. Two classification routes are detailed in what follows. The first, illustrated in Figure 3, manifests an extension of the classification method of Figure 2. This extension centers around considering the surface production route. It is noted that the production method is influential to the quality of the resulting surface, and the order-disorders nature of the roughness features. The connection between production rates and layout pattern of the surface prompted introduction of yet another surface classification based on 'proposed function' or behavioral traits (Figure 4).

Figure 3. Classification of surfaces in reference to processing routes. Surfaces resulting from conventional manufacturing methods (e.g. casting, turning, etc.,) show random, or stochastic, topographical features. Deliberate ordering, or shaping, of roughness features renders a surface an "engineered" construct.

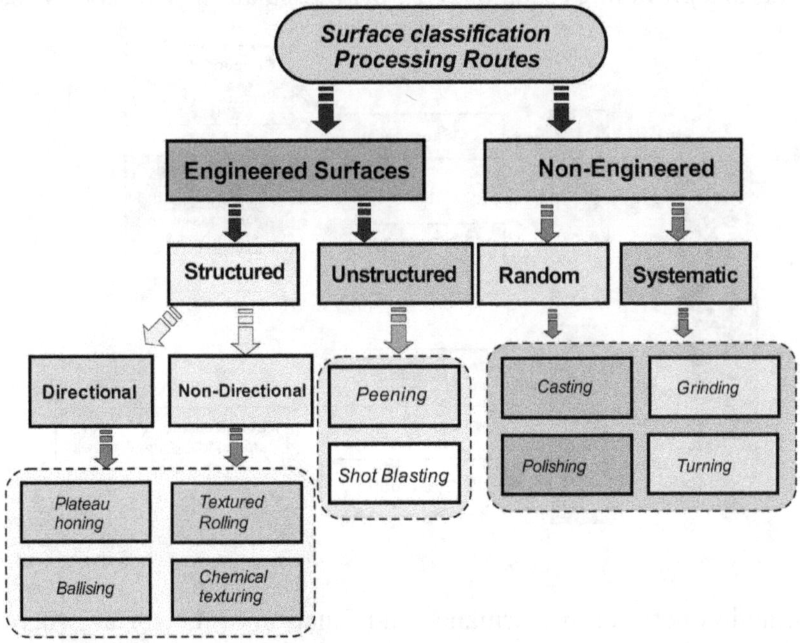

An interesting distinction between surfaces within the classification of figure 4, with respect to properties, is the purpose of the roughness. Accordingly, the major difference between engineered and non-engineered surfaces is the objective of generating a particular roughness pattern. In nonengineered surfaces, roughness features are not optimized. This is so because the main objective is to provide an overall geometry for an object. That is, in this case, a non-engineered surface falls within the classical view of surfaces as enclosures. On the other hand, an engineered surface comprises textural features purposely designed to facilitate a property or a "function". We notice here that the definition of function for engineered surfaces overlaps purpose and properties (and not necessarily performance). That is, in engineered surfaces, function implies the purpose of the design or successful fulfillment of a task. This immediately leads to a narrow definition of texture, within the context of function, as geometries that are optimized to achieve a functional task.

Figure 4. Classification of surfaces according to the properties of the surface layer. This classification route is based on whether the layers are "treated" to withstand a derogatory condition that dominates during operation. Thus, an "engineered" surface within this view is one of which topographical layers are optimised to resist failure due to a prevailing condition (e.g., wear, residual stresses, corrosion, etc.,)

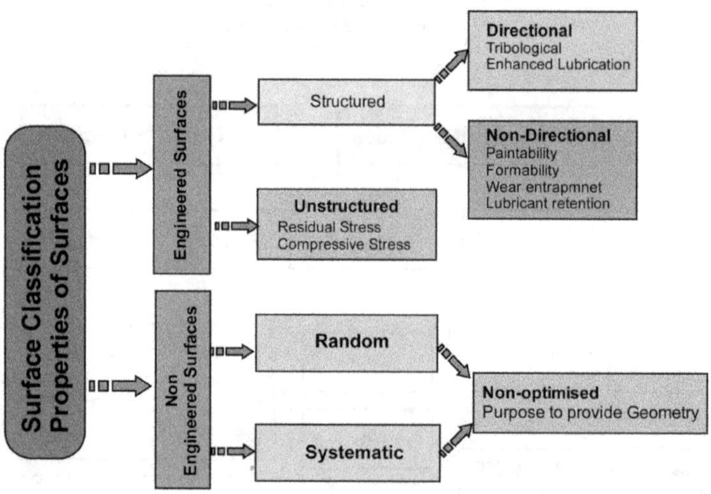

Distinction between performance and function of the surface (here with taken to imply functional purpose of the system) has its roots in attempts to evaluate 3-D surface topography. Two-dimensional surface topography has been traditionally the major focus of analysis (Blunt and Jiang 2003, Whitehouse and Archard 1970, Jiang and Whitehouse 2012). However, starting the late 1970s it became clear that other operational and design factors are as equally important to the quality of functional performance over surface (see figure 5).

Figure 5. Generalized effect of surface integrity on part functionality. The figure illustrates the effect of altering the state of surface and sub-surface layers on functionality. Note that functionality here implies preserving the structural integrity of the component under a dominant detrimental operation condition (e.g., wear, corrosion, etc.,)

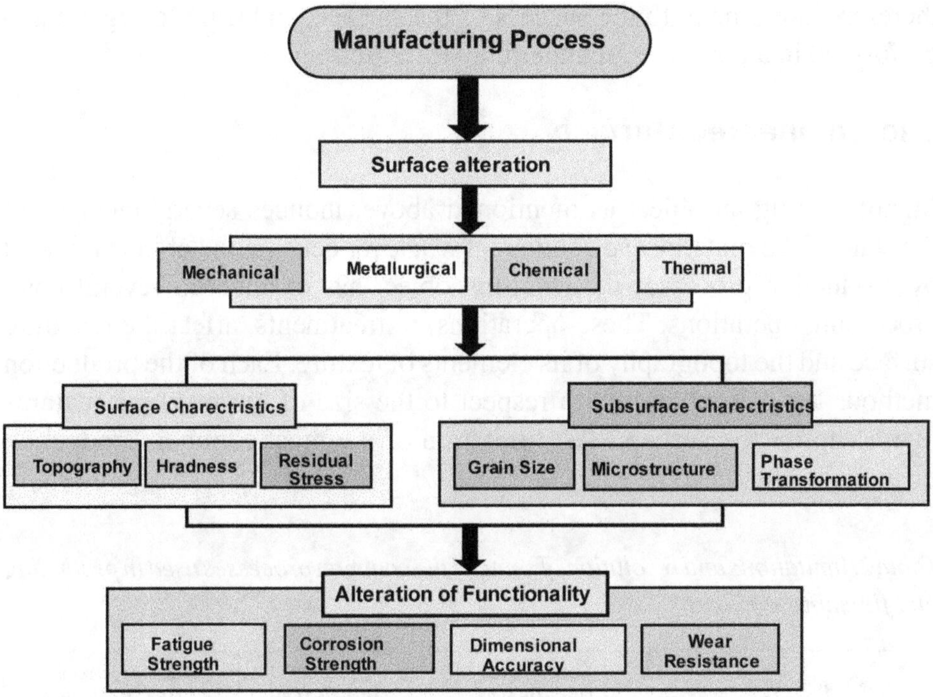

Such a realization has led to the recognition of functional environment (alternatively termed as constraints) in which surfaces are employed is of fundamental importance to the success of a surface (Stout 1993, Whitehouse and Archard 1969, Whitehouse and Phillips 1982). As a result, investigative efforts, from that point on, were inclusive of the effect the operating environment has on performance.

The environment of operation, in most cases, mandates changes to the raw produced surfaces. These changes, in turn, can alter the initial estate of the material used in production, and if not accounted for, predictions about quality of function may fall short. Metallic surfaces, for example, regardless of the original manufacturing method, can undergo surface modifications aiming at enhancing surface integrity. These may include heat treatment, coatings, chemical treatment, or mechanical treatment (figure 5). The treatments which constitute the so-called surface engineering, aim to modify the upper layers of the surface to induce a specific property (e.g. wear resistance, scratch resistance, etc.). In doing so, the resulting surface layers, will differ from those obtained directly after manufacturing (say by turning), and for sure are different than the original state of the pre-machine stock. The induced changes, therefore, are crucial to the success of the surface, and if not designed and performed in a controlled manner may result in failure.

The Engineered Surface

Manufacturing an object as mentioned above, induces several changes of the state of the material. To counteract whatever detrimental effects induced by production processes the resulting object has to undergo several post-processing operations. These operations, or treatments, affect the resulting surface and the topography of its elements of texture. Each of the production methods has a limitation with respect to the size of the resulting features that comprise the surface. The limitation of the most common production methods is illustrated in Table 2

Table 2. limitations and resolution of some of the common processes used in producing and finishing surfaces.

Creation Process	Conventional (10-3-10-4) m	Precision (10-4-10-7) m	Nano Processing (10-7-10-10) m
Removal	Turning milling Boring	Grinding honing lapping	Diamond machining etching lithography
Addition	Painting galvanizing plastic coating	Plating Mirroring deposition	Plasma atomic placement vapor deposition
Forming	Rolling bending forging	Extrusion peening molding	Other precision molding

It is noted that in order to reduce the size of the features is sequence of finishing operations (post the creation processes) is needed. In order to achieve final resolution that fits an advanced functional performance profile, the sequence of operations from production to finish becomes more complex. In such a case, more cumulative damage may be sustained by subsurface layers. This may drastically reduce the quality of performance. Realizing the possibility of reducing the quality of service performance due to multiple production-finishing cycles have forced the development of alternative routes of controlling texture of surfaces.

New approaches center on controlling the physical properties of the surface through specifying processing conditions necessity to yield the desired topography, and the corresponding mechanical properties of the various operative layers. The success of that approach, however, hinges on in-depth understanding of the effects of each process on the material being processed. Understanding the effects of processes may be facilitated by linking the topography of the surface to the physical-mechanical properties. The following step to linking topography and properties is to correlate the findings to the particular machining or fabrication process. The final step, is to link the findings to the integrity of the surface after a specified period of operation.

Establishing a relationship between production steps and the resulting surface, and then after to the actual performance (success), of the surface represented a shift in the view of surfaces. In effect, a translated the surface, and its textural features, from being a bi-product of the shape defining operation to an integral part of the object (Figure 6)

Figure 6. Shift in view of the surface and its texture

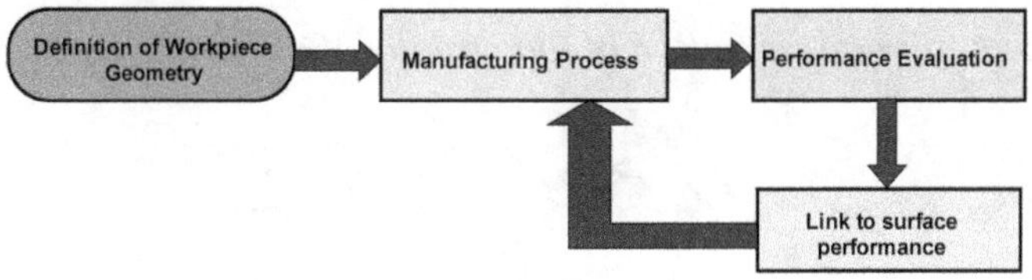

Relating performance to the state of the surface further characterizes the surface as an important integral function parts of the system. Therefore, it mandates optimization of the elements of texture as a viable means of achieving it preplanned level of success. In essence, therefore, relating production processes to performance and surface geometry calls for a surface-by-design paradigm. Such a paradigm, comprises deliberate design, and implementation, of both geometry and topography of textural elements within a surface to suit the target application. This fundamental shift in view of the surface necessitates the availability of comprehensive performance data sets. These may include component specification, for parametric relationships between production machines and product performance, and data concerning functional testing products. In addition, such a view requires comprehensive validation and standardization not only of the data, but also, testing methods and data collection protocols. The later requirements may not necessarily conform to conventional procedures. However, in order to understand these requirements, first there is a need to examine conventional practice as applied to examination and characterization of surface texturing elements and their topography.

METROLOGY AND ROUGHNESS

Surface metrology is the science of measuring small-scale geometrical features of surfaces (Greenwood 1984, Whitehouse and Phillips 1978). The importance of that science drives from the close relationship between surface texture and performance of surfaces in tribology. Measuring the roughness has become more of a priority with the growing needs to understand the interaction between surfaces during operation. In our modern times, there is an emphasis on establishing standardized parameter sets which describe, and somehow predicts, suitability of the surface for a particular application.

A conventional rough surface, figure 7, manifests peaks and valleys. These results from the interaction of the solids during surface generation processes.

Figure 7. Schematic representation of topography of a real surface.

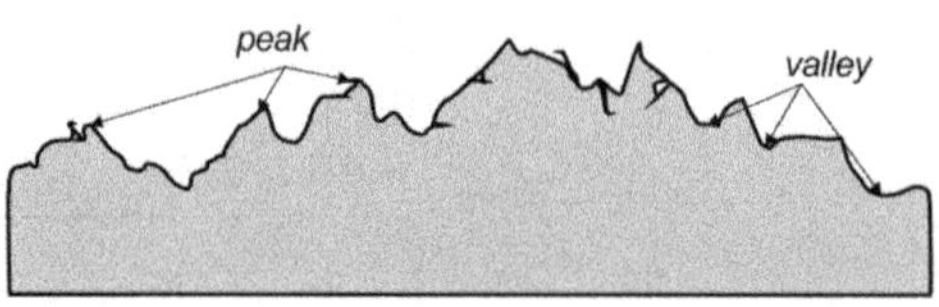

Analyzing features over surface is rooted in stochastic analysis, with the work of Rice (1945) on random noise signals forming the foundation. His work showed that within a random noise signal, the heights of the maxima along with the frequency of excursions about a particular reference datum is predictable once a so-called correlation function is evaluated. This correlation function is a property of the signal and is given by,

$$R(x) = E\left\{Z(x')Z(x' + x)\right\} \quad (1)$$

where Z(x) is the height of the surface profile and location (x) and E {.} is the expectation operator. Longuet-Higgins (1957 a, b) generalized the analysis of Rice from a one-dimensional signal to a two-dimensional surface. He further was able to obtain the distribution of the surface slope and of the heights of surface maxima (summits), and the mean length of contours of a given height per surface area. It was Nayak (1971, 1973) who extended the work of Longuet-Higgins to solid surfaces. Students reading your profile of the machine solid surface as random noise, Nayak obtained properties of interest in the study of contact between solid surfaces. In particular, he was able to find the distribution of summit curvatures. An important contribution of Nayak to characterization of surface roughness was clarification of terminology. His distinction between a "Peak" as a point on the profile that is higher than its two neighbors and a "summit" as it points on a 2-D surface higher than all its neighbors, has become the standard in the field.

Although that the analysis of roughness has its mathematical roots in the analysis of random noise signals, there has always been attempts to improve or to develop alternative routes for analysis. The foundation of the analysis has always been that assumption of Rice concerning the existence of a correlation function. Rice has assumed that such a function can be expanded near the origin in terms of the well-known Taylor series to yield and expression containing only even terms (Stout and Blunt 2001) viz,

$$R9x) = R(0) + \frac{x^2}{2!} R''(0) + \frac{x^4}{4!} R^{iv}(0) + ... \quad (2)$$

The expression resulting in equation (two) friends on the original form of the correlation function being free of singularities at the origin. This condition restricts the expressions valid for describing the function. An example is the expression used by Whitehouse and Archard (1978). These authors used an alternative random process theory that proposed the correlation function of the exponential form:

$$R(x) = R(0) \exp\left(\frac{-|x|}{\beta}\right) \quad (3)$$

Such an expression, however, has a discontinuity in its slope at the origin and cannot be expanded in a Taylor series. As a consequence, only "odd" derivatives would exist. There have been many attempts to develop a theory that is adequate for analysis of roughness (Sayles and Thomas 1978, 1979, Whitehouse and Phillips 1982, Scott. 2004, Jiang et al, 2007), that the self-contained and takes into account some of the potential mathematical pitfalls and sensitivities.

MEASUREMENT AND CHARACTERIZATION OF TOPOGRAPHY

Characterization of topographical features is a complex process. It is a demanding branch of surface metrology. It translates into a set of parameters that define, and standardize the state of the surface. Therefore it assists in certifying the state of the surface.

Characterization, and parameterization, depend on the advances and quality of measuring devices. Traditionally measurement of surface topography resulted in a 2-D profile. This was a consequence of the stylus-based instruments dominant starting the 1930s. At that time, the surface probing instrument used a simple tactile method comprising the stylus connected to a beam, transducer, amplifier, and meter (De Chiffre et al, 2003). This gave a profile of heights in terms of analog electrical signals. The simple idea of transforming heights into analog signals represented the basis of all instruments to follow for quite some time. Thus, the measurements resulted in heights translated into an electrical signal. This signal was further analyzed to yield descriptive parameters. Naturally, because of the animal nature of the signal, the number of possible obtainable parameters was limited. More parameters begin possible later on with the advent of digital measuring devices.

Two dimensional parameters at some shortcomings especially that the interaction of services is essentially volumetric. That is, any contact situation the volume of the roughness elements on one surface interacts with the one in the roughness on the counter surface. This interaction manifests either in direct contact or in the roughness of one surface filling the voids between the elements of roughness on the counter surface. In all, compliance, contact, and eventually released motion are three-dimensional phenomena. Therefore the ability to characterize, or predict, the behavior of the surface, describing its elements of roughness in 3-D based parameters became a must. However, as mentioned earlier, the ability to develop descriptive parameters close related to the existence, or not, of instrumentation that allow adequate probing of the surface. It also depends on the development of analytical methods capable of deducing the basic features of measurement data. Both of these factors, existence of an analytical base and development of suitable surface probes, are intertwined in a sense. The advancement of one, advances the other.

The advent of instruments that are capable of probing the volumetric makeup of surface topography led to the feeding of 2-D parameters in favor of 3-D based parameters. Thus, the description of surfaces moved from a so-called profile based meters (typically referred to as the R-parameter family) to the so-called areal parameters (typically referred to as the S-parameter family). The necessity of concealing areas (volumetric) aspects of surface topography is rooted in the shift, and advances, in production capabilities. Two main reasons are frequently quoted for such a push, advances in nanotechnology and futuristic needs for alternate precision surfaces (Whitehouse 1999).

Advances in nanotechnology has revealed some of the technical drawbacks of 2-D measurements. The drawbacks are especially apparent when measurement and qualification of commercial micro-and-Nano metric scale manufactured components is needed. Miniaturization of topographical features emphasizes the importance of the surface and its properties. The level, and path of the interaction of the mechanical, chemical, and tribological properties, and not necessarily the value of the property, becomes crucial to successful surfaces. Moreover, because of the minute size, accurate measurements at a comparative scale becomes more crucial, and directly links to function.

Other precision requirements of services are also demanding when it comes to measurement and characterization. Necessary for many modern devices, these surfaces may be theoretically flat, or their roughness may be of atomic magnitude. The surfaces also would require complex shapes with precision of up to one part of 108. Measuring, then analyzing these surfaces also demands different specifications and parameterization. Profile to areal characterization which reflects a stochastic two structures and simple to complex free-form shapes is therefore a practical necessity (Blunt and Jiang, 2003).

The shift from profile to areal parameters, in principle, stands on the stochastic analytical base. Such a shift started along the lines of the simple statistical descriptions of peek-to-pit heights and surface slope. A considerable shift from this simple description to place in the early 1990s. This was motivated by the needs reflected by modern applications of surfaces. Conventional surfaces, those which essentially defined the shape of an object, required characterization that guarantees consistency of production. There is no emphasis, however, on diagnostics nor on prediction of quality of performance.

Areal description of surface features, however, caters to the comprehensive needs of modern surface operation. Thus, areal characterization methods consider fundamental and functional features within a surface. It further accesses shape and direction of that shift, estimates feature attributes and differentiates between connected and isolated features. As such, areal surface characterization is not simply an extension from profile to areal case but is a genuine effort towards to characterization of aerial features (Townsend et al, 2017).

Irregularities of the surface have traditionally been separated into three groups: roughness, waviness, and errors of form (Table 3).

Table 3. Definition of Surface irregularities

Component	Definition
Roughness	The texture generated by the material removal mechanism (e.g., tool marks)
Waviness	Shape deviations due to imperfect operation of the machine tool
Errors of form	Errors generated of the machine tool due to distortion effects (e.g. gravity, thermal expansion, etc.).

Areal parameters that describe the roughness (texture) of the surface are developed according to their geometrical properties. They comprise two main classes: “field” and “feature” parameters

Field parameters are statistical in origin. They classify averages, deviations, extremes and specific features from scale-limited continuous surface. Featured parameters, on the other hand, manifest a Novell concept. The definition uses subsets of pre-defined features from a scale-limited continuous surface using segmentation-based pattern recognition (Blateyron, 2013). The two parameter sets, field and feature, form the core parts of modern technical specifications for surface texture.

Field Parameter Set

The field parameter set comprises two subsets: and S-parameter subset and a V-parameter subset. The S-parameters subset, Figure 8, depends on heights and spacing frequency to describe amplitude and spatial information. The V-parameters set, Figure 9, provides volumetric information that is based on the Abbott-Firestone curve.

Figure 8. The S parameter set

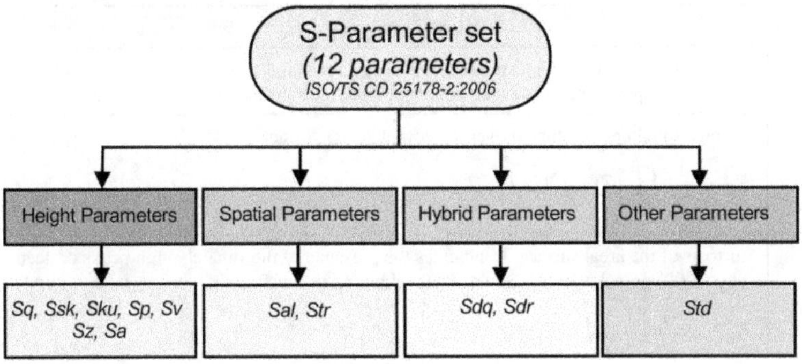

Figure 9. V-Parameter set

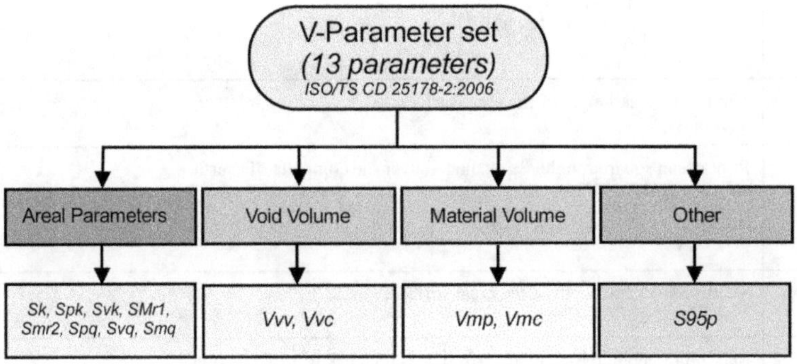

As seen in figure the S-parameter set comprises 12 parameters (definitions are given in table 4). This parameter set as being further divided into four basic types: height, spacing, hybrid, and miscellaneous. The V-parameter set (figure 9) assesses functional topographical features of the surface. They analyze material volume and void volume of a scale limited surface. Thus, the material ratio-curve of the surface is divided into three height zones: a peak, a core, and a Valley. The peak zone corresponds to initial running-in wear. The core zone corresponds to wear throughout the lifetime of the components, and the valley zone to lubricant retention under heavy wear conditions.

Table 4. Areal ISO Parameters

Height Parameters	
Flatness	Areal flatness deviation. The measure of surface deviation from perfectly flat. It is the distance between two parallel planes obtained by applying a Chebyshev fit to the surface data.
Sa	Average roughness evaluated over the complete 3D surface. $S_a = \frac{1}{A}\iint_A \lvert Z(x,y)\rvert dxdy$
Sku	Kurtosis of the areal surface. It indicates the presence of inordinately high peaks or deep valleys (Sku>3.00) or lack thereof (Sku<3.00) making up the surface. $S_{Ku} = \frac{1}{S_a^4}\left[\frac{1}{A}\iint_A \left(Z(x,y)\right)^4 dxdy\right]$
Ssk	Skewness of the areal surface. This represents the degree of symmetry of the surface heights about the mean plane. The sign of Ssk indicates the predominance of peaks (Ssk>0) or valley structures (Ssk<0) comprising the surface. $S_{Sk} = \frac{1}{S_a^3}\left[\frac{1}{A}\iint_A \left(Z(x,y)\right)^3 dxdy\right]$
Sp	Maximum peak height of the areal surface. $S_p = \max(Z(x,y))$
Sq	Root Mean square roughness evaluated over the complete 3D surface. $S_q = \sqrt{\frac{1}{A}\iint_A Z^2(x,y)dxdy}$
Sv	Maximum Valley depth of the areal surface. $S_v = \min(Z(x,y))$
Sz	Maximum height of the areal surface. It is the peak to valley height. $S_z = S_p + S_v$
Functional Parameters	
Smq	The material ratio at which the line-fits of the two characteristic linear regions of the material probability curve intersect.
Spq	The root-mean-square average of the height deviations in the peak or plateau portion of the Material Probability plot.
Svq	The root-mean-square average of the height deviations in the valley portion of the Material Probability plot.
Sxp	Peak Extreme Height. A measure of the difference in heights on the surface from the areal material ratio value of "p" and the areal material ratio of "q". According to the ISO standard, the default value for p is 2.5% and the default value for q is 50%.
Areal ISO Functional (Material Ratio) Parameters	
Sk	Core Roughness Depth. A measure of the "core" roughness (peak-to-valley) of the surface with the predominant peaks and valleys removed. This is a measure of the nominal roughness (peak-to-valley) and may be used to replace parameters such as Sz when anomalous peaks or valleys may adversely affect the measurement.
Spk	Reduced Peak Height. The area above the region of the material ratio curve which delimits the core roughness. A measure of the peak height above the core roughness. During a running in operation, Spk is the nominal height of the material that may be removed. A large Spk implies a peak dominant surface.

continues on following page

Table 4. Continued

Height Parameters	
Svk	Reduced Valley Depth. A measure of the valley depth below the core roughness. Svk impacts a surfaces ability to retain lubricant and trap debris.
Areal ISO Functional (Volume) Parameters	
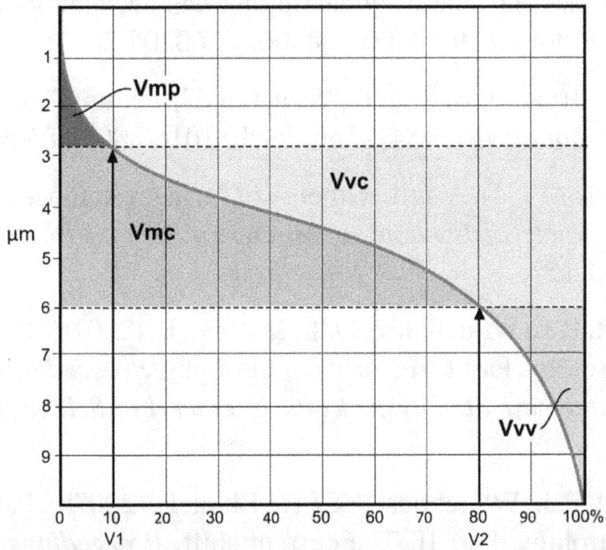	
V1	The material ratio control setting where Vmp and Vmc meet. Default value is 10% (ISO 25178-3:2008); however, it is user adjustable.
V2	The material ratio control setting where Vmc and Vvv meet. Default value is 80% (ISO 25178-3:2008); however, it is user adjustable.
Vmc	Core Material Volume. The volume of material bound by the surface texture between heights of "V1" and "V2". The default values of V1 and V2 are 10% and 80% respectively (ISO 25178-2:2012).
Vmp	Peak Material Volume. The volume of material bound by the surface texture at a height of V1 to the highest peak (Tp = 0%). The default value of V1 is 10% (ISO 25178-2:2012).
Vvc	Core Void Volume. The void volume enclosed from V1 to V2 of surface material ratio and normalized to the unit sampling area.
Vvv	Valley Void Volume. The volume of space bound by surface texture at a height of V2 to the lowest valley (Tp = 100%). The default value of V2 is 80% (ISO 25178-2:2012)

REFERENCES

Blateyron, F. (2013). The areal field parameters. In *Characterisation of areal surface texture* (pp. 15–43). Berlin: Springer. doi:10.1007/978-3-642-36458-7_2

Blunt, L., & Jiang, X. (2003). *Advanced techniques for assessment surface topography: development of a basis for 3D surface texture standards "surfstand"*. Elsevier.

De Chiffre, L., Kunzmann, H., Peggs, G. N., & Lucca, D. A. (2003). Surfaces in precision engineering, microengineering and nanotechnology. *CIRP Annals*, *52*(2), 561–577. doi:10.1016/S0007-8506(07)60204-2

Evans, C. J., & Bryan, J. B. (1999). "Structured", "textured" or "engineered" surfaces. *CIRP Annals*, *48*(2), 541–556. doi:10.1016/S0007-8506(07)63233-8

Greenwood, J. A. (1984). A unified theory of surface roughness. *Proceedings of the Royal Society of London. A. Mathematical and Physical Sciences*, *393*(1804), 133-157.

Jiang, X., Scott, P. J., Whitehouse, D. J., & Blunt, L. (2007a). Paradigm shifts in surface metrology. Part I. Historical philosophy. *Proceedings of the Royal Society A: Mathematical, Physical and Engineering Sciences*, *463*(2085), 2049-2070.

Jiang, X., Scott, P. J., Whitehouse, D. J., & Blunt, L. (2007b). Paradigm shifts in surface metrology. Part II. The current shift. *Proceedings of the Royal Society A: Mathematical, Physical and Engineering Sciences*, *463*(2085), 2071-2099.

Jiang, X. J., & Whitehouse, D. J. (2012). Technological shifts in surface metrology. *CIRP Annals*, *61*(2), 815–836. doi:10.1016/j.cirp.2012.05.009

Longuet-Higgins, M. S. (1957a). Statistical properties of an isotropic random surface. *Philosophical Transactions of the Royal Society of London. Series A, Mathematical and Physical Sciences*, *250*(975), 157–174.

Longuet-Higgins, M. S(1957b). The statistical analysis of a random, moving surface. *Philosophical Transactions of the Royal Society of London. Series A, Mathematical and Physical Sciences*, *249*(966), 321–387.

Nayak, P. R. (1971). *Random process model of rough surfaces*. Academic Press.

Nayak, P. R. (1973a, November). Some aspects of surface roughness measurement. *Wear*, *26*(2), 165–174. doi:10.1016/0043-1648(73)90132-4

Nayak, P. R. (1973b, December). Random process model of rough surfaces in plastic contact. *Wear*, *26*(3), 305–333. doi:10.1016/0043-1648(73)90185-3

Rice, S. O. (1945). Mathematical analysis of random noise. *The Bell System Technical Journal*, *24*(1), 46–156. doi:10.1002/j.1538-7305.1945.tb00453.x

Sayles, R. S., & Thomas, T. R. (1978). Surface topography as a nonstationary random process. *Nature*, *271*(5644), 431–434. doi:10.1038/271431a0

Sayles, R. S., & Thomas, T. R. (1979). Closure to "Discussion of 'Measurements of the Statistical Microgeometry of Engineering Surfaces' (1978, ASME J. Lubr. Technol., 101, p. 418). *Journal of Lubrication Technology*, *101*(4), 418–418. doi:10.1115/1.3453386

Scott, P. J. (2004). Pattern analysis and metrology: the extraction of stable features from observable measurements. *Proceedings of the Royal Society of London. Series A: Mathematical, Physical and Engineering Sciences*, *460*(2050), 2845-2864. 10.1098/rspa.2004.1291

Stout, K. J. (1993). The development of methods for the characterisation of roughness in three dimensions. Commission of the European Communities.

Stout, K. J. (1997). A review of nano-technology. *Korean J. Precis. Eng*, *14*, 29–51.

Stout, K. J., & Blunt, L. (2000). *Three dimensional surface topography*. Elsevier.

Stout, K. J., & Blunt, L. (1995a). Nanometres to micrometres: Three-dimensional surface measurement in bio-engineering. *Surface and Coatings Technology*, *71*(2), 69–81. doi:10.1016/0257-8972(95)80023-9

Stout, K. J., & Blunt, L. (2001). A contribution to the debate on surface classifications—Random, systematic, unstructured, structured and engineered. *International Journal of Machine Tools & Manufacture*, *41*(13-14), 2039–2044. doi:10.1016/S0890-6955(01)00069-4

Stout, K. J., & Blunt, L. A. (1995b, February). Application of 3-D topography to bio-engineering. *International Journal of Machine Tools & Manufacture*, *35*(2), 219–229. doi:10.1016/0890-6955(94)P2376-Q

Townsend, A., Pagani, L., Scott, P., & Blunt, L. (2017). Areal surface texture data extraction from X-ray computed tomography reconstructions of metal additively manufactured parts. *Precision Engineering*, *48*, 254–264. doi:10.1016/j.precisioneng.2016.12.008

Whitehouse, D. J. (1999). Identification of two functionally different classes of maxima in random surface waveforms. *Proceedings of the Institution of Mechanical Engineers. Part C, Journal of Mechanical Engineering Science*, *213*(3), 303–306. doi:10.1243/0954406991522662

Whitehouse, D. J., & Archard, J. F. (1969, November). The properties of random surfaces in contact. In *Surface Mechanics, Proceedings of the ASME Annual winter meeting* (pp. 16-20). Academic Press.

Whitehouse, D. J., & Archard, J. F. (1970). The properties of random surfaces of significance in their contact. *Proceedings of the Royal Society of London. Series A, Mathematical and Physical Sciences*, *316*(1524), 97–121. doi:10.1098/rspa.1970.0068

Whitehouse, D. J., & Phillips, M. J. (1978). Discrete properties of random surfaces. *Philosophical Transactions of the Royal Society of London. Series A, Mathematical and Physical Sciences*, *290*(1369), 267–298. doi:10.1098/rsta.1978.0084

Whitehouse, D. J., & Phillips, M. J. (1982). Two-dimensional properties of random surfaces. *Philosophical Transactions of the Royal Society of London. Series A, Mathematical and Physical Sciences*, *305*(1490), 441–468. doi:10.1098/rsta.1982.0043

Chapter 4
Surface Engineering Techniques

ABSTRACT

This chapter reviews both the economic and technological landscapes of surface engineering. It is shown that surface engineering aims to modify the properties of surface and subsurface layers of an object to meet a specific performance goal. The focus is on the enormous growth in the field in light of emerging technologies that enable manipulation of matter at virtually any scale. The presentation emphasizes the diversity of the subject and its relation to desired performance of objects in tribological situations. The chapter also defines the traits of an ideal tribological surface and briefly discuses obstacles that hinder materialization of the envisioned ideal surface construct.

INTRODUCTION

Properties of materials may be classified as those pertaining to the bulk and those that result from the characteristics of the surface. As seen in previous chapters the function of modern objects depends, to a great extent, on surface characteristics. Moreover, if those characteristics do not meet functional demands of the object, the structural integrity of the bulk material may be compromised. It is often the case that the ideal surface properties for a particular application are not satisfied by using certain material. To meet functional constraints, deliberate, often calculated, alteration of the properties of the surface has to take place. Alteration is realized through several routes

DOI: 10.4018/978-1-7998-1647-8.ch004

and may include changing surface topography, heat treating of surface and subsurface layers, alteration of the chemical composition of the surface, altering the microstructure of subsurface layers, etc.

The process of deliberate change, or customization, of surface and subsurface properties is known as "surface engineering". Such a process stands core of preventing catastrophic failure of engineering components. Although the scientific analysis of such a practice intensified around the early 1980s, the practice of customizing of surface properties has been around since antiquity.

Surface engineering has been practiced since the dawn of age when man learned to make and use tools, figure 1 (Frere, 1800)

Figure 1. A typical Acheulean handaxe; this example is from the Douro valley, Zamora, Spain. The small chips on the edge are from reworking

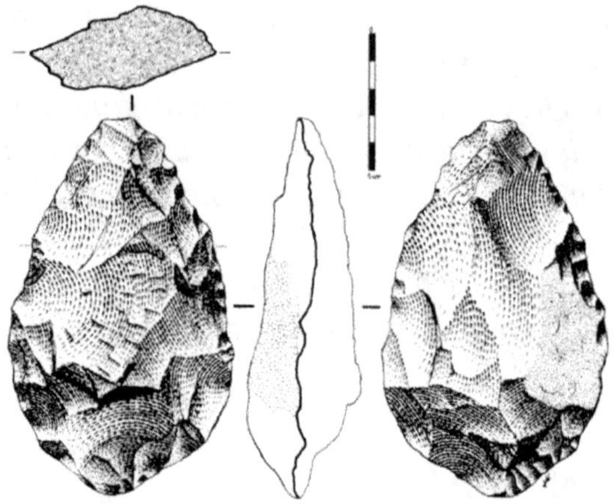

The fundamental techniques of surface modification comprise adsorption of small or large molecules onto surfaces through incorporation of a layer forming additive in a fluid layer above the surface, deliberate roughening (i.e. engineering of texture), or morphological modification of the surface. It is interesting that these techniques, in principle, have been practiced since antiquity. In ancient Egypt black patinated copper alloy objects has been produced as early as the middle Kingdom Period (2040 to 1782 BC). Production continued until disappearing after the fall of the Roman Empire (Giumlia-Mair. 1996, 2005). Studies on objects from the Egyptian Museum in Cairo (Rehren, et al. 2013), showed that the black patina was induced by thermal treatment of the surface and by application of animal glue as a surface coating (Johnson et al., 2013). Debated evidence of smelting iron from meteorites, have also been identified in tombs of Egyptian royalty (Mohamed and Darweesh, 2012). Surface engineering of sword blades is perhaps a well-known practice. It is synonymous with Damascus steel (Kennedy, D. M., 2008), and samurai-sword steels (Ranganathan and Srinivasan, 2006).

Bell (1990), reports that surface engineering has been practiced in Europe for several thousand years. The first written reference to thermal quench-hardening for treatment the weapons was referred to in 880 BC in Homer's Odyssey (Bell, 1990). Carburizing case hardening, which is an of austenitic thermochemical process, was documented in Latin by the German monk Theophilus in his "schedule diversarum artium". This 11th century manuscript clearly mentions the word "temper", which is a direct translation from the Latin "temperament". The usage of this word is equivalent to the modern usage of "quench and self-temper",which is necessity for adequate mass transfer of carbon and nitrogen from the nitrogenous compounds used at the time-urine in the case of Theophilus, to enhance surface hardness and hardenability. Hardening of steel through using organic material was common practice for quite sometimes. In medieval times many references mentioned a wide variety of organic nitrogenous compounds used in hardening of weapons and agriculture tools. The practices of nitriding by using organic compounds were not confined to the Middle East nor to Europe. It perhaps originated in Asia. Archaeological mythological studies in China (Scott, 1907-a), revealed that in the early Chinese dynasty the strength of steel swords was improved by extraction of soybean proteins.

Hardening of weapons in 17th century England was performed using burnt hoofs and horns, bag felt, old bird leather and tartar all mixed together with urine. The scientific study of this practice was initially undertaken by Scott (Scott, 1907-b), who reported of the influence of temperature, time and different materials on case hardening. His results showed that no carbon penetration takes place on heating mild steel bars in burnt leather for 4 hours at 700°; a penetration to 0.13 mm. found after similar treatment at 800°, whilst at 900° the penetration of carbon reaches 1.58 mm. At 1000° the penetration is found to be more than twice that obtained at 900° for equal times, but becomes irregular, and free cementite is found near the outside of the samples. Scott also reported that heating mild steel in a mixture of barium carbonate and wood charcoal facilitates the most rapid penetration for a short time, burnt leather coming next, whilst wood charcoal gives the least. When the heating is sufðciently prolonged, however, all used mixtures gave approximately the same results. With anthracite and hard coke (substances practically free from nitrogen) very slight carbon penetration was obtained.

Inducing carbon penetration and nitriding using natural methods is the foundation technology for many modern surface engineering techniques. It is noted that the goal, with respect to the surface, is the same across techniques meanwhile the means of inducing the planned change are different. Almost all of the traditional methods relied on heat generated from burning some organic material. Extensive research, into electrical phenomena in the late 1800s and early 1900s provided the scientific basis for new surface engineering techniques. Many of the technologies conceived at that era later reached commercial maturity.

The efforts of many physicists in the early 1900s (Thomson, 1909, Crowther, 1946, Bell, 1982) led to the discovery of electron discharge phenomena which formed the basis of plasma assisted thermo-chemical and coating technologies (Bason, 1928). Bason recognized the potential of using the glow discharge for heat treatment around 1928. That potential was materialized by Berghaus some four years later when he filed consecutive patents in England and Germany in 1940, 1941 consecutively (Berghaus,1932,1939). The patents filed by Berghaus facilitated commercial exploitation of plasma thermo-mechanical processing. Sputtering by ionic bombardment was also recognized in the 19th century. However, the practical applications of that phenomena emerged many years later towards the end of the 1930s. Stimulated emission, the phenomena responsible for laser processing, was also postulated as a novel type of radiation by Einstein in 1917. That was slated phenomena was capitalized upon by Maiman (1960) to make the first LASER in 1960. His device comprised a Ruby crystal irradiated by a Xenon flash lamp which gave a coherent single wavelength beam of red light. Surface treatment technology continues to be strong to this day. There are many advances in surface treatments especially with the advent of nanotechnology. Surface engineering continues to be an integrative part of advanced manufacturing as well. Many of the printed metallic parts have to undergo surface treatment for conditioning prior to operation.

SURFACE ENGINEERING TECHNOLOGIES

Surfaces engineering can improve the reliability, durability, and lifespan of many mechanical components. In doing so, it reduces manufacturing times, and reduces waste. Therefore, it can contribute to the reduction of anthropogenic emissions which are the root cause of global warming. Benefit to the environment, although indirectly, has contributed to the growth of the technology and its adoption by many industries, table 1. This reflects on the size of the global market of surface engineering and its various niches

Table 1. Traditional and advanced applications of surface Engineering

Processes	Types	Properties	Applications
Chemical Vapour Deposition (CVD) Physical Vapor Deposition (PVD) Ion- plating Ion implantation	Diamond film Diamond like Carbon (DLC) Titanium Nitride Titanium Carbide	Hardness Wear resistance Shock resistance Chemical resistance High Temperature Resistance Low friction	Cutting tools and Dies Drills, Mills, Punches Tapping, Reaming Extruders Injection moulding Ball Bearings Rolling mills
Weld facing	Silicon Nitride	Toughness	Mining equipment
Cladding			
			Broaching
Thermal spray	WC-Co		Forging, Quarrying
	Tungsten carbide		
Plasma spray	Aluminum oxide	Diffusion barrier	
	Silicon carbide	Impact resistance	Pump housings
	Chromium carbide	Fatigue resistance	Aircraft components
Laser surfacing	Chromium oxide	Corrosion resistance	Agricultural parts
Dipping		Surface finish	Electrical contactors
Electroplating		Electrical insulation	Engine parts

The surface engineering industry is practically split into sections. The first is surface treatment whereas the second is surface coating (figure 2). Surface treatment entails changing the texture, density, or the microstructure of the surface and immediate surface layers of an object (figure 2-b). Surface coating, meanwhile, is a process by which a cover is applied to the surface (figure 2-b). The material of that cover usually is one with properties suitable for a targeted application (e.g. diamond coatings tools, nonstick coatings for surfaces, etc.). Owing to the diversity of applications, variety of methods and usages (tables 2 a-c), and a wide spectrum of customer segments, the business of surface engineering has grown immensely. Currently that sector of global economy comprises several specialized enterprises each including numerous small and medium-size ventures.

Figure 2. The two sections of surface Engineering

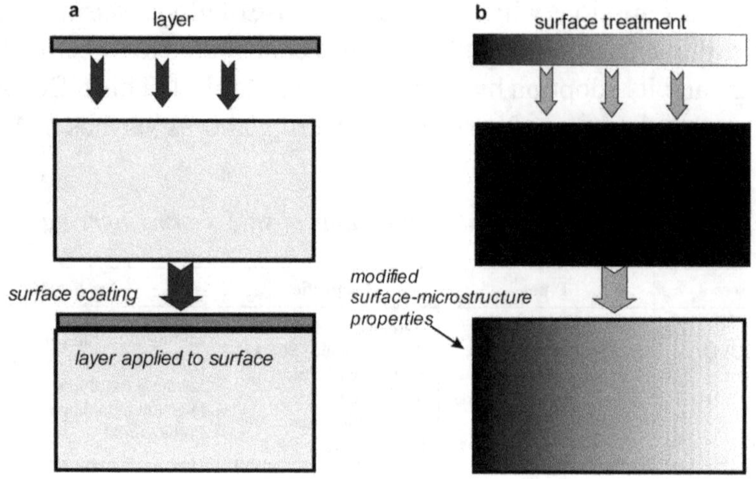

SIZE OF GLOBAL SURFACE ENGINEERING MARKET

The economic landscape drastically changed within the past two decades or so. In particular, surface engineering has emerged to the forefront due to vested interest of various stakeholders. In addition, redefined global priorities (e.g. sustainability, energy savings, natural resource management, etc.) have driven market needs. This resulted in enormous growth of global market needs. While no exact statistics concerning the actual size of the global surface engineering market exist, it is possible to envision the extent of that market through several financial estimates of selected constituents. The emerging view confirms that the size of the surface engineering global market is constantly expanding. The global paints and coatings market, for example, was worth USD 141.5 bn in 2015 and is predicted to hit the USD 164.0 bn threshold by 2021. The metal finish components of the market, for example, was USD 64.2 bn in 2014, and is expected to grow at CAGR of 6.5% by 2020.

Growth in the surface engineering market is triggered by international demand and is not confined to a certain geographical region. The Asia-Pacific region is predicted to grow at a CAGR of 6.5% until 2020. This region accounted for more than 40% of the market demand in 2015. Noting that this area has two of the upcoming powerhouses in Manufacturing, namely China and India, is easy to understand the stimulus for growth of surface engineering in that region. This is especially visible with the rate of setup of automotive and electronics industrial firms within these two countries. In Europe the expected growth is limited compared to that of the Asia-Pacific region. Compared to the CAGR of 6.5 percent for the later, surface engineering in Europe is expected to grow at a CAGR of 2% until 2020, and to remain the second largest market with around 23% share of the global market.

Table 2a. Processes and methods of the two main sections of surface engineering technology

Surface Treatment Technologies
• Mechanical finishing is often used as a pre-treatment of the product, to ensure effectiveness of further procedure. Processes like grinding, polishing, peening, blasting and deburring are used • Thermal treatment is the alternating use of heat and cold to change the properties, like hardness, of metal surfaces. This method includes processes like flame hardening, laser treatment and electron beam treatment • Laser Texturing aims to reduce wear and friction. The laser beams generate three-dimensional patterns and change the material inner structure only at an extremely thin surface layer. This makes the material almost frictionless and less prone to wear. • Flame hardening is a process where metal surfaces, such as steel, are heated using a high-temperature flame and then quenched. This results in a hard surface that is wear and corrosion resistant. The original working surface layer of the material, in this method, is modified without changing its composition. • Electron beam surface technologies are thermal high-speed heat treatment processes in solid and/or liquid state. The method is used for surface annealing, hardening, tempering or surface re-melting, alloying, dispersing, or cladding. • Diffusion hardening is a process designed to increase the hardness of steel by exposing the surface to elements which diffuse into the surface. Typical processes are carburizing, aluminizing, nitriding and boriding • Carburization is a heat treatment process in which iron or steel absorbs carbon while the metal is heated in the presence of a carbon-bearing material, such as charcoal or carbon monoxide. The process aims to increase the hardness of the material. Increase in hardness depends on duration of the process and its temperature. Longer carburizing times and higher temperatures typically increase the depth of carbon diffusion. When the iron or steel is cooled rapidly by quenching, the higher carbon content on the outer surface becomes hard due to local phase transformation while the core remains soft and tough. • Nitriding (aluminizing, and chromizing) are high-temperature diffusion processes, in which elements are assisted to diffuse into the surface layer. The diffusion of interstitials (carbon, nitrogen, and boron) makes the surface hard and, due to substitutional (chromium, aluminum) elements, results in the formation of a corrosion, and oxidation-resistant surface. • Ion implantation aims to modify the surface regarding electrical conductivity properties and is often used in the semiconductor industry. This method includes nitrogen and carbon implantation
Surface Coating Technologies
• Chemical processes add thin films of sulphide or oxide by chemical reactions. The method is used for metal coloring, corrosion protection and priming • Plating is a process to cover a surface with a metallic layer, e.g. gold, zinc and copper to increase corrosion and wear protection, reduce friction or for decorative use • Physical vapor deposition (PVD) adds a thin layer to the surface. Therefore, the coating is first transferred into vapor state by physical processes and secondly condensed on the surface. • Chemical vapor deposition (CVD) involves the formation of a thin solid film on a surface by a chemical reaction within a vacuum. The surface is exposed to the coating in gaseous state and is often used in the semiconductor industry • Thermal spraying is a method to add a melted or heated coating to a surface without heating the surface itself. Typical coatings include metals, ceramics.

Table 2b. Stakeholders of surface engineering techniques by industry

Material	Function or role	R& D	Process	Coating Type	Applications
Metals	Corrosion	Quality	Thermal	Nano	Textiles
Ceramics	Wear	Design	PVD/CVD	Micro	Sport
Polymers	Friction	Testing	Sputtering	Thin	Marine and offshore
Powder form	Tribo corrosion	Innovation	Plasma	Thick	Machinery
Rod form	Adhesion	Colour and texture	Dipping	Multi-layer	Building Industry
Atomic form	Oxidation	Thermochemical	Mechanical	Multi process	Renewable energy (PV)
Gaseous form	Fracture	Electrochemical	Heat Treatment	Hard	Transport
Liquid form	Optical	Paints & Organic coatings	Ion Implantation	Soft	Aerospace
Plasma	Fatigue	Characterisation	Laser & Electron Beam	Transparent	Tools
	Thermal barrier	Surface modelling	Electrochemical	Insulator	Biomedical
	Diffusion barrier	Mechanics	Atmospheric & Vacuum	Conductor	Cutlery & kitchenware
		Cost reduction		Tough	Electronic
		Optical	Anodising	Hard	Mining
		Bio medical		Abradable	Military
		Sports & motor racing		Weight reduction	Porosity

Generally speaking, surface treatment processes are considered less innovative than the processes of surface coatings. This reflects on the size of the market niche occupied by each of the approaches. The market for surface treatment is less diverse than that for the coating sector. This situation reflects on research activity in the field. Most activities are focused on coating techniques. As a consequence, the coating segment shows more innovation and broader applications. Coating is advantageous in many aspects.

Protective coatings may shield surface and prevent, or delay, the effects of detrimental surface phenomena (e.g. whether conditions, war, rust, etc.). Coatings may induce desirable effects on surfaces (e.g. hydrophobicity, reflectivity, etc.). Being that diverse and versatile, the coating sector is predicted to continue to grow. This predicted that the market for coating will grow to USD 10.7 bn in 2022 (figure 3).

Table 2c.

Process	Costs	Value	Applications	Functions	Flexibility
Plasma spraying	Expensive to setup. High power requirement Water/Air cooling, high material consumption.	Used for high value parts aero-engines and exhaust, wide range of coatings and thicknesses. High/rapid deposition rates.	Wide range of thick coatings. Atmospheric Small and large components. Broad range of substrates	Wear, salvage, Corrosion Erosion, Abrasion, Hardness, Friction coatings, thermal barriers etc.	Versatile, Mobile, Rapid pre-and post- treatment, Thick and thin coatings, line of site.
Electroplating	Relatively cheap to set up but environmentally unpopular, requires skilled staff.	Broad range of coatings, low waste Can be used for metallic and non metallic surfaces. Broad range of materials can be deposited.	Decorative, wear and wide range of coatings, thick and thin, Need conductive substrates.	Decorative, wear, conductive coatings,	Limited coating hardness, Sample sizes, non-mobile, non-line of site process. Limited to certain substrates, no heat treatment of substrates
CVD & PVD	Expensive to set up but relatively clean, fully automatic, requires skilled operators, expensive and slow pre cleaning	Excellent life of components, used for high value but relatively small components, not suitable for salvage,	Cutting tips and tools, implants, engine parts, high hardness, wear, impact resistant coatings	Thin coatings, high bond strength, low porosity coatings,	Restricted to small components, requires expert operators, wide range of coatings and multi-layers, not good for re-salvage work.
Sherardizing	Relatively cheap process but limited mainly to steels.	High corrosion resistant process where zinc is used on threads of nuts and bolts,	Steel components suchas metal casings, nuts and bolts where clogging of threads does not occur.	Thin uniform coating of zinc applied at approximately 370oC in rotating drum.	Relatively fast process, component size is limited by drum size. Cementation process
Carburizing	Relatively cheap to develop but limited to steels in face centered cubic crystal form.	Extremely high hardness and wear resistant properties developed to appreciable surface depths. High temperatures limit the process and cause metal distortions.	Gas, solid or liquid processes make it versatile for different substrates	Wear parts, high hardness, limited to metals, namely steels, good depth of penetration.	Samples can be masked if no heat treatment is required in certain areas of the surface, post heat treatment can be conducted on components.

Figure 3. Predicted growth in the smart coating sector

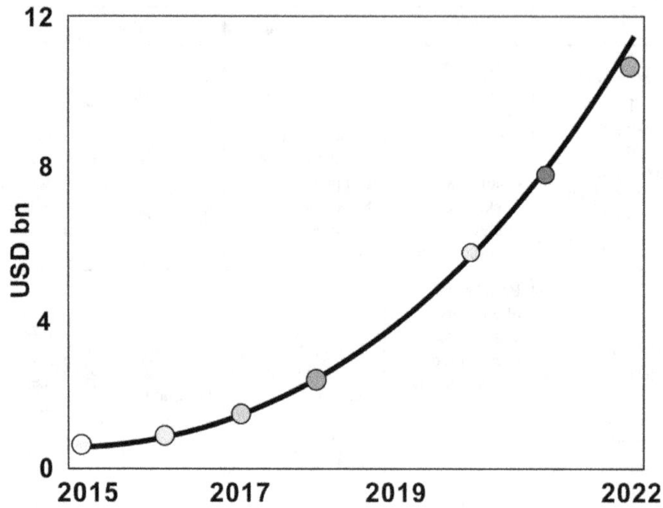

Most of the predicted growth, it is noted, is expected to take place because of maturation of research focusing on smart coatings into commercial products. That is, the growth in the coating sector is triggered by products that capitalize on functionalization of surfaces through specialized coatings. Functionalization, figure 4, increases the value of the product due to specialized properties induced on the surface. In contrast, surface treatment cannot provide the level of multi-functionality and adaptability provided by surface coatings.

A functional surface is one that has been deliberately engineered to interact with its operation environment in a targeted way. It is a surface that has the embedded ability to sense and respond to a specific stimulus or constraint in its environment. The stimuli can be heat, electricity or the presence of water, etc. Functional surfaces are created by manipulating chemical composition, the physical structure, or the electrical, optical, transport or magnetic characteristics at the surface of the material. This manipulation facilitates achieving specific behaviors (e.g., repelling or attracting water, conducting or insulating electricity or heat, or detecting and responding to specific chemicals). The science of "smart surfaces" is confined to the interface. That is, it considers the joint boundary between two surfaces or between the surface of a material and its environment. The interaction at the surface may be controlled, or guided, by changing the properties of the base material, adding a smart coating, or manipulating the surface of one or both adjoining materials at the interface.

Table 3. Some aspects of Functionalization of smart surface

Chemical: Surfaces can be altered to chemically repel or attract water or other substances or to react with another substance in specific ways.
Physical: Changing the morphology of a surface can improve adhesion between surfaces, make a surface less hospitable for bacteria or other small organisms or serve a broad range of other purposes.
Electrical: Surfaces or coatings with specific electrical properties have a wide variety of uses for medical devices, as well as consumer and industrial products.
Thermal: Thermal properties can be manipulated to provide better insulating or conducting behaviors.
Optical: Optical properties can be manipulated to control the amount of light that passes though materials.

MODIFICATION OF SURFACE TEXTURE

Modification of surface texture is a subject of interest to many industries. Many technologies evolved to meet the individual demands of industry. Differences in precision, quality of resulting surfaces, minimum size of the resulting textural feature, etc. do exist between the various technologies (figure 5). The difference in the quality of the outcome, governing dynamics, or any generalized attributes may form the basis of a classification for techniques used to engineer a surface. Thus, one may develop a classification based on chemical aspects of the process, size and accuracy, cost, etc. One feasible classification is to distinguish between techniques according to the direction of material within the surface during the modification process itself. Broadly speaking, therefore, texturing methods may fall under one of the four categories shown in figure 4.

Figure 4. Attributes of modern surfaces

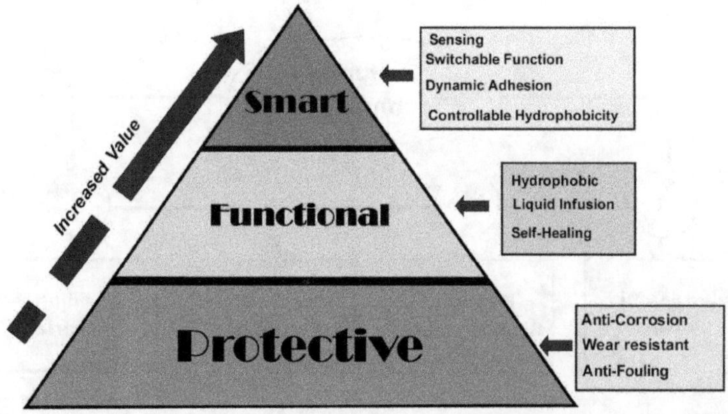

Figure 5. Ranges of Surface-engineered layer thickness obtained by the different processes

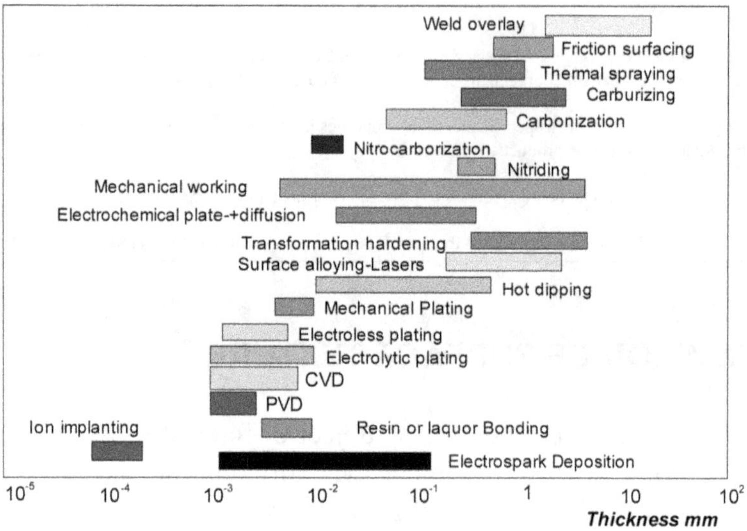

For some cases, modification of a surface may take place by adding a thin layer of a wear resistant coating, or ultra-lubricating film. In such instance modification is achieved by adding material, not originally present, to the surface. Carving texture out of the surface layer (e.g. by machining, Ion beam etching, or laser ablation) involves removal of original material from the surface (i.e. subtraction of material). Other modification methods may involve shifting of original material within the surface to enhance mechanical properties or to introduce intentional deformation (e.g in shot peening).

Figure 6. Main classification categories of surface engineering techniques

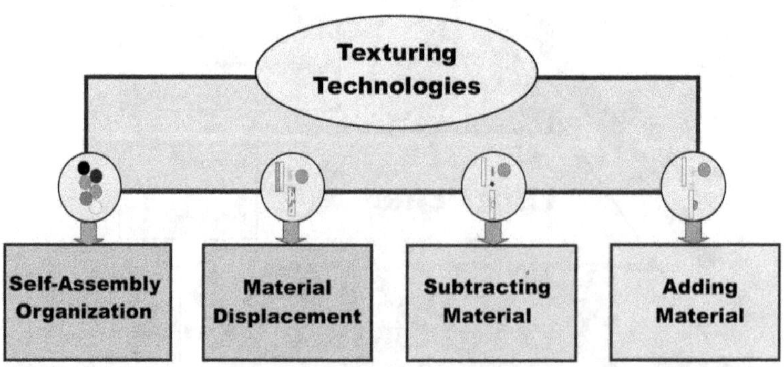

Here the mechanism involves displacement of surface-layer constituents to manipulate properties. Finally, in some cases, modification of texture by manipulating the organization or bonding between molecules (e.g. through radiation) takes place for modifications of a surface. In such case, surface engineering takes place through inducing self-organization of the target interfacial layers. Table 4 presents a comprehensive review of the available texturing/surface modification techniques categorized according to the definitions of figure 6.

The data presented imply that for each conceptual pathway to texturization, several technological routes enable material manipulation. These routes provide unprecedented opportunities to manipulate surface material, virtually, across any scale, phase, thickness, or compositional threshold. The rapid evolution of material manipulation technologies, however, is not paralleled by textural-design or conceptualization capabilities. Namely, viable design algorithms leading to true designer surfaces are still lacking. In addition, even at the conceptual level, a viable strategy to address basic technical challenges brought forward by futuristic needs haven't taken shape yet.

Table 4. Summary of available material manipulation techniques for texture engineering

Adding material		
	Conversion coatings	Chromating (Achtsnick, et. al. 2004)
		Phosphating (Achtsnick, et. al. 2004)
	Deposition coatings	Patterned chemical deposition (CVD) (Aspinwall et., al,1992)
Chemical processes		autocatalytic plating (Bartz et, al, 2000, Becker and Ludema 1999)
		Electrolytic coating,
		Anodising (Achtsnick, et. al. 2004)
		Electro-deposition (Blau, Qu, and Riester 2005, Bowden et al 1997)
	precipitation coating (Brinksmeier, at al. 2007)	
Physical deposition	Inkjet printing (Bulatov et al, 1997, Choi at al., 2004, Costa and Hutchings 2007)	
	Patterned curing (Choi at al., 2004)	
	Focused Ion Beam (Costa and Hutchings 2007)	
Removing material		

continues on following page

Table 4. Continued

High temperature	Laser methods	Laser texturing (LT) (Costa and Hutchings 2008, Daniel et al, 2000, Datta 1995)
		Masked excimer Laser (De Soete et. al., 1997, Defourny and Jodogne 1992, Deng et al.,2000)
		CNC focused lasers (Dornfeld at. Al, 2006, Dos Santos et al, 2006, Dumitru et. al., 2000)
		Femto-second lasers (Fadeai-Tehrani and Imanian 2004, Fletcher et al.,2000, Flucke et al., 2007)
	Electrical discharge machining (EDM)	Electrical discharge texturing (EDT) (Fu et al., 2001, Geiger et al., 2001)
		Micro EDM (Achtsnick, et. al. 2004)
	Ion beam texturing (IBT) (Gordon et al, 2005, Gutierrez-Mora et al., 2005)	
Chemical etching	Masking methods	Chemical texturing (Achtsnick, et. al. 2004, Heuberger et al., 2002, Hua et al., 2006, Hupp and Hart 2005)
		Electrochemical texturing (Ishibashi 1993, Kalvala et al., 1999)
	Non-masking methods	Self-assembling (Langford et al.,2002, Li and Chai 2001, Lonardo et al., 2004, Achtsnick, et. al. 2004)
		Maskless-electrochemical texturing (Lonardo et al., 2004, Mailis et al., 1999)
		Anisotropic etching (Meijer, 2004., Melcher et al., 1983)
Mechanical	CNC ultrasonic machining (Achtsnick, et. al. 2004)	
	Mechanical honing (Mott and Evans 1999, Mott et al., 1999)	
	Grinding (Nguyen and Butler 2005)	
	Free abrasive machining (Okamoto et al., 1999, Pawelski et al., 1994)	
	Micro-cutting (Pettersson, 2005, Reverberi et al., 2006, Achtsnick, et. al. 2004)	
	Patterned erosion (Reverberi et al., 2006)	
Moving material		
Mechanical	Shot blasting (Defourny and Jodogne 1992)	

continues on following page

Table 4. Continued

	Embossing	Patterned embossing tools (Sheu et. al., 1998, Slikkerveer and Veld 1999, Snell and Coombs, 2000)
		Vibro-rolling (Snell and Coombs, 2000)
Chemical	Molecular migration (Song et al., 1999)	
Self-forming		
	UV contraction (Srinivasarao 2001)	
Localized diffusion	Chemical conversion, thermochemical treatments (Achtsnick, et. al. 2004)	
	Printing + heating (Achtsnick, et. al. 2004)	
hard and soft phases	Thermal implantation of hard phases (Steinhoff et. al., 1999, Stutzmann et. al, 2000)	
	Embedding of soft phases (Tian and Chao 1996)	

Adding Material Technologies

Adding material technologies comprise texturing methods that are mostly related to the deposition of patterned coatings. Localized deposition is normally controlled by locally alteration of chemical or physical properties of the surfaces before deposition of the coating, which either promotes or inhibits the deposition in those areas.

Chemical processes involve printing on the surface with inks that can inhibit or promote some chemical or electrochemical reaction, in order to generate a patterned coating. However, when metallic substrates and/or coatings are involved, the texturing method can expose the substrate material, so that a galvanic pair can form on the surface. Therefore, in most applications, it is vital to have a process that produces a very thin coating, followed by the masking step, and then the final coating that will be textured.

In another approach, features with special characteristics are deposited onto the surface. For example, wear-resistant reliefs composed of ceramic particles can be deposited by inkjet printing. The range of materials that can be inkjet printed on a surface includes sol–gel materials, conducting polymers, structural polymers, organic transistors, ceramics and metals. The current resolution of the technique is believed to be approximately 15 μm.

Focused ion beam technique (FIB) can also be used for local deposition of material, with excellent resolution (<100 nm), however, the process is very slow. Thermoset polymers cannot be made more plastic through heating, as is currently done for the embossing of thermoplastic polymers. However, when a textured coating of a thermoset polymer is desired, a liquid resin can be deposited on the surface, and a tool with the desired texturing then used to emboss the resin. The tool must be used when the resin is still plastic, which occurs during curing (Song et al., 1999).

Nano-surface textures can be produced by spin coating a colloidal nanoparticle solution onto a flat substrate. Its main advantage is the flexibility of the process in terms of the material, size and density of the nano-texture produced, because a large variety of nanoparticles are commercially available (Zimmer at al., 2000).

Self-organizing effects during the growth of thin polycrystalline films can lead to coatings with columnar microstructures and surface corrugations in the 100 nm range (Deng et al.,2000). The cleaning of the surface prior to the coating is crucial and it involves fairly sophisticated deposition equipment. The selection of the deposition parameters must be very carefully chosen to ensure columnar growth. TiAlN and TiO2 have been successfully deposited with nano-sized diamond-like relief on silicon surfaces, but the growth mode has not been detected to be dependent upon the substrate material (Wuhrer and Yeung 2004).

Removing Material Technologies

Chemical Texturing

In this group of methods, chemical reagents are selectively applied to a surface to remove material in specific regions, in order to produce texturing. Material is removed from selected areas by immersing it in a chemical etchant. Etching only in selected areas can be achieved either by masking, or by self-assembling of solid particles and/or or liquids to form a mask. Material removal results from a microscopic electrochemical cell action, equivalent to that responsible for corrosion or chemical dissolution of a metal. The resolution of the features that compose the texture depends strongly on the method used to mask the surface.

Depending on the printing method, texturing can also be applied to surfaces with irregular shapes. Also, there are few restrictions related to the shape and distribution of the features that compose the surface texture. The overall speed of chemical texturing methods is restricted by the speed of both the printing method and etching process. Normally, these methods change the surface topography, but they do not change the mechanical properties of the textured region. The main equipment costs are related to the printing method. Figure 7, shows an example of a texture obtained on copper by using photochemical etching (PCE) and the corresponding pseudo-color map of the surface. The copper surface can be etched in different conditions: changing the rate of reagent diffusion, the temperature of the eroding bath and the geometry of the masks protecting parts of the substrate.

Figure 7. PCE patterned surface.

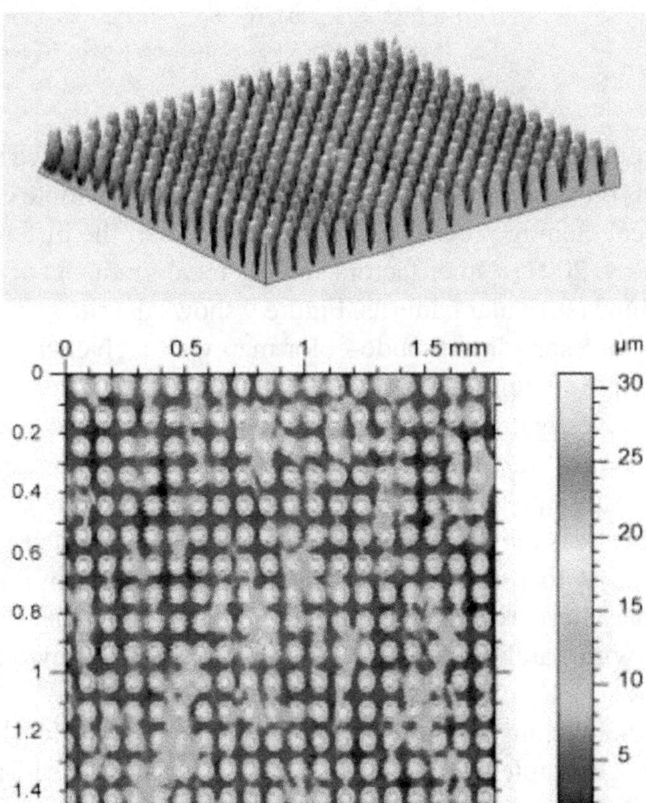

Figure 8. Desired attributes in modern surfaces as indicated by industrial stake holders

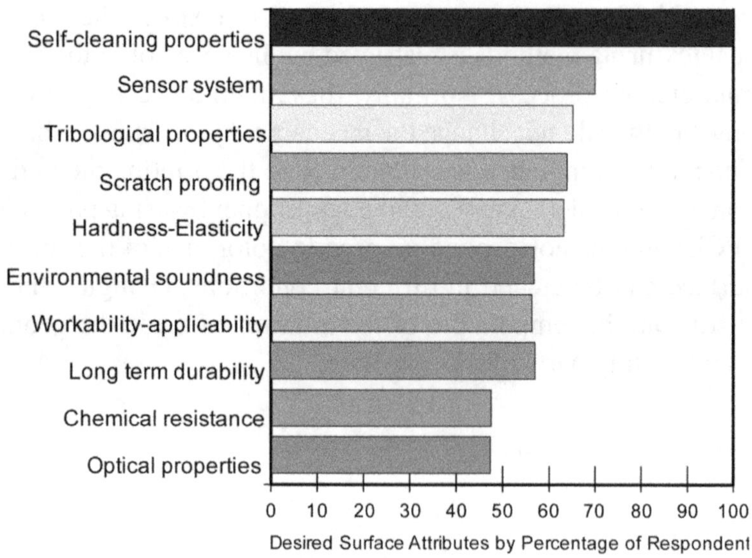

Change from a reaction-limited etching to a diffusion-limited etching can be observed as the eroding bath temperature increases. A Monte Carlo model of the chemical etching process can take into account the different erosion conditions (Roy, 2007). Other factors such as local gradients of the etching media determine particular patterns. Figure 9 shows the surface of an etched hole and the corresponding pseudo-color map with a characteristic pattern due to convective streams in the liquid etcher.

Costa also investigated other possible masking methods to be used before etching for chemical texturing. The use of inkjet printing for masking steel surfaces, with subsequent etching and ink stripping, proved to be a versatile, fast and highly feasible technique for texturing steel surfaces (Costa and Hutchings 2008). Other surfaces could potentially be textured by this technique using adequate alternative etchants. Figure 9 show the inkjet printing of a steel surface with parallel gaps and in multi-steps with chevron-like gaps, respectively.

Some special etching methods that do not require prior masking can be envisaged. For example, methods involving simultaneous self-assembling of particles (Bowden et al 1997), droplets, or bubbles (Steinhoff et. al., 1999), followed by etching or methods involving the formation of locally etched regions created by acid droplets (Gutierrez-Mora et al., 2005), could considerably reduce texturing time and process costs. These methods generally rely on the hydrophobicity and hydrophilicity of surfaces, which allow small particles or droplets to arrange themselves, forming patterns. The assembly process depends on the differences in surface energy between the various components.

Mechanical Methods

Material can also be removed from a surface by mechanical action. A cutter could be used to remove material from selected areas of a sample. If the surface is masked with an erosion-resistant mask, material can be removed by the action of a high-velocity stream of abrasive particles. If no masking is applied, the topography can be controlled by conventional surface finishing methods, such as mechanical honing, grinding, polishing, etc. In abrasive ultrasonic machining (USM), if a CNC table moves the sample, different patterns can be created on the surface.

Figure 9. Major attributes of an ideal tribo-surface in the futuristic sense. The surface is considered a smart entity that manages the complexity arising from interaction of function, geometry, and material

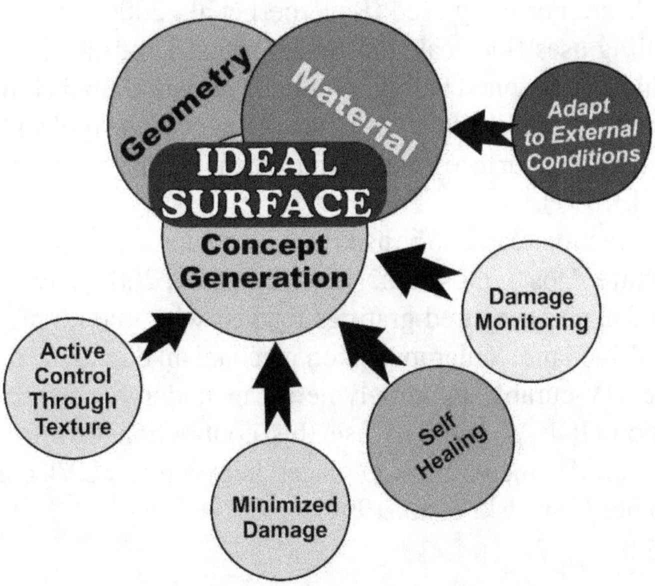

Moving Material Technologies

In this group, plastic deformation of surfaces is used to form the surface topography. Some increase in surface hardness can be obtained by strain hardening, but it is generally less pronounced than that obtained by high-temperature material removal methods. These methods are generally restricted to ductile surfaces, such as thermoplastic polymeric materials or metals of high ductility, but very hard embossing tools can also be used to emboss ceramic surfaces.

In shot blasting (SB), hard shot particles impact the surface at high velocity to cause impact damage and roughening of the surface. The change in the surface structure is attributable to plastic deformation. It is difficult to precisely control the texture because of the large number of variables inherent in the process, such as the size, shape, velocity and hardness of the shot. Also, shot blasting can only form random textures.

To allow for the ease of embossing of thermoplastic materials, the surface to be textured must be heated until it becomes plastic. The embossing tools do not need to be very hard, which lowers the tooling costs. This is a simple and versatile way to create well-resolved structures on thermoplastic substrates. The ductility of the substrate permits it to be embossed into regular, micrometric-sized structures (Tian and Chao 1996). For harder materials, micro-structured diamond tools are normally used (Reverberi et al., 2006).

Vibro-rolling uses fine-scale plastic deformation of metals. A hard tool, normally a highly hardened ball or a rounded diamond tip, is forced into the surface. Complicated kinematics of the movement of the tool and the sample produces a uniform surface microrelief, with regularly shaped asperities (Bulatov et al, 1997).

Chemical reactions can also be used to move material and therefore produce a surface texture. Dos Santos et al. (Dumitru et. al., 2000) have reported the fabrication of nano-structured gratings formed with mass transport arising from a light-driven molecular migration mechanism due to the use of a laser beam. Some UV-curable photopolymers can undergo contraction during polymerization. It is possible to use this contraction to form a textured polymer surface if a metal mask is placed between the UV source and the liquid monomer (Pawelski et al., 1994).

Self-Forming Technologies

Once produced, a surface texture may not have sufficient mechanical resistance to survive high loads. Severe stresses, especially the shear stress, may wear away surface textures within the first few cycles. This takes place because the uniformity of wear properties of the textured surface at all points. Raised areas of the surface, which support larger portion of applied loads, wear more rapidly than the surrounding material. Research (Flucke et al., 2007) suggest that the creation of a surface structure that is maintained over a long functional period requires a method that capitalizes on wear itself to produce the required surface structure. These methods are called 'self-forming by wear'. They use localized wear-resistant regions on a workpiece surface; these regions can be achieved by having a local gradient in mechanical properties, which ensures that wear occurs at different rates in different regions, leading to the development of geometrical structures. A gradient in mechanical properties can be created by using known metallurgical treatments. Shot blasting and subsequent polishing prior to the use of the surface wears the substrate at a rate higher than that of the hard regions, and therefore produces a surface texture (Gutierrez-Mora et al., 2005).

The Ideal Tribological Construct

Surface engineering is cross-disciplinary. It utilizes appropriate technologies to optimize surface properties and meet design constraints for pre-defined applications, all at minimized cost. In this sense, the field is a conduit of ideas, technologies, and expertise that bridges the needs of stakeholders, at the application end, the skills of surface technologists, and visions of researchers. Stake holders, by virtue of being at the forefront of consumer needs, may very well require a complex array of attributes within their particular surfaces. Planning forward necessitates forecasting future surface technical needs. Accordingly, many measures were taken within the past twenty years to address this requirement. Efforts aim at systematically quizzing stakeholders about future surface needs, therefore, enhancing ability to forecast the needs of, and necessary level of interaction between, of the various constituents of the field.

Figure 8 provides an example of the output. The data graphs categorized responses to an industry-wide questionnaire forwarded to German manufacturers. Bars in the plot represent the percentage of responses among all quizzed stakeholders that anticipate a need for a given surface attribute within products. Percentages reflect priority of manufacturers, and therefore aid in predicting futuristic views of surfaces and their function.

In order for an operational engineered designer-surface, in the futuristic sense, comes into existence. The top four priorities, self-cleaning, sensory, tribological performance, and scratch resistance are interdependent, and their co-existence within a surface reflects recognition of the surface as a "smart system" rather than an "encasing" component. This view requires new insights into surface design. Focusing on tribo-surfaces, a logical first step toward a new insight is to define what constitutes an ideal smart tribo-surface. Once attributes of such a construct are defined the scope of the design process and its philosophy should be easier to devise. Identification of ideal attributes is not simple due to the complexity of the factors involved and their level of interference. The process is further complicated by the view of a surface as a smart system, which is not only capable of satisfactorily functioning at an initial point, but also is capable of self-monitoring and-control of performance.

The question of what is an ideal tribological surface although straight forward, has no simple answer. This is because of the complexity of the factors involved and their level of interconnectedness. The answer is further complicated by the view of a surface as a smart system that not only is able to satisfactorily function at an initial point, but also is capable of self-monitoring and of controlling it performance. Keeping in mind the limitations, we may compose a "generic list" that highlights the major attributes favored to exist within a tribo-surface. Friction being a surface phenomenon is associated with energy losses, material damage and mass loss. Therefore, initial iteration yields those attributes directly related to generalized function and to structural integrity during operation. Keeping these factors as the basis of our requirement list, we arrive at the attributes shown in figure 6. These are: minimized damage, control of tribological response, maintaining a preset damage limit, adaptability to changes in external rubbing conditions, self-healing-ability, and self-condition monitoring.

The attributes shown reflect complex interaction between material choice, geometry, and manufacturing techniques. The later factor, as previously mentioned is currently well defined and advanced. As such, successful design and implementation of an engineered surface within the modern view requires advanced design paradigms that emphasize management of the complexity arising from interaction between functional attributes and performance constraints. Figure 9 presents a thematic map that lists the major activities within a so-called "Engineered Surface technology pipeline". The map illustrates the sequence of activities that should take place. The map shows that the crucial step is idea generation. Again, the difficulty arises from limited experience with holistic design of surfaces where multiple functional demands, which initially may be conflicting, have to exist and despite conflicts the surface has to function efficiently as a coherent unit.

Figure 10. Sequence of activities within the Engineered Surface Pipeline. The map shows the interplay between concept generation, standardization, and implementation of a functional-designer surface.

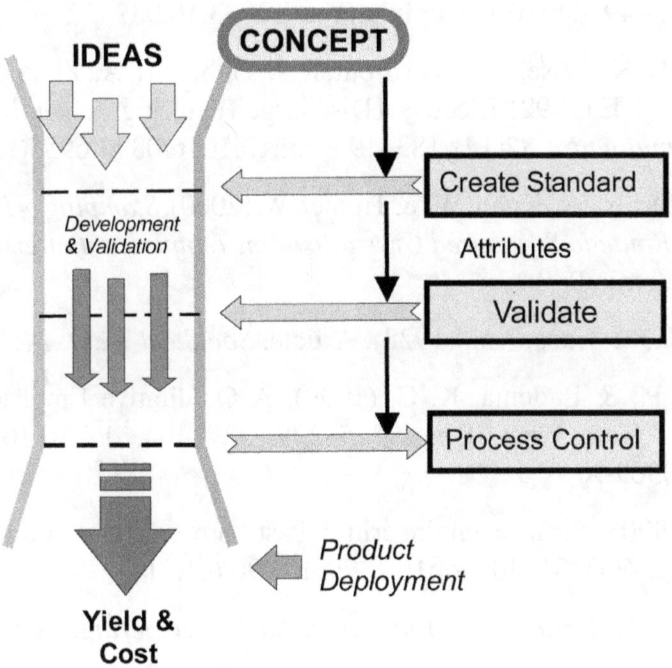

Currently there exist many methods and technologies to implement textures from nano to micron size on virtually any material. To date, however, there is no agreement on the optimal values to be implemented given a particular surface. More importantly, a methodology for generation of specialized functional textures for tribological surfaces is virtually non-existent. Design of texture is still viewed as "black art" to the effect that currently there exists a gap between available enabling texturing technologies and conceptual texture design paradigms. Crucially absent is a holistic texture-generation methodology that merges form and topology of texture subject to function constraints. Such a paradigm while not mature within the realm of human surface engineering, is advanced in natural surface constructs. Consequently, currently many researchers look for natural systems for mimicry or inspiration.

REFERENCES

Achtsnick, M., Drabbe, J., Hoogstrate, A. M., & Karpuchewski, B. (2004). Erosion Behaviour and Pattern Transfer Accuracy of Protecting Masks for Micro- Abrasive Blasting. *Journal of Materials Processing Technology*, *149*(1-3), 43–49. doi:10.1016/j.jmatprotec.2003.10.037

Aspinwall, D. K., Wise, M. L. H., Stout, K. J., Goh, T. H. A., Zhao, F. L., & El-Menshawy, M. F. (1992). Electrical Discharge Texturing. *Journal of Machine Tools & Manufacture*, *32*(12), 183–193. doi:10.1016/0890-6955(92)90077-T

Bartz, M., Terfort, A., Knoll, W., & Tremel, W. (2000). *Stamping of Monomeric SAMs as a Route to Structured Crystallisation Templates: Patterned Titania Films*. Academic Press.

Bason, G. F., & Trans, Am. (1928). Article. *Soc Steel Treat*, *14*, 932–934.

Becker, E. P., & Ludema, K. C. (1999). A Qualitative Empirical Model of Cylinder Bore Wear. *Wear*, *225–229*, 387–404. doi:10.1016/S0043-1648(98)00369-X

Bell, T. (1990). Surface engineering: Past, present, and future. *Surface Engineering*, *6*(1), 31–40. doi:10.1179ur.1990.6.1.31

Bell, T. (1982). *Dearnley in Plasma surface engineering*. Senlis, France: CETIM.

Berghaus, B. German Patent, DPR, 668629 1932

Berghaus, B. German Patent, DPR, 851560, 1939

Blau, P. J., Qu, J., & Riester, L. (2005) Effects of Laser Surface Texturing on the Micro- structure and Frictional Characteristics of Zirconia Surfaces Under Liquid and Solid Lubrication. *World Tribology Congress*, 311–312. 10.1115/WTC2005-63197

Bowden, N., Terfort, A., Carbeck, J., & Whitesides, G. M. (1997). Self-Assembly of Mesoscale Objects into Ordered Two-Dimensional Arrays. *Science*, *276*(5310), 232–235. doi:10.1126cience.276.5310.233

Brinksmeier, E., Glabe, R., Riemer, O., & Twardy, S. (2007). Precision Cutting Processes for the Manufacture of Micro Molds. *Proceedings of the 2nd International Conference on New Forming Technology*, 1–13.

Bulatov, V. P., Krasny, V. A., & Schneider, Y. G. (1997). Basics of Machining Methods to Yield Wear- and Fretting-Resistive Surfaces, Having Regular Roughness Patterns. *Wear*, *208*(1-2), 132–137. doi:10.1016/S0043-1648(96)07403-0

Choi, K. H., Meijer, J., Masuzawa, T., & Kim, D. (2004). Excimer Laser Machining for 3D Microstructure. *Journal of Materials Processing Technology*, *149*(1-3), 561–566. doi:10.1016/j.jmatprotec.2004.03.005

Costa, H. L., & Hutchings, I. M. (2007). Hydrodynamic Lubrication of Textured Steel Surfaces Under Reciprocating Sliding Conditions. *Tribology International*, *40*(8), 1227–1238. doi:10.1016/j.triboint.2007.01.014

Costa, H. L., & Hutchings, I. M. (2008). Inkjet Printing for Patterning Engineering Surfaces. *Proceedings of 24th Conference on Digital Fabrication*, 256–259.

Crowther, J. A. (1946). *Iron, electrons and ionizing radiations* (7th ed.). London: Edward Arnold.

Daniel, J. H., Moore, D. F., & Walker, J. F. (2000). Focused Ion Beams and Silicon-on- Insulator: A Novel Approach to MEMS. *Smart Materials and Structures*, *9*(3), 284–290. doi:10.1088/0964-1726/9/3/306

Datta, M. (1995). Fabrication of an Array of Precision Nozzles by Through-Mask Electrochemical Machining. *Journal of the Electrochemical Society*, *142*(11), 3801–3806. doi:10.1149/1.2048416

De Soete, D., Pans, D., & Steinhoff, K. (1997). *EBT Technology and its Applications*. Academic Press.

Defourny, J., & Jodogne, P. (1992). Lasertex, A Concept of Roughness—From the Rolling-Mill to the Coated Sheet. *Revue de Métallurgie*, *89*(5), 439–446. doi:10.1051/metal/199289050439

Deng, T., Wu, H., Brittain, S. T., & Whitesides, G. M. (2000). Prototyping of Masks, Masters and Stamps/Molds for Soft Lithography Using an Ofðce Printer and Photographic Reduction. *Analytical Chemistry*, *72*(14), 3176–3180. doi:10.1021/ac991343m

Dornfeld, D., Min, S., & Takeuchi, Y. (2006). Recent Advances in Mechanical Micro- machining. *Annals of the CIRP*, *55*(2), 745–768. doi:10.1016/j.cirp.2006.10.006

Dos Santos, D. S., Cardoso, M. R., Leite, F. L., Aroca, R. F., Mattoso, L. H. C., Oliveira, O. N., & Mendonca, C. R. (2006). The Role of Azopolymer/Dendrimer Layer-by-Layer Film Architecture in Photo induced Birefringence and the Formation of Sur- face-Relief Gratings. *Langmuir*, *22*(14), 6177–6180. doi:10.1021/la060399q

Dumitru, G., Romano, V., Weber, H. P., Haefke, H., Gerbig, Y., & Pñuger, E. (2000). Laser Micro structuring of Steel Surfaces for Tribological Applications. *Applied Physics. A, Materials Science & Processing*, *70*(4), 485–487. doi:10.1007003390051073

Einstein, A. (1906). Article. *Annalen der Physik*, *19*, 371–379.

Einstein, A. (1917). Article. *Physikalische Zeitschrift*, *18*, 121.

Fadeai-Tehrani, A., & Imanian, E. (2004). A New Etchant for the Chemical Machining of St304. *Journal of Materials Processing Technology*, *149*, 404–408.

Fletcher, D. I., Kapoor, A., Steinhoff, K., & Schuleit, N. (2000). Wear BEhaviour and Surface form Evaluation of a Novel Titanium Carbide Implanted Surface Under Lubricated Conditions. *Proceedings of the Institution of Mechanical Engineers*, 597–610.

Flucke,, C., Glabe, R., & Brinksmeier, E. (2007). Manufacturing of Moulds for the Replication of Prismatic Microstructures by a Novel Diamond Cutting Process. *Industrial Diamond Review*, *67*, 25–30.

Frere, J. (1800). Wikisource-logo.svg "Account of Flint Weapons Discovered at Hoxne in Suffolk." *Archaeologia*, *13*, 204–205. doi:10.1017/S0261340900024267

Fu, Y. H., Ye, Y. X., Zhang, Y. K., & Cai, L. (2001). The Technology of Laser Honing Applied in Distinctively Improving the Lubrication of Frictional Units. *Key Engineering Materials*, *202–203*, 265–270. doi:10.4028/www.scientific.net/KEM.202-203.265

Geiger, M., Kleiner, M., Eckstein, R., Tiesler, N., & Engel, U. (2001). *Microforming*. Academic Press.

Giumlia-Mair, A. (2005, September 1). On surface analysis and archaeometallurgy. *Nuclear Instruments & Methods in Physics Research. Section B, Beam Interactions with Materials and Atoms*, *239*(1-2), 35–43. doi:10.1016/j.nimb.2005.06.178

Giumlia-Mair, A. R. (1996). Das Krokodil und Amenemhat III. aus el-Faiyum. Hmti km-Exemplare aus dem mittleren Reich. *Antike Welt*, *27*(4), 313–321.

Gordon, N., Starosvetsky, D., & Ein-Eli, Y. (2005). Negative Potential Dissolution (NPD)-Advanced and Rapid Texturing Method of As-Cut Silicon. *Electrochimica Acta*, *50*(27), 5313–5321. doi:10.1016/j.electacta.2005.03.008

Gutierrez-Mora, F., Erdemir, A., Goretta, K. C., Dominguez-Rodriguez, A., & Routbort, J. L. (2005). Dry and Oil-Lubricated Sliding Wear of Si3N4 and Si3N4/BN Fibrous Monoliths. *Tribology Letters*, *18*(2), 231–237. doi:10.100711249-004-2747-6

Heuberger, M., Feldman, K., Zach, M., & Spencer, N. D. (2002). Dynamic Control of Friction via Surface Structuring. In D. Dowson (Ed.), *Boundary and Mixed Lubrication: Science and Applications* (pp. 67–73). Elsevier. doi:10.1016/S0167-8922(02)80008-6

Hua, M., Tam, H. Y., Ma, H. Y., & Mok, C. K. (2006). Patterned PVD TiN Spot Coatings on M2 Steel: Tribological Behaviors Under Different Sliding Speeds. *Wear*, *260*(11–12), 1153–1165. doi:10.1016/j.wear.2005.07.014

Hupp, S. J., & Hart, D. P. (2005). Quantifying the Effect of Lubricant Elasticity on Micro textured Surfaces. *Iron and Steel Engineer*, *74*(9), 36–40.

Ishibashi, K. (1993). *Methods for Treating the Surface of a Solid Body*. Patent US 5,198,073.

Johnson, D., Tyldesley, J., Lowe, T., Withers, P. J., & Grady, M. M. (2013, June). Analysis of a prehistoric Egyptian iron bead with implications for the use and perception of meteorite iron in ancient Egypt. *Meteoritics & Planetary Science*, *48*(6), 997–1006. doi:10.1111/maps.12120

Kalvala, P. R., Beesabathina, D. P., & Wiedemann, K. E. (1999). *Process for Forming a Wear-Resistant Coating that Minimizes Debris*. US Patent 5,955,145.

Kennedy, D. (n.d.). Exploitation of surface engineering technology research. *Proceedings of International Conference, Matrib Materials, Tribology and Processes*.

Langford, R. M., Dale, G., Hopkins, P. J., Ewen, P. J. S., & Petford-Long, A. K. (2002). Focused Ion Beam Micromachining of Three-Dimensional Structures and Three- Dimensional Reconstruction to Assess their Shape. *Journal of Micromechanics and Microengineering*, *12*(2), 111–114. doi:10.1088/0960-1317/12/2/303

Li, L., & Chai, D. M. (2001). The Research and Application of Honing Technology. *Key Engineering Materials*, *202–203*, 385–388. doi:10.4028/www.scientific.net/KEM.202-203.385

Lonardo, P. M., De Chiffre, L., & Bruzzone, A. A. G. (2004). Characterisation of Functional Surfaces. *Proceedings of 2nd International Conference on Tribology in Manufacturing Processes*, 39–62.

Mailis, S., Zergioti, I., Koundourakis, G., Ikiades, A., Patentalaki, A., Papakonstantinou, P., ... Fotakis, C. (1999). Etching and Printing of Diffractive Optical Microstructures by a Femtosecond Excimer Laser. *Applied Optics*, *38*(11), 2303–2308. doi:10.1364/AO.38.002301

Maiman, T. H. (1960, June 1). Optical and microwave-optical experiments in ruby. *Physical Review Letters*, *4*(11), 564–566. doi:10.1103/PhysRevLett.4.564

Meijer, J. (2004). Laser Beam Machining (LBM), State of the Art and New Opportunities. *Journal of Materials Processing Technology*, *149*(1-3), 2–17. doi:10.1016/j.jmatprotec.2004.02.003

Melcher, R. L., Romankiw, L. T., & Von Gutfeld, R. J. (1983). *Method for Maskless Chemical Machining*. Patent US 4,379,022.

Mohamed, W., & Darweesh, S. (2012, February). Ancient Egyptian black-patinated copper alloys. *Archaeometry*, *54*(1), 175–192. doi:10.1111/j.1475-4754.2011.00602.x

Mott, M., & Evans, J. R. G. (1999). Zirconia/Alumina Functionally Graded Material Made by Ceramic Ink Jet Printing. *Materials Science and Engineering A*, *271*(1-2), 344–352. doi:10.1016/S0921-5093(99)00266-X

Mott, M., Song, J. H., & Evans, J. R. G. (1999). Microengineering of Ceramics by Direct Inkjet Printing. *Journal of the American Ceramic Society*, *82*(7), 1653–1658. doi:10.1111/j.1151-2916.1999.tb01982.x

Nguyen, T. A., & Butler, D. L. (2005). Simulation of Precision Grinding Process. Part 1. Generation of the Grinding Wheel Surface. *International Journal of Machine Tools & Manufacture*, *45*(11), 1321–1328. doi:10.1016/j.ijmachtools.2005.01.005

Okamoto, T., Mori, M., Karasawa, T., Hayakawa, S., Seo, I., & Sato, H. (1999). Ultraviolet-cured Polymer Microlens Arrays. *Applied Optics*, *38*(14), 2991–2996. doi:10.1364/AO.38.002991

Pawelski, O., Rasp, W., Zwick, W., Nettelbeck, H. J., & Steinhoff, K. (1994). The Iñuence of Different Work-Roll Texturing Systems on the Development of Surface Structure in the Temper Rolling Process of Steel Sheet used in the Automotive Industry. *Journal of Materials Processing Technology*, *45*(1-4), 215–222. doi:10.1016/0924-0136(94)90343-3

Pettersson, U. (2005). *Surfaces Designed for High and Low Friction* (PhD Thesis). Uppsala University.

Ranganathan, S., & Srinivasan, S. (2006, June 1). A tale of Wootz steel. *Resonance.*, *11*(6), 67–77. doi:10.1007/BF02838885

Rehren, T., Belgya, T., Jambon, A., Káli, G., Kasztovszky, Z., Kis, Z., ... Pigott, V. C. (2013, December 1). 5,000 years old Egyptian iron beads made from hammered meteoritic iron. *Journal of Archaeological Science*, *40*(12), 4785–4792. doi:10.1016/j.jas.2013.06.002

Reverberi, A., Del Borghi, A., Briano, C., Bruzzone, A., & Massei, M. E. A. (2006) Mass Transfer Effects in Copper Surface Chemical Etching Through Masks. Summer Computer Simulation Conference SCSC2006, Calgary, Canada.

Roy, S. (2007). Fabrication of Micro- and Nano-Structured Materials Using Mask-Less Processes. *Journal of Physics. D, Applied Physics*, *40*(22), R413–R426. doi:10.1088/0022-3727/40/22/R02

Scott, G. S. (1907). Article. *Mechanical Engineering (New York, N.Y.)*, *20*, 452–454.

Scott, G. S. (1907). *Iron and Steel Inst.* Scpt.

Sheu, S., Hector, L. G., & Richmond, O. (1998) Tool Surface Topographies for Con- trolling Friction and Wear in Metal-Forming Processes. *Journal of Tribology— Transactions of the ASME, 120*(3), 517–527.

Slikkerveer, P.J., & Veld, F.H. (1999). Model for Patterned Erosion. *Wear, 233–235*, 377–386.

Snell, D., & Coombs, A. (2000). Novel Coating Technology for Non-Oriented Electrical Steels. *Journal of Magnetism and Magnetic Materials, 215-216*, 133–135. doi:10.1016/S0304-8853(00)00095-0

Song, J. H., Edirisinghe, M. J., & Evans, J. R. G. (1999). Formulation and Multilayer Jet Printing of Ceramic Inks. *Journal of the American Ceramic Society*, *82*(12), 3374–3380. doi:10.1111/j.1151-2916.1999.tb02253.x

Srinivasarao, M., Collings, D., Philips, A., & Patel, S. (2001). Three-Dimensionally Ordered Array of Air Bubbles in a Polymer Film. *Science*, *292*(5514), 79–83. doi:10.1126cience.1057887

Steinhoff, K., Kapoor, A., & Guillon, N. (1999). Controlled Wear as Mechanism for the Design of Geometrically Deðned Nanometric Surface Structures on Forming Tools. *Advanced Technology of Plasticity*, *1*, 265–270.

Stutzmann, N., Tervoort, T. A., Bastiaansen, K., & Smith, P. (2000). Patterning of Polymer-Supported Metal Films by Microcutting. *Nature*, *407*(6804), 613–616. doi:10.1038/35036545

Thomson, J. J. (1909). Article. *Philosophical Magazine*, *18*(6), 441.

Tian, H., & Chao, A. M. (1996). Optimization of Disk Surface Texturing and Lubrication on CSS Performance. *IEEE Transactions on Magnetics*, *32*(5), 3666–3668. doi:10.1109/20.538722

Wilson. (1900). Article. *Phil Mag., 49*(5), 505.

Wuhrer, R., & Yeung, W. Y. (2004). Grain Reðnement with Increasing Magnetron Discharge Power in Sputter Deposition of Nanostructured Titanium Alumi- nium Nitride Coatings. *Scripta Materialia*, *50*(6), 813–818. doi:10.1016/j.scriptamat.2003.12.022

Zimmer, K., Braun, A., & Bigl, F. (2000). Combination of Different Processing Methods for the Fabrication of 3D Polymer Structures by Excimer Laser Machining. *Applied Surface Science*, *154–155*, 601–604. doi:10.1016/S0169-4332(99)00423-7

Chapter 5
Texturing for Technological Surfaces

ABSTRACT

This chapter explores methods of engineering the elements of the roughness of a tribological surface and the effect of that process on performance. Texture and topography have a crucial influence on the tribological performance of surfaces. The presentation focuses on the dominant technology of laser texturing. The authors review the principles of the texturing process, its effects on friction and wear, and then discuss existing technical difficulties. It is shown that despite being practiced for a long time, and being promising, laser texturing still lacks standardization. To date, there is no agreement on a standardized design methodology that yields optimized textural features that meet the requirements of a prespecified application.

INTRODUCTION

Texture and topography have a crucial influence on the tribological performance of surfaces. The previous chapters detailed the classification and characterization of topography. It was shown that the roughness and texture of the surface, along with their distribution, are directly linked to the surface loadbearing capacity. This has been pointed out by Abbott and Firestone (Abbott and Firestone 1933) in the classic work. Such an influence is to be expected on physical basis since texture is at the forefront of any contact or sliding taking place between two complying solids.

DOI: 10.4018/978-1-7998-1647-8.ch005

The connection between roughness distribution and loadbearing, as detailed in chapter 3, formed the basis of the so-called areal performance parameters (i.e. the so-called Sk-family). These parameters are supposed to predict the suitability of the surface for tribological events (including anticipation of weigh-in and lubricant retention). The relation between roughness distribution, and its geometry, to loadbearing and friction performance, however, is more rooted in friction laws as formulated by Amonotons-Coloumb and others (Amontons 1699, Bélidor, 1782, Coulomb, 1801).

When two surfaces are brought into contact under the influence of the normal force, the roughness asperities on both bodies will deform. This deformation is proportional to the portion of the load befalling each of the asperities on each surface. The low share of asperities, in turn, is proportional to the height of the individual roughness features. The longest asperity establishes contact first, shorter ones follow, and the process continues until the normal load is totally supported by the roughness. Accordingly, the true area of contact would be a small fraction of the apparent (nominal) area, and the formation of the asperities will be either elastic, the elasto-plastic, or totally plastic (see chapter 2). At this point, it is instructive to recall the friction laws of Amonotons. The two simple laws imply that control of friction may be attempted through two approaches. The first is to control the normal active load which causes initial compliance contacting monies and maintain their contact. Whereas the second, is to control the real area of contact that develops due to asperity interaction (i.e. to control roughness distribution and geometry). It is seen, therefore, that although the state of roughness of the surface is not a direct factor in these two laws, roughness (distribution and topography) indirectly contributes to the evolution of the friction process. Roughness and texture features additionally are subject to wear. Wear, in turn, changes the distribution of heights of individual features. Change of heights in turn affects friction. In particular it affects the dynamic load bearing and the instantaneous true area of contact. Wear effects are dynamic and require active control.

The evolution of the real area of contact in relation to the geometry and texture contacting couple have been the subject of numerous works in tribology literature. Modern models have the roots in 17th and 18th century literature. Inspired by the sketches of da Vinci, Amontons (Amontons 1699) conducted several friction tests in the course of designing his heat engine "moulin s feu". The observations of his experiments were formulated into the friction rules described earlier. Amonotons further remarked that the friction force developing between two sliding bodies depends on the complex relationship between normal pressure, time, and the sliding velocity. Seemingly simple, Amonotons remark signifies the complexity of friction and associated effects.

A contemporary of Amonotons, Antoine Parent, published the results of some friction experiments (Bélidor, 1782). In his work, Parent was the first to indicate that the action of sliding surface does not coincide with the direction of the normal that surface. He called the maximum angle inclination of this action the angle of equilibrium. He further showed that the tangent of this angle is numerically equal to the coefficient of friction.

Belidor (1782), investigated the problem of dry sliding friction between two solid surfaces. He considered that the actual surfaces of solids are uneven, and that they consist of hemi-spherical peaks and valleys. Friction, accordingly, is prescribed by the force which is needed to move the hemi-spheres up and over each other. Thus, two surfaces in contact will interlock through the asperities and the work needed for one surface to climb over the other would be equivalent to the energy loss due to friction. Such a model manifests the endless consideration of the effect of surface topography on the friction process. Coulomb (1801) investigated in more detail the effect of surface finish on lubricated friction and also on dry interaction between surfaces. He experimented with a wide range of loads and speeds. He considered that the fibers covering wood surfaces intermeshed as bristles of two brushes. Thus, to investigate the force developing between the surfaces due to friction, it is necessary to study the instantaneous position of the bristle at the time of sliding. In order to pinpoint this position, there is a need to know the distribution of heights of the individual bristles (i.e. the surface roughness features).

A combination of the bristles and hemi-spherical asperities evolved into a hemi-spherical capped roughness features as that used in 20th century contact models such as the well-known Greenwood Williamson model (Greenwood and Williamson, 1966). The model describes the evolution of the contact conditions and their dependence on surface roughness and feature height distribution. It was shown that if the real area of contact will be proportional to the hardness of the softest material. A complex situation takes place, however, in case of elasto-plastic deformation or plastic deformation with work hardening. For these conditions, the statistical distribution of roughness heights has to be considered. The greenwood-Williamson model, therefore, classifies the types of contact in terms of a so-called plasticity index, ψ. The value of that index distinguishes contact-deformation regimes. This parameter is given by,

$$\Psi = \frac{E'}{H}\sqrt{\frac{\sigma}{\beta}} \tag{1}$$

where E` is the so-called composite modulus, H is the hardness of the softest material, σ, is the standard deviation of the roughness, and β is the radius of the asperity tips (assumed hemi-spherical). Accordingly, the transition in deformation of the contact regime from plastic to elastic, passing by the elasto-plastic regime, is determined by the interaction of the composite stiffness of the sliding pair, which results from the mechanical properties of the pair, and the average aspect ratio of the individual roughness features. That is, the nature of information of the surface roughness, and its extent, depend on the level of interaction between the geometry and form of roughness features, on the one hand, and the mechanical properties of the surface layer on the other.

A remarkable feature equation (1) is its implication for the energy dissipated in friction. Namely, whether the friction process with consumer larger or smaller portion of the energy used to move the sliding pair will depend on the level of interaction with geometry, statistics of the roughness features, and the mechanical properties of the contacting pair. Therefore, the Greenwood-Williamson model brings forward the effect of roughness on the sliding process and devices pathway control that cross through controlling the degree of randomness in the statistical distribution of the roughness features.

In lubricated contacts, Friction-Induced Energy Losses, FIEL, in a rubbing system has two contributions. The first is a result of friction between the micro-topography at the interface between the contacting bodies. The second is a consequence of the friction between the lubricants, if present, with the interface. The magnitude of the second component increases upon using a lubricant with high viscosity (which is necessary to support high frictional loads). Similar to dry rubbing, roughness controls evolution of friction and damage. One on the early investigations into the effect of roughness on lubrication was that of Salama in 1948 (Salama, 1950). Salama have shown that the performance of parallel-faced thrust bearings is sensitive to the changes in the surface macro-roughness profile. He demonstrated that film lubrication performance is related to the hydrodynamic effect of the oil pockets shaped by the surface profile. Further work (Hamilton et al., 1966, Anno et al., 1968, Anno et al., 1969, Moore 1969, Lo and Wilson 1999) on lubrication provided tools relating surface roughness to lubricating film thickness as a guide to mark the transitions among various lubrication regimes. Investigation of lubricated sliding under extreme operating conditions (high load, high temperature, and high speed), smooth isotropic surfaces may encounter scuffing and seizure (diesel piston ring-liner interface, metal extrusion process). Solutions to these problems point in the direction of using of surface textures (lines, grooves, and discrete surface dimples) to provide lubricant reservoirs and trapping wear debris to avoid damage and failures. The afore mentioned studies, for both dry and lubricated friction, confirm that controlling the geometry of surface micro-texture (i.e, its customization to meet specific operational conditions) is an efficient method to control energy losses and enhance the structural integrity of rubbing pairs.

CHARACTERIZATION OF TEXTURING

Texturing in frictional surfaces have traditionally evolved in the context of solving problems arising in lubrication. Many lubricated surfaces exhibit failure manifested in scuffing and seizure. In engines, for example, retaining lubrication oil in expansion and compression strokes was problematic. It induced scuffing and seizure. One remedy was to devise a new surface-texture pattern to be embossed on cylinder liners to improve quality of lubrication and entrapment of wear debris. The idea is to create grooves of controlled geometries capable of retaining lubrication oil, and entrapping the wear debris resulting from rubbing. In addition to lubrication, the volume of the material in between the grooves, will act as a raised island that "cushions" the counter body. This reduces the actual contact area and thereby enhances the quality of the friction process. Additionally, if the "cushioning islands" are structured appropriately, wear may be also reduced. It is often the case that the texturizing process aims to alter the film thickness of the lubricating fluid to achieve complete separation between the rubbing surfaces at an early stage (i.e. in a shorter distance upon entry of lubricant between the two rubbing surfaces). Two possible configurations, protrusions or dimples represent the building block of texturing (figure 1). These can have endless foot prints (ellipse, square, circle, chevrons, etc.,). An added difficulty, arising from the available choices, is combining suitable chemical or metallurgical treatment to the textured surface.

There are several techniques to emboss textures (e.g., multistep honing, helical-slide honing, controlled thin layer deposition (Willis, 1986, Bolander and Sadeghi,2007, Borghi et al., 2008, Brizmer and Kligerman, 2012), and laser texturing (Etsion et al. 1999, Eichstät et al., 2011, Etsion and Halperin 2002, Galda, 2009). In honing, the goal of texturing is to create an array of micro-channels and plateaus on the target surface. The plateaus provide raised cushions (islands) for the counter face surface to slide on, and in the meantime, they reduce the contact area between the siding solids. This results in reducing the friction forces between the sliding bodies. The micro-channels aid in reducing oil consumption in lubricated friction by keeping remnants of the lubricant to replenish the interface continuously. Controlling the precision of surface texture generation by honing is currently a pressing problem. This is due to the multitude of parameters that influence conventional texturing technologies (around four hundred parameters involved in texturing by honing (Borghi et al., 2008)). Among the currently available texturing technologies, texturing by means of a laser beam is the most advanced and is considered by many as a promising enabling technology (Gerbig, et al. 2004, Golloch, et al., 2004, Gonzalo, e al., 2012, Halperin, et al., 1997, Hu and Hu 2012, Hu and Ding 2012). This is because the Laser is extremely fast and allows for short processing times. It is also environmentally friendly (i.e, green and sustainable). More importantly, however, laser texturing offers excellent control of the shape and size of the texture features.

Figure 1. Possible elements of texture, protrusions (left-hand side), dimples (right-hand side).

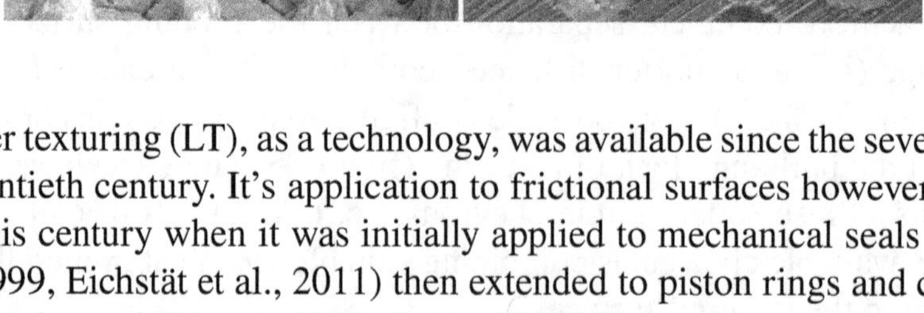

Laser texturing (LT), as a technology, was available since the seventies of the twentieth century. It's application to frictional surfaces however, began early this century when it was initially applied to mechanical seals (Etsion et al. 1999, Eichstät et al., 2011) then extended to piston rings and cylinder bores (Etsion and Halperin 2002, Galda, 2009). The process involves creation of an array of micro-dimples on the target surface using a material ablation process with a pulsating laser beam. The "dimple" acts as a "plateau" and the channel between two dimples acts as a "groove". The grooves retain remnants of the lubrication oil during sliding and thereby they can replenish the lubrication film in subsequent sliding cycles. The plateaus, meanwhile, provide raised cushions (islands), for the surface to slide on, so that the area of friction reduces without sacrificing the load carrying capacity.

Texturing is also conditionally effective in enhancing hydrodynamic lubrication between parallel surfaces in relative sliding compared to untextured surfaces that cannot provide any significant hydrodynamic load carrying capacity. The operating mechanism in this case is illustrated in figure 2 which shows two parallel sliding surfaces with the lower one moving at a sliding velocity U.

Figure 2. Schematic description of parallel sliding surfaces: (a) A single protrusion, and (b) the hydrodynamic pressure distribution over the single protrusion.

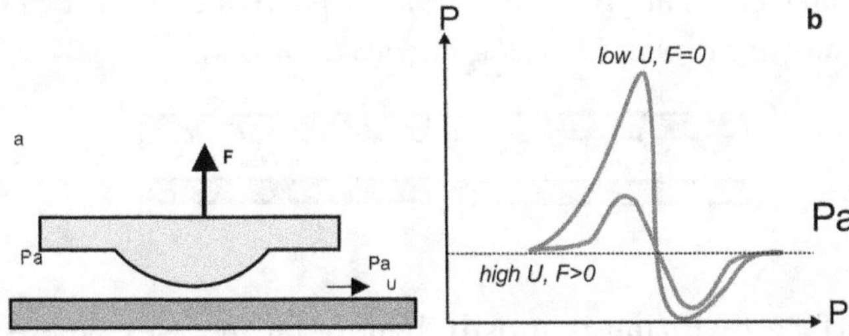

The texturing of the surface in figure 2-a is represented in by a single protruding asperity attached to the upper surface. The ambient pressure surrounding the two surfaces is Pa. In the absence of any texturing the relative sliding velocity results in viscous shear only without any effect on the pressure between the two flat surfaces. Therefore, in this case, the load carrying capacity is zero (i.e., F = 0). The introduction of the protruding asperity changes the local film thickness into a converging diverging one. This, generates a hydrodynamic pressure distribution figure 2-b. Due to the relative velocity U the pressure increases above Pa in the converging portion and decreases below Pa in the diverging portion of the clearance. At low velocities the maximum pressure is smaller than the absolute value of the cavitation pressure Pc and the pressure distributions is anti-symmetric about Pa. Integrating the pressure distribution along its axial span results in zero load capacity since the above and below ambient pressures cancel each other. As the velocity U increases the pressure distribution becomes asymmetric about Pa since the minimum pressure value is bounded from below by the cavitation pressure Pc while the maximum pressure is not limited. The integration of the pressure now results in a net positive value F > 0. When the textured surface contains a large number of protruding asperities the total load carrying capacity is the sum of their individual contributions. Exactly the same effect with an asymmetric pressure distribution as shown in figure 2-b can be obtained with indented dimples instead of protruding asperities. However, in this case the diverging portion of the clearance precedes the converging one. It should be noted here that because of the micro-scale of the asperities or dimples (see figure 3) the total load capacity of textured parallel surfaces is relatively small compared to the load capacity that can be generated in conventional hydrodynamic slider bearings.

Figure 3. Schematic profiles of two types of texturing commonly implemented in technological surfaces. The figure illustrates semicircular textures that protrude away from the surface or sinks into the surface (dimples).

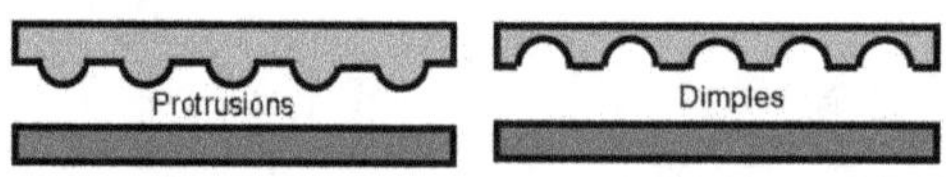

Hence, surface texturing is mostly beneficial in cases where conformal mating surfaces at very small uniform clearances are required, for example, in various sealing applications. Also, as shown in figure 3, the dimples configuration is a better choice for surface texturing compared to that of protrusions. This is because the increase in the real area of contact, compared to protrusions, when the surfaces are brought into contact. This results in the smaller average contact pressure in the dimples and lower wear. An additional advantage of using dimples-configuration is minimizing separation between the mating surfaces. This enhances sealing and considerably minimizes leakage.

DESCRIPTIVE METRICS FOR DIMPLED SURFACES

The makeup of any surface entails features of many different scales. Such features comprise the surface texture when the surface is engineered to meet a predetermined functional requirement, such as enhanced lubrication or reduced frictional response, the sizes and distribution of the surface features will vary according to the design intent.

The distribution of the surface features, along with the type, will also determine the method of describing the performance metrics. Traditional surfaces (surfaces that result from conventional manufacturing processes such as grinding, turning etc.,) the organization of the surface features (height, intervals, size etc.,) is stochastic in essence. Description of surface metrics in such a case stems from signal processing techniques (e.g., spectral analysis, auto correlation functions, root mean square height etc.,). The reason being that on a fundamental level, the stochastic elements resulting from the surface generation process dominate the texture. When, on the other hand, a surface contains dominant deterministic patterns (such as tessellations, axially symmetrical patterns, rotationally symmetric patterns) a stochastic description of the surface features becomes meaningless (Gerbig, et al. 2004).

The conventional approach for describing the compositional features of a surface invokes statistical techniques. Chapter three detailed the stochastic nature of describing the roughness of a surface and the various parameters used for characterization. Consequently, the distinctive property of any texture descriptor, for a given stochastic surface, is that this descriptor is a statistical variable (e.g., mean, maximum, root mean square, Skew etc.). The value of a characteristic texture feature in this frame of reference is the expected value of the relevant statistical variable (s) or, in some cases, is the relationship between values of two statistical variables. By invoking statistics, one can account for individual variations between all textural features of the same kind present in the studied surface. For example, one can describe the roughness of the surface by accounting for the variation in heights of all the asperities present on the surface. The roughness feature in this case, Ra, will be the statistical average of the heights of all summits present on the surface. In the case of deterministic-surfaces the quasi-invariance in the size of surface features renders statistical analysis meaningless. This is because in the limit, the value of the statistical variables, conventionally used to express surface features, will converge to the size of the described feature. In other words, the statistical description of surface features of a deterministic surface will converge to the scales and dimensions of the building elements of the texture. Such a fundamentally important result emerges from the extensive work done in studying performance and design of LTS (Etsion et al. 1999, Eichstät et al., 2011, Etsion and Halperin 2002, Galda, 2009, Gerbig, et al. 2004, Golloch, et al., 2004).

Texturing of surfaces entail placing a unit element of primitive geometrical shape (cone, hemisphere, ellipse, etc.) on the surface and repeating that element in an array over the target area. It follows that in order to describe a surface of such deterministic structure, there is a need to describe the geometry of the feature, its location within the array, and the relation of the array to the entire target surfaced. That is, one need parameters that describe the geometry of the element of texture and parameters that describe the topology of the features in relation to the textured surface (figure 4).

Continuous investigation of frictional performance resulted in identifying four ratios between the various factors listed in table 1. These ratios, listed in table 2, directly relate to efficiency of the frictional process and to the quality of lubrication (Gonzalo, e al., 2012, Halperin, et al., 1997, Hu and Hu 2012, Hu and Ding 2012, Huang, et al. 2012, Kovalchenko et al., 2005). They characterize the extent of interaction between topology and feature geometry and the influence of that interaction on surface performance. Therefore, these ratios indicate the quality of performance of a texturized surface. Performance indicators have been the subject of many investigations

Figure 4. Major geometrical and topological attributes of textured-deterministic surfaces

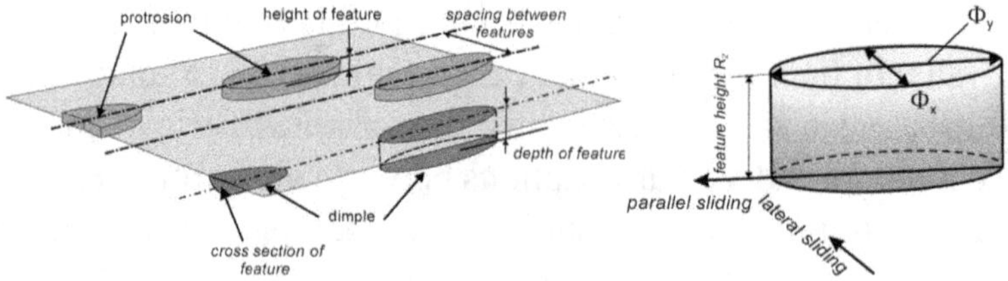

(Liu et al., 2010, Marian, et al., 2011, Mann and Zum Gahr, 2012, Podgornik, et al., 2012, Qiu 1986, Rechenberg, 2003, Ronen et al., 2001., Ryk et al., 2002, Santochi and Vignale1982, Scaraggi 2012,Shinkarenko et al., 2009., Sun, 2011, Xiaolei et al., 2009, Yamakiri et al., 2011, Yin, et al., 2012). Many attempted to define optimal values, or domains of optimal values, for each of these ratios. Reports in literature, however, provide conflicting information. Reasons for the conflict are many; yet, the main contribution has its roots in two factors: the lack of standardization of testing methods, and the absence in uniformity of reporting the recommended values. Inconsistencies span a wide range of constituents such as dimple or protrusion, feature shape, size, and spacing, used tester, and fabrication techniques. It is not clear so far if the source of such discrepancies could be reduced or at best eliminated in the near future. One serious difficulty is the almost infinite possible choices of feature geometries that may be chosen for texturing.

Table 1 provides a simple classification of the parameters involved in describing a deterministic surface textured surface in terms of relation to function of the surface.

The absence of common consensus about standardized testing or characterization procedures makes it impossible to cover the entire range of potential operation conditions.

Table 1. Classification of Texture attributes in relation to texture function

Function	Textural Attribute
Quality of Lubrication	Subsurface alterations Feature cross section Edge sharpness
Feature Geometry	Feature aspect ratio Feature slenderness ratio
Feature Orientation	Parallel to sliding direction Anti-parallel to sliding direction
Topology	Centerline-to-centerline spacing Width of textured area Are fraction of texturing

CRITERIA FOR EVALUATION PERFORMANCE

In principal, setting feasibility criteria for a surface should stem from in-depth analysis of the perceived function of the target surface, its loading and its operating environment. The requirements of surfaces that sustain higher loads but have to maintain high wear resistance in dry conditions differ from demands on surfaces operating in lubricated contacts. There are many constraints on the structure of the later class of surfaces. The designer in this case should decide on the type of lubrication to be maintained, what type of lubricant, permissible COF, and at which distance from the entrance the lubricant should establish maximum separation. These demands are functions of both surface texture, flow conditions and lubricant physical properties. In all the process is not simple and a tree of decisions may result in any given situation. When considering deterministic, or textured, surfaces therefore, deciding on a feasible bio-analogue is not as simple. In addition, the lack of cumulative design experience with deterministic surface features adds another layer of difficulty to the design process.

Table 2. Definition of common primitive geometrical attributes for Textured Surfaces. The height of the dimple (h), Feature base diameter (Φ), and the center line-to-centerline spacing between features λ.

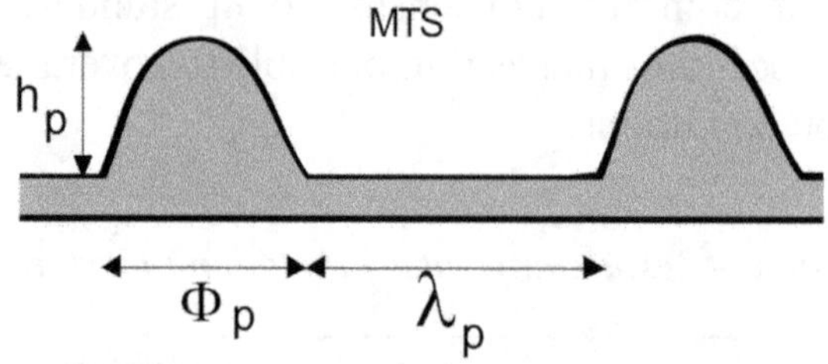

Ratio	Definition	
TAR	$\frac{\text{Area of Feature}}{\text{Total Surface Area}}$	$\frac{A_f}{S_A}$
FSR	$\frac{\text{Height of feature}}{\text{Length of Axis in direction of motion}}$	$\frac{h_p}{\Phi_P}$
FAR	$\frac{\text{Length of feature}}{\text{width of feature}}$	$\frac{\Phi_P}{\Phi_L}$
SAR	$\frac{\text{Centerline - to - Centerline spacing of features}}{\text{feature height}}$	$\frac{\lambda_P + \Phi_P}{h_P}$

The primary purpose of the additional criteria, in this work, is to rank the predicted performance of the snakeskin-inspired surfaces under lubricated contact conditions. One adequate ranking metric is ranking according to the COF developed in lubricated contact conditions. In this context, we adopt the model of Siripuram and Stephens (Siripuram and Stephens 2004) which allows for computing the COF for lubricated micron-sized textured surfaces. Accordingly, the COF in case of lubricated sliding of textures takes the form,

$$\mu_f = \left[\frac{\nu U}{W}\right]\left[\frac{\left(TAR\right)^2}{b} + \frac{\left(1-\left(TAR\right)^2\right)}{\left(h_f + b\right)}\right] \tag{2}$$

for raised fibrils and,

$$\mu_p = \left[\frac{\nu U}{W}\right]\left[\frac{(TAR)^2}{(h_d + b)} + \frac{\left(1 - (TAR)^2\right)}{b}\right] \quad (3)$$

for dimples or pores. Figure 5 illustrates the definitions of the geometrical parameters involved in equations three and four.

Figure 5. Elements of the model equations developed by Siripuram and Stephens (2004) a-parameter description in case of protrusions, and b parameter description in the case of dimples.

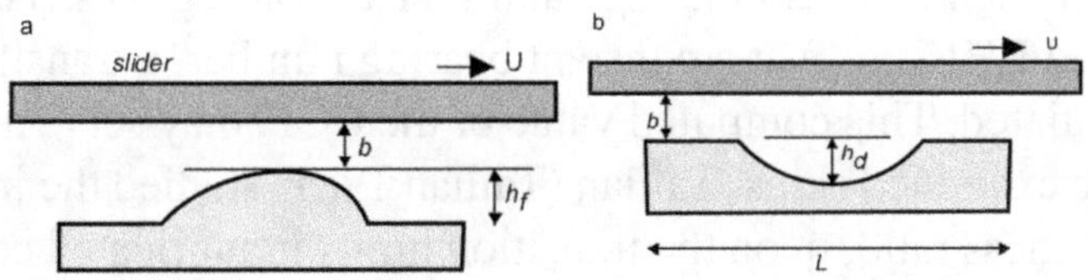

With b being the thickness of the lubrication film that develops between the rubbing surfaces. For ease of comparison and computation, we recast the film thickness in equation (3) in a non-dimensional form by defining a new parameter,η, that denotes the ratio between the thickness of the lubricant film, b, to the surface feature height (hs). Equations one and two can be recast in terms of this new ratio as,

$$\mu_f = \left[\frac{\nu U}{W}\right]\left[\frac{\left(1 + TAR^2\right)}{h_f(1+\eta)}\right] \quad (4)$$

$$\mu_p = \left[\frac{\nu U}{W}\right]\left[\frac{1}{h_d(1+\eta)} + \frac{\left(1 + (TAR)^2\right)}{(1+\eta)}\right] \quad (5)$$

The height of the surface feature hs is equivalent to the roughness parameter Ra. As such, we rewrite equation five as:

$$\mu_f = \left[\frac{\nu U}{W}\right]\left[\frac{\left[R_a^s\left(1+\left(TAR\right)^2\right)\right]+1}{\left(\eta+1\right)}\right] \cdots\cdots \eta \geq 0 \tag{6}$$

Equation 6 indicates that given a constant bearing number (νU/W), the COF represents the reaction of the fluid to the geometrical and metrological parameters of the surface features. Further, when the film thickness ratio, η, is constant, the COF manifests the interaction between the geometrical parameters and the bearing number (which represents external operation conditions). As such, if one defines the limits of the film thickness at the transition between different damage and lubrication regimes (i.e., between boundary and EHL, etc.,) at a constant bearing number, a transitional COF may be calculated. This computed value of the COF, may serve as a ranking index for the examined skins. Tallian (Tallian 1981) studied the influence of the film thickness ratio, η, on the transition limits in lubricated contacts. He identified five domains that fuly describe the possible operational conditions in Elasto-Hydrodynamic Lubrication (EHL). Table 3, provides a summary of the identified critical transition limits.

Table 3. values of the film thickness ratio, η, at the transition limits adopted in the current work

η	Effect
η < 1	Surface smearing or deformation accompanied by wear
1 <η <1.5	Possible surface distress
1.5 <η <3	Surface glazing is possible
η > 3	Minimal wear is possible
η > 4	Full separation of surfaces by EHL is expected.

CURRENT UNDERSTANDING AND TECHNICAL CHALLENGES

In principal, surface engineering through texturing exploits the two main possible routes for mitigation of friction, namely reduction of the true area of contact and alteration of the material properties of the mechanically affected zone in friction. Thus far, application of texturing within tribological contexts considered only lubricated contacts. Success was reported in mechanical seals, metal forming, and magnetic hard disks (Etsion I and Burstein L 1996, McNickel and Etsion 2004, Feldman et al., 2007, Raeymaekers et al., 2007). Successful application of the concept to dry contacts has not been reported yet. Despite instances of success, systematic investigation of textured surface performance across all lubrication regimes is still lacking. Hsu (Stephen et al., 2014) attributes the deficiency in cross-regime studies to the nature of friction at the scale of the texturing features.

The tribological events at the scale of the texturing features (micro-scale) may include transient frictional responses, asperity collisions, micro-scratching, and harmonic oscillation (among other complex factors). Exact frictional measurements are not always possible under these conditions, whence available conclusions are based on "averaged" results. As such, the mechanistic details of texture-surface interaction mostly remain obscure.

The vast majority of studies on texturized surfaces considered circular dimples fabricated by laser texturing. Other geometries (squares, chevrons, ellipses, etc.,) were not studied as frequently. Results show that successful application, at least where circular dimples are considered, is reported for conformal contacts at low contact pressures. The benefits achieved in this case include early transition to a full hydro-dynamic lubrication regime (Kovalchenko et al., 2005), lubricant reservoir under high shear contact pressure for retention of a lubricant (Vilhena, 2011), and entrapment of wear particles (Wang et al., 2009, Wang et al., 2010). So, it seems that circular dimples appear to be suitable under low contact pressures. This excludes surfaces that operate under medium to high contact pressures (e.g., gears, rolling element bearings) where the lubrication mechanism differs from that of seals or journal bearings. Keeping in mind that the shape of the textural element affects the lubrication regime, the inadequacy of available information about the viability of texturing becomes apparent. In particular, a comprehensive knowledge base that covers texture performance at all possible combinations of loads and speed categories is not in existence.

Many researchers started to investigate the influence of various shapes and orientations on quality of friction. Wang et al (2009, 2010) investigated the effect of triangle-shaped textures on the performance of a lubricated point contact, where a range of equilateral triangular dimples (with a range of side lengths, depths, and area densities) were laser textured on the steel flat discs. Unidirectional ball-on-flat sliding tests were conducted under a constant load and a series of speeds and full lubrication conditions. Modest conclusions were achieved from the enormous number of experiments, where the optimal friction reduction was found for dimple depth of 15 μm and area density of 10%. The conclusion about dimple orientation confirmed the findings of Yu (2010, 2011).

Shen and Khonsari (2013) found that dimple internal structure has a profound influence on the load-carrying capacity for HD lubrication. While keeping the shape of the dimples on the surface circular, they produced dimples with three different internal structural shapes: rectangle, oblique triangle, and isosceles triangle. A torsional rheometer was used in their experiment, where load-carrying capacity of the parallel surfaces was determined. The lower disc surfaces were textured using laser techniques. The diameters of the dimples were 800 and 6000 μm, which were much larger than those used in most other studies. The depths of dimples were ~30 μm and the area density varied from 14.2 to 33.2%. The results showed that the smaller dimples had limited effect on the load-carrying capacity of the contacts, whereas the larger rectangular cross-section dimples showed more increase on the load-carrying capacity than the triangular dimples. However, the dimples showed little effect on the frictional torque between the contacts.

Yuan et al (2011) studied the effect of microgrooves and their orientations on sliding surfaces and found that the orientation of microgrooves significantly influences the friction performance of the contacts. Depending on the contact pressure, grooves perpendicular to the motion at 7 μm depth were found to reduce the friction by 38.2% under low pressure, whereas grooves with depth of 19 μm parallel to the motion achieved a better effect at relatively high contact pressure. It was also concluded that, benefitting from a microgroove, requires the features to be arranged at angles other than parallel or perpendicular to the direction of motion (i.e., the textural matrix should not be orthogonal). Table 4 summarizes current state of knowledge of performance at the different load-speed combinations.

Depending on the configuration of the features, the materials involved, the alignment of the sliding surfaces, the type of relative motion, the speed of relative motion, the lubricant properties, and the design of the component, the effects of texturing have ranged from beneficial to undetectable to negative. For the majority of applications, the proper alignment of the mating surfaces is a must for achieving the beneficial effect of textures. Periodic changes in the sliding speed or load changes can disrupt lubricant films and prevent the texturing from reaching its potential beneficial effects. Textural geometry and the orientation of features with respect to the direction of relative motion affects the efficiency of texturing in reduction of friction. Proper selection of texture shapes, sizes, spacing, and orientation influences the ability of textures to enhance the performance in lubricated contacts.

Table 4. Summary of the major findings of research on man-textured surfaces

Regime	Main Findings	Applications
High speed low load conditions	Under steady state conditions, the load is fully supported by the hydrodynamic film pressure. The surfaces are separated by a continuous fluid film and the thickness of the film is controlled by the contact geometry, speed, load, and the viscosity of the lubricant. friction reduction mechanisms (a) enhanced hydrodynamic lubrication by early entry into the hydrodynamic regime (b) Possible cavitational lift effects for some systems. The hydrodynamic lift depends on the size, depth, and number of dimples in the contact area. The number of dimples in the contact area can be quantified by the area ratio defined by the area of dimpled surface over the total area in the contact.	Seals, Thrust, Bearings
High speed medium load conditions	The textures working in the low load-high speed regime, did not achieve an elasto-hydrodynamic regime, resulting in increased friction as compared to the baseline case. There are contradictory reports on the effects of textures on friction under elasto-hydrodynamic lubrication (EHL) regime. When the apparent contact pressure exceeds 180MPa, friction actually increases. This is due to the edge stresses induced by the elastic contact around the dimple edge.	Sliding Bearings, Tappets, cam rings and liners
Low speed high load conditions	This is the boundary lubrication regime where the load is supported by the asperity contacts under low speed, high load conditions. Under such conditions, local plastic deformation of asperities and wear usually occur. Studies in this regime are characterized by contradictory reports concerning friction reduction by introducing parallel grooves in relatively poorly lubricated systems such as titanium alloys to trap the wear particles, hence significantly reducing the third-body abrasion effects. Other studies report achieving increased anti-galling by surface texturing. Additional reports claim some benefits under very limited conditions using different geometries but also reported instances where friction actually increased.	Gears, hydraulic motors, transmissions

SIZING AND FEATURE SHAPES

Friction reduction when laser texturing is applied depends on the regime of sliding (dry or lubricated). Different operation conditions, even within the same regime, trigger distinct mechanisms. In mixed lubrication, three scenarios may take place. The first is a quasi-hydrodynamic lubrication regime which establishes minimum elastic contacts. Support of the load in this case takes effect through equal surface contacts and fluid film pressure. That is, a combination of fluid pressure and load bearing asperity contacts. In this regime, large shallow dimples and grove lines are effective. Upon increasing surface contact, the so-called edge stress acts as artificial roughness. This causes an increase in friction. Reduction of friction, in this case, will depend on the ability of the textural features to create hydrodynamic pressure.

In a boundary lubrication regime, when the load is fully supported by the surface contact, negative dimples, regardless of geometry will introduce edge stresses. Edge stresses, in turn, will increase friction. To reduce friction, texture has to be able to induce a hydrostatic force when the dimple is passed over. In addition, the normal reaction force has to be high enough to decrease the load and enable the formation of a continuous lubricant film.

The existence of several possible mechanisms that depend on the applied loads and sliding speeds suggest that a unique textural solution is not applicable. Friction reduction in this case is impacted by operation conditions and the dimensional requirements for the textural features will depend on the particular situation. Table 5 provides a summary of most used criteria for dimensioning the features.

Dimensional requirements have been extensively researched, and modelled, under several sliding regimes. Shinkarenko and co-investigators (2009) presented a theoretical model to evaluate the feasibility of using Laser Texturing in the form of spherical micro-dimples for soft elasto-hydrodynamic lubrication. They performed an extensive parametric investigation to identify key parameters of the problem. The parameters identified were the aspect ratio of the dimples which defines the average slope of the converging film thickness zone of each dimple. This parameter is known to be the main factor affecting load capacity in slider bearings. Shinkarenko deduced the optimal range of this parameter to fall within the range 0.06<DSR< 0.08. Further, the Dimple Area Ratio (equivalent to TSAR) was identified as 0.05 <SAR<0.5 with SAR equal to 0.2 being the recommended value.

Costa and Hutchings (2007) extensively investigated the influence of surface topography on lubricant film thickness. They measured film thickness using a capacitance technique in reciprocating sliding of patterned plane steel surfaces. The study employed circular patterns (pockets), grooves, and chevrons. They also varied the sliding orientation relative to the texture. The results of Costa and Hutchings concluded that a TAR of 0.11 seems to achieve the maximum film thickness when circular pockets are used for texturing. The same authors also found that maximum film thickness, for a sample with a TSAR ratio of 0.06, was achieved when the DSR was about 0.07. Consequently, they rationalized that such a value is optimal for texturing.

Table 5. Summary of surface texturing effects depending on texture geometry (h: structure height, w- structural width and f: area coverage with texture) and lubrication conditions

	Dry Sliding	Boundary Lubrication	Mixed Lubrication	Full Film Lubrication	
				HD	EHL
Main effects	Entrapment of wear debris	Friction increase	Entrapment of wear debris	Friction reduction	Increase of film thickness
	Reduction of contact area	edge of dimple wear by tensile cracking	Secondary oil effect	Increase of load support	Reduction of film thickness
	Formation of tribolayers	dimple worn off	Friction reduction (strong effect)	Occasionally friction increase	Friction reduction
	Effect on stress distribution	if texture can generate hydrostatic reaction force, wear disappears and friction reduction	Increase of film thickness		Friction increase
	Reduction in stiction		Shift in Stribeck curves		-Increase wear
h	Mostly below 20 μm	2 to 15 μm	1 to 10 μm	~6 to 10% of feature width	Small below 500 nm
w	Mostly below 60 μm	Depends on material properties	mostly 5 to 100 μm but depends on contact width	< bearing width	< elastic contact width (generally below 5 μm)
f (area coverage)	10 to 70%		7 to 12%	10 to 30%	10 to 20%
Shapes	Pockets (mostly) sinusoidal patterns	Pockets, grooves	closed shapes (pockets, chevrons, grooves (less efficient)	closed shapes (pockets; chevrons, grooves (less efficient)	Pockets grooves
Directionality	non-parallel to sliding	N/A	Perpendicular to sliding	Perpendicular to sliding	Perpendicular to sliding
Potential texturing techniques	LST, DLIP (used)	PCT, MECT, IPT	LST, PCT, MC, MECT (used)	LST, PCT (used)	DLIP
	MECT, PCT, IPT, MC (potential)		IPT (potential)	MECT, IPT, MC (potential)	EBL

Results of Petterson and Jacobson (2003) Kovalchenko et al. (2004), and Costa and Hutchings (2007) support the preposition that texturing of surfaces does not always effectively influence friction. They found that, although surface texturing reduces friction under boundary lubrication conditions, friction is likely to be insensitive to texturing when a full film exists between the sliding surfaces surface texturing has little or no effect on friction. Most of the damage sustained in rubbing therefore takes place before building up of the film thickness. That is where texturing is efficient.

Table 6. Summary of recommended performance indicators extracted from open literature

TAR	DSR	SAR	Reference
	0.06-0.08	0.05-0.5	Shinkarenko et al (2005)
0.025-0.1	0.07	0.11	Costa and Hutchins (2007)
	0.05		Dongsheng et al (2010)
		0.05-0.2	Ronen et al (2001)
0.25	0.02-0.25		Halperin et al (1997)
0.05	0.01-0.02		Wang et al (2009)
0.03-0.12			Yu and Zhou (2010)
0.6			Gonzalo et al (2012)
0.075-0.2	0.03-0.08		Galda et al (2009)
0.085			Hu, Hu (2012)
0.2-0.4	0.03-0.1		Yin et al (2012)

In principle, deliberate surface texturing is powerful. Full potential of texturing, however, hasn't been realized thus far. A major obstacle is lack of details about the interference of geometry and topology in lubricated contacts (conformal and non-conformal). By far the problem of optimal choice of geometry and topology still dominates. The benefits of texturing haven't extended to dry friction although enhancing friction in this area has the potential to reduce the viscosity of the lubricant used and therefore, further reducing friction losses. The technology is reaching a critical joint, not because of the inadequacy of texturing enabling technologies, but because of the need of fundamental information that can drive texturing design. One direction that potentially can vitalize the field is natural functional surfaces.

A structured surface for enhanced tribo-performance should possess several advanced features. One principal feature is the ability to tune the frictional response upon rubbing. That is, the surface should be able to adapt its' frictional profile in response to sensed changes in sliding conditions (e.g. changes in texture of the mating surface, variation in contact pressure, etc.). The tuning requirement may stem from geometry of the surface topographical building blocks, their distribution and placement within the surface, presence of embedded sensory, or a combination of all these factors. One of the difficulties in engineering such a surface, despite the availability of several enabling technologies, is the current limited understanding of the interaction between deterministic surface textures and frictional response. This, in turn, is due to the relatively recent history of deterministic surface texturing in human engineering.

While the technical world lacks diverse examples of functional-self adapting tribo-surfaces, our surroundings contain an abundance of examples of hierarchically structured naturally occurring surfaces capable of delivering super functionality. These may provide inspiration for surface designers. The richness and diversity of the examples provided by natural surfaces are worthy of study to extract viable solutions for surface design problems encountered in the technical world. This is particularly feasible since the natural world obeys the same physical laws that govern the behaviour of engineering systems. As such, any extracted design rules should, in principle, be valid across both realms: the natural and the technological.

REFERENCES

Abbott, E. J., & Firestone, F. A. (1933). Specifying surface quality: A method based on accurate measurement and comparison. *Mechanical Engineering (New York, N.Y.)*, *55*, 569–572.

Amontons, G. (1699). De la resistance cause'e dans les machines [About resistance and force in machines]. *Mem l'Acedemie R A*, 257–282.

Anno, J. N., Wallowit, J. A., & Allen, C. M. (1969). Load support and leakage from microasperity-lubricated face seals ASME J. *Lubr. Technol.*, *91*(4), 726–731. doi:10.1115/1.3555030

Anno, J. N., Walowit, J. A., & Allen, C. M. (1968). Micro-asperity lubrication Trans. *ASME F*, *90*, 351–355.

Bélidor, B. F. (1782). *Architecture hydraulique, ou l'Art de conduire, d'élever et de ménager les eaux pour les différents besoins de la vie* (Vols. 1–2). Paris: Jombert.

Bolander, N. W., & Sadeghi, F. (2007). Deterministic modeling of honed cylinder liner friction. *Tribology Transactions*, *50*(2), 248–256. doi:10.1080/10402000701309471

Borghi, A., Gualtieri, E., Marchetto, D., Moretti, L., & Valeri, S. (2008). Tribological effects of surface texturing on nitriding steel for high-performance engine applications. *Wear*, *265*(7-8), 1046–1051. doi:10.1016/j.wear.2008.02.011

Brizmer, V., & Kligerman, Y. (2012). A laser surface textured journal bearing. *ASME J of Trib.*, *134*(3), 031702. doi:10.1115/1.4006511

Costa, H. L., & Hutchings, I. M. (2007). Hydrodynamic lubrication of textured steel surfaces under reciprocating sliding conditions. *Tribology International*, *40*(8), 1227–1238. doi:10.1016/j.triboint.2007.01.014

Coulomb, C. A. (1801). Experiences destinees a determiner la coherence des fluides et les lois de leur resistance dans les movements tre lents. Paris: Academic Press.

Eichstät, J., & Römer, G. R. B. E. (2011). Towards friction control using laser-induced periodic Surface Structures. *Physics Procedia*, *12*, 7–15. doi:10.1016/j.phpro.2011.03.099

Etsion, I., & Burstein, L. (1996). A model for mechanical seals with regular microsurface structure. *Tribology Transactions*, *39*(3), 677–683. doi:10.1080/10402009608983582

Etsion, I., & Halperin, G. A. (2002). Laser surface textured hydrostatic mechanical seal. *Tribology Transactions*, *45*(3), 430–434. doi:10.1080/10402000208982570

Etsion, I., Kligerman, Y., & Halperin, G. (1999). Analytical and experimental investigation of laser-textured mechanical seal faces. *Tribology Transactions*, *42*(3), 511–516. doi:10.1080/10402009908982248

Feldman, Y., Kligerman, Y., & Etsion, I. (2007). Stiffness and efficiency optimization of a hydrostatic laser surface textured gas seal. *Journal of Tribology*, *129*(2), 407–410. doi:10.1115/1.2540120

Galda, L., Pawlus, P., & Sep, J. (2009). Dimples shape and distribution effect on characteristics of Stribeck curve. *Tribology International*, *42*(10), 1505–1512. doi:10.1016/j.triboint.2009.06.001

Gerbig, Y., Haefke, H., Bruneau, S., Hermann, J., & Sentis, M. (2004). Femtosecond Laser Ablation of Cemented Carbides: Properties and Tribological Applications. *Applied Physics. A, Materials Science & Processing*, *79*(3), 629–632. doi:10.100700339-004-2675-1

Golloch, R., Merker, G. P., Kessen, U., & Brinkmann, S. (2004). Benefits of Laser-Structured Cylinder Liners for Internal Combustion Engines. *Proceedings of the 14th Int. Colloquium Tribology*, 321–328.

Gonzalo, B, G., Backhaus, K., & Knoll, G. (2012). Numerical analysis of laser-textured piston-rings in the hydrodynamic lubrication regime. *J. Trib.*, *134*. Doi/ doi:10.1115/1.4007347

Greenwood, J. A., & Williamson, J. P. (1966). Contact of nominally flat surfaces. *Proceedings of the Royal Society of London. Series A, Mathematical and Physical Sciences*, *295*(1442), 300–319. doi:10.1098/rspa.1966.0242

Halperin, G., Greenberg, Y., & Etsion, I. (1997). Increasing mechanical seals life with laser-textured seal faces. *15th International Conference on Fluid Sealing, BHR*, 3–11.

Hamilton, D. B., Walowit, J. A., & Allen, C. M. (1966). A theory of lubrication by micro-irregularities Trans. *ASME D*, *88*(1), 177–185. doi:10.1115/1.3645799

Hu, N., & Ding, Q. (2012). Effective solution for the tribological problems of Ti-6Al-4V: Combination of laser surface texturing and solid lubricant film. *Surface and Coatings Technology*, *206*(24), 5060–5066. doi:10.1016/j.surfcoat.2012.06.014

Hu, T., & Hu, L. (2012). The study of tribological properties of laser-textured surface of 2024 aluminium alloy under boundary lubrication. *Lubrication Science*, *24*(2), 84–93. doi:10.1002/ls.1165

Huang, Z., Liu, H., Shen, Z., Li, P., Hu, Y., Liu, H., ... Wang, X. (2012). Process parameters analysis on surface texturing under laser shock peening. *Chinese Journal of Lasers*, *39*, 107–113.

Kovalchenko, A., Ajayi, O., Erdemir, A., Fenske, G., & Etsion, I. (2004). The effect of laser texturing of steel surfaces and speed-load parameters on the transition of lubrication regime from boundary to hydrodynamic. *Tribology Transactions*, *47*(2), 299–307. doi:10.1080/05698190490440902

Kovalchenko, A., Ajayi, O., Erdemir, A., Fenske, G., & Etsion, I. (2005). The effect of laser surface texturing on transitions in lubrication regimes during unidirectional sliding contact. *Tribology International*, *38*(3), 219–225. doi:10.1016/j.triboint.2004.08.004

Liu, H., Han, H., Xue, Y., & Li, J. (2010). Influence of laser surface texturing distribution patterns on the hydrodynamic lubrication. *Advanced Materials Research*, *97-101*, 1429–1432. doi:10.4028/www.scientific.net/AMR.97-101.1429

Lo, S., & Wilson, W. R. D. (1999). A theoretical model of micro-pool lubrication in metal forming Trans. ASME. *Journal of Tribology*, *121*(4), 731–738. doi:10.1115/1.2834129

Mann, M., & Zum Gahr, K. (2012). Effect of a contact area texturing under unidirectional sliding load dependent on the viscosity of the liquid lubricant. *Tribologie und Schmierungstechnik*, *59*, 35–40.

Marian, V. G., Gabriel, D., Knoll, G., & Filippone, S. (2011). Theoretical and experimental analysis of a laser textured thrust bearing. *Tribology Letters*, *44*(3), 335–343. doi:10.100711249-011-9857-8

McNickel, A., & Etsion, I. (2004). Near-contact laser surface textured dry gas seals. *Journal of Tribology*, *126*(4), 788–794. doi:10.1115/1.1792695

Moore, D. F. (1969). A history of research on surface texture effects. *Wear*, *13*(6), 381–412. doi:10.1016/0043-1648(69)90020-9

Parent, A. (1722). Nouvelle statique avec frottemens et sans frottemens ou Regles pour calculer les frottemens des machines dans l'état d'équilibre. *Mémoires de l'Académie Royale des Sciences, 1704*, 173-197.

Pettersson, U., & Jacobson, S. (2003). Influence of Surface Texture on Boundary LubricatedSliding Contacts. *Tribology International*, *36*(11), 857–864. doi:10.1016/S0301-679X(03)00104-X

Podgornik, B., Vilhena, L. M., Sedlaček, M., Rek, Z., & Žun, I. (2012). Effectiveness and design of surface texturing for different lubrication regimes. *Meccanica*, *47*(7), 1613–1622. doi:10.100711012-012-9540-7

Qiu Willis, E. (1986). Surface finish in relation to cylinder liners. *Wear*, *109*(1-4), 351–366. doi:10.1016/0043-1648(86)90278-4

Raeymaekers, B., Etsion, I., & Talke, F. E. (2007). A model for magnetic tape/guide friction reduction by laser surface texturing. *Tribology Letters*, *28*(1), 9–17. doi:10.100711249-007-9242-9

Rechenberg, I. (2003). *Tribological Characteristics of Sandfish.* In Nature as Engineer and Teacher: Learning for Technology from Biological Systems, Shanghai, China.

Ronen, A., Etsion, I., & Kligerman, Y. (2001). Friction-reducing surface texturing in reciprocating automotive components. *Tribology Transactions*, *44*(3), 359–366. doi:10.1080/10402000108982468

Ryk, G., Kligerman, Y., & Etsion, I. (2002). Experimental investigation of laser surface texturing for reciprocating automotive components. *Tribology Transactions*, *45*(4), 444–449. doi:10.1080/10402000208982572

Salama, M. E. (1950). The effect of macro-roughness on the performance of parallel thrust bearings. *Proceedings - Institution of Mechanical Engineers*, *163*(1), 149–161. doi:10.1243/PIME_PROC_1950_163_019_02

Santochi, M., Vignale, M., & Giusti, F. (1982). A study on the functional properties of a honed surface. *Ann. CIRP*, *31*(1), 431–434. doi:10.1016/S0007-8506(07)63342-3

Scaraggi, M. (2012). *Textured Surface Hydrodynamic Lubrication: Discussion. Trib. Lett.* doi:10.100711249-012-0025

Shen, C., & Khonsari, M. M. (2013). Effect of dimple's internal structure on hydrodynamic lubrication. *Tribology Letters*, *52*(3), 415–430. doi:10.100711249-013-0225-8

Shinkarenko, A., Kligerman, Y., & Etsion, I. (2009). The effect of surface texturing in soft elasto-hydrodynamic lubrication. *Tribology International*, *42*(2), 284–292. doi:10.1016/j.triboint.2008.06.008

Siripuram, R. B., & Stephens, L. S. (2004). Effect of deterministic asperity geometry on hydrodynamic lubrication. *Journal of Tribology*, *126*(3), 527–534. doi:10.1115/1.1715104

Stephen, M. (2014). Friction reduction using discrete surface textures: Principle and design. *Journal of Physics. D, Applied Physics*, *47*(33), 335307. doi:10.1088/0022-3727/47/33/335307

Sun, Yuan. S.H. (2011). Effect of micro-dimple shapes on tribological properties of specimen surfaces. Huanan Ligong Daxue Xuebao/Journal of South China University of Technology. *Nature and Science*, *39*(1), 106–110, 123.

Tallian, T. E. (1981). Rolling bearing life modifying factors for film thickness, surface roughness, and friction. *Journal of Lubrication Technology.*, *103*(4), 509–516. doi:10.1115/1.3251724

Vilhena, L. M., Podgornik, B., Vizintin, J., & Mozina, J. (2011). Influence of texturing parameters and contact conditions on tribological behavior of laser textured surfaces. *Meccanica*, *46*(3), 567–575. doi:10.100711012-010-9316-x

Wang, X., Liu, W., Zhou, F., & Zhu, D. (2009). Preliminary investigation of the effect of dimple size on friction contacts. *Tribology International*, *42*(7), 1118–1123. doi:10.1016/j.triboint.2009.03.012

Wang, X., Yu, H., & Huang, W. (2010). Surface texture design for different circumstances. *The 1st International Brazilian Conf. on Tribology-TriboBr-2010.*

Willis, E. (1986). Surface Finish in relation to cylinder liners. *Wear*, *109*(1-4), 351–366. doi:10.1016/0043-1648(86)90278-4

Xiaolei, W., Wei, L., & Fei, Z. (2009). Preliminary investigation of the effect of dimple size on friction in line contacts. *Tribology International*, *42*(7), 1118–1123. doi:10.1016/j.triboint.2009.03.012

Yamakiri, H., Sasaki, S., Kurita, T., & Kasashima, N. (2011). Effects of laser surface texturing on friction behavior of silicon nitride under lubrication with water. *Tribology International*, *44*(5), 579–584. doi:10.1016/j.triboint.2010.11.002

Yin, B., Li, X., Fu, Y., & Yun, W. (2012). Effect of laser textured dimples on the lubrication performance of cylinder liner in diesel engine. *Lubrication Science*, *24*(7), 293–312. doi:10.1002/ls.1185

Yu, H., Deng, H., Huang, W., & Wang, X. (2011). The effect of dimple shapes on friction of parallel surfaces. *Proceedings of the Institution of Mechanical Engineers. Part J, Journal of Engineering Tribology*, *225*(8), 693–703. doi:10.1177/1350650111406045

Yu, H., Wang, X., & Zhou, F. (2010). Geometric shape effects of surface texture on the generation of hydrodynamic contacting surfaces. *Tribology Letters*, *37*, 123–130. doi:10.100711249-009-9497-4

Yuan, S., Huang, W., & Wang, X. (2011). Orientation effects of micro-grooves on sliding surfaces. *Tribology International*, *44*(9), 1047–1054. doi:10.1016/j.triboint.2011.04.007

Chapter 6
Nature:
The Design Mentor

ABSTRACT

This chapter discusses the historical origins of the concept of borrowing from nature and coining the science of bio-mimetics. The material also surveys examples of tribological systems in nature. Generation of design in natural systems (geometry, pattern, form, and texture) is shown to be holistic in essence. It synchronizes simple interaction of design constituents and efficient performance. Such an approach yields deterministic design outputs that while conceptually simple are of minimized energy footprint. Natural engineering, it is shown, seeks trans-disciplinary technically viable alternatives to our technological practices. These alternatives, given functional constraints, require minimum effort to construct and economize effort while functioning.

INTRODUCTION

Bio-mimetics, Bio-inspiration, Natural engineering, Bionics, etc., stand for widely spreading trends in the technological world. A tool common among these trends is scouring technical solutions available in the natural world then; transfer them to the human engineering domain. Although the previous terms are of recent origin, the underlying concept is deeply rooted in antiquity. Indeed, inspiration by nature dates back to primitive man. Human fascination with nature is rooted in the harmonic expression of form, function, and

DOI: 10.4018/978-1-7998-1647-8.ch006

proportion often observed in our surroundings. Throughout history, humans sought to identify a perceived "heavenly order" manifested as beauty, order, or optimality of function.

There are many examples in nature that reflect optimized "smart" function. One intriguing example is that of bees (Hales, 2001). Humans have recognized the brilliance of constructing beehives since at least two millennia, especially when the relation between the hexagonal shape of the unit cell and the economy of material (wax in this case) is considered. The earliest claim of such knowledge appeared as early as 36 B.C., in the writings of Marcus Terentius Varro within his book on agriculture. He supported the claim that the hexagonal structure was explained by the isoperimetric property of the hexagonal honeycomb:[1]

"Does not the chamber in the comb have six angles. The geometricians prove that this hexagon inscribed in a circular figure encloses the greatest amount of space."

Toth (Tóth, 1964) alluded to the same property and explained that the three-dimensional problem of shaping the ends of the hexagonal cells to interlock with the ends of the cells on the other side is quite complex. Interestingly, according to his calculations, he concluded that the material in the bees' idealized three-dimensional structure can be reduced by a fraction of one percent. In 1999 Hales (2001) proved that a honeycomb of regular hexagons is the most efficient way to partition the plane into unit areas (using the least amount of average amount of material). In the world of the beehives, this statement confirms that bees have practically solved a recondite problem, and have made their cells of the proper shape to hold the greatest possible amount of honey, with the least possible consumption of precious wax in their construction.

Branch compliance presents a major problem for arboreal animals that travel and feed in the forest canopy. Branches taper towards their ends, but the narrowest gaps between trees are situated between the thin terminal branches of adjacent tree crowns. Efficient travel through the rainforest canopy requires animals to minimize the deviations from their travel path, unless such deviations are more energetically efficient than direct travel (although non-energetic factors such as the risk of falls may also influence gap-crossing decisions). In theory, compliant terminal branches may act as external springs adding momentum to jumps and leaps across gaps, but most researchers have suggested that compliant supports increase the cost of arboreal locomotion. Alexander (1991) showed that quadrupedal monkeys lose energy during takeoff and landing as branches bend under their weight and jumping lemurs generally lose contact with branches before they recoil, expending energy to deform supports but failing to exploit the elastic energy to aid momentum. However, field observations suggest that some primates can use branch compliance to their advantage: galagos and pygmy chimpanzees (Pan paniscus) use branches as catapults (Cant, 1987) and various species pump branches prior to leaps to make use of the propulsive force of branch recoil (Cant, 1994). Recent studies of orangutans offer increasing evidence that many aspects of their gap-crossing behaviour use elastic energy stored in compliant branches (Alexander, 1991, Cant, 1987, Cant, 1994, Demes et al.,1995, Markham and Groves, 1990), even though their large body mass might be expected to render arboreal gap crossings energetically costly and a safety risk. The best documented example of orangutans using support compliance is tree sway, in which individuals oscillate thin vertical tree trunks with increasing magnitude until a support on the opposite side of the gap may be reached (Demes et al.,1995, Markham and Groves, 1990, Thorpe and, Crompton 2005, Thorpe and Crompton, 2006)

Many species of snakes use constriction—the act of applying pressure via loops of their trunk—to subdue and kill their prey. Constriction is costly and snakes must therefore constrict their prey just long enough to ensure death. Constriction is an energetically costly and potentially dangerous activity. The act of striking and constricting prey exposes snakes to retaliatory attacks from the prey and other predators (Mehta and Burghardt 2008). In boas, constriction demands a nearly sevenfold increase in aerobic metabolism above basal levels (Murphy and Henderson, 1997) and the duration of a constriction event can be extensive (9–16 min (Murphy and Henderson, 1997, Canjani et al., 2003)). Thus, it would be greatly advantageous for constricting snakes to accurately and precisely determine when prey is incapacitated and no longer capable of retaliation or escape. Boas have the remarkable ability to detect a heartbeat in their prey and, based on this signal, they modify the pressure and duration of constriction accordingly.

The term, Biomimetics has a Greek origin: "bios" meaning life and "mimesis" meaning to imitate. It follows that the domain of biomimetics comprises the study and imitation, where possible, of nature's methods, designs, and processes. In its essence, this rapidly growing field, is but a rigorous "contemporary" reformulation of our ongoing curiosity about "the ways of nature".

Our natural surroundings hold a vast reservoir of ideas. Not all ideas, of course, avail themselves to feasible mimicry in the human world. In such a case, the focus of a student of nature is to comprehend the intricacies of how nature implements such an idea to deduce a design rule fit for the human engineer. These design rules are based on correlating the structure to the function of the particular bio-analogue. Such a process, termed as bio-inspired or natural-inspired design is one of the building blocks of human interaction with nature. Further, this process, a human engineer being inspired by the study of the construction of a natural species, constitutes the premise of viewing nature as an "engineer" and the objects in or surroundings as "designs". Whence, one can speak of the correlation of structure, function, and form in natural constructs as elements of design-in-nature. In this sense, the role of "nature as a mentor" design is not alien to the human intellect. Despite that, however, the philosophy of design generation in nature fundamentally differs from that in the technical world.

In nature, design for function strictly dominates. Underpinning this paradigm is the general principle of "separation of function". Simple single-cell organisms, for example, have to provide all their function in one cell, whereas higher animals and plants have many different kinds of cells for special purposes. Separation of function enhances survivability of species in a fiercely competitive world that promotes economy of living material. Natural designs, also, portray a complimentary relation between form and function. The pursuit of a single purpose requires careful considerations of the imperatives of balancing economy of design (with regards to materials and effort) and efficiency of function. This often results in designs that reflect elegance, and for the keen observer manifests a dimension of beauty. This beauty, in the most part, stems from form following function rather than conscious aesthetic intentions.

The imperatives of design in nature are severe. They pertain to reproduction (conservation of the species), growth, and survivability. There are also limitations on the kind of materials used, the sorts of mechanisms, and structures, which can be developed. This reflects on the degree of complexity of the outputs. Complexity, in itself, doesn't appear to be costly in natural designs. In the human design domain, where cost is balanced against performance, design outputs, have relatively less complicated shapes to accommodate ease of manufacturing rather than optimal fit to the intended task. Simplicity generally is considered a virtue in human engineering. Meeting complex tasks, for example, with few and simple parts, makes for a strong aesthetic appeal.

Perhaps a major difference between human and natural engineering is the dependence on energy in the former, while depending on information within the later. Such a difference highlights the essence of change that bio-inspiration can bring to human engineering. That is, the establishment of a reproduction based, rather than a manufacturing-based design culture were designs that facilitate recycling, sustainability, and remanufacturing without reformulation of chemistry are generated. In all, the interaction with nature is hoped to bring a cultural change in human engineering where "co-existing with" rather than "taming" nature is possible.

BIO-MIMICRY: A HISTORICAL TIMELINE

Although of recent origin as a discipline, bio-inspiration as a design paradigm is quite ancient. Indeed, the concept of technology transfer between life forms and human engineering extends deeper in history. The history of heavier than air flight is but one of the many examples that supports such a notion. The myth of Pegasus and the legend of 'Icarus and Daedalus' are glimpsing that manifest success (conceptually at least). An embodiment of the myth and legend took place as early as 875 A.D. That is when Abbas Ibn Firnas (810–887 A.D.), also known as Abbas Qasim Ibn Firnas, a Muslim Andalusian polymath (also inventor, engineer, aviator, physician, Arabic poet, and Andalusian musician) had an attempt at aviation using the concept of Icarus (i.e., fixing bird like wings to his body). He attempted to glide down the tower of Cordoba using his contraption. 2 Further study of the history of the subject reveals more successful instances where life form technology was extracted then employed to advance human engineering (and in some cases to solve pressing technical problems faced by humans). More famous are the designs of Leonardo da Vinci which were based on his studies of bird flight. Although never implemented, they formed the basis for many designs to follow. Clement Ader, perhaps following the footsteps of da Vinci, built several steam-powered aircrafts using a wing design based on that of the bat. Although reportedly achieving an altitude of some 300 m he encountered problems controlling his vehicle.

Borrowing from the technology of living forms, is not confined to aviation. Biological structures have been instrumental in inspiring architects and advancing building technology throughout history. Evidence for bio-inspiration may be traced to Egyptian temples. Moreover, there are suggestions that Gustav Eiffel based the design of his well-known tower on the taper of a tulip stem. The "La Sagrada familia church" (figure 2) is one famous example in modern architecture. Modern marvels such as the Shinkansen Bullet train of the West Japan railway also made use of technology present in life forms. In its original design the train produced large thunder claps upon emerging from a tunnel. The effects were felt at a distance of more than 500 meters away. The problem was solved when the front end of the train was modelled after the beak of kingfishers. Kingfishers dive from air into water to catch fish. The shape of their beak causes minimal splash of water upon entry.

Figure 1. Jacob Peter Gowy's The Flight of Icarus. La caída de Ícaro, óleo sobre lienzo, 195 x 180 cm, Madrid, Museo del Prado This is a faithful photographic reproduction of a two-dimensional, public domain work of art. The work of art itself is in the public domain for the following reason: This work is in the public domain in the United States, and those countries with a copyright term of life of the author plus 100 years or less.

Figure 2. View of the Passion Façade of the Sagrada Familia Basilica in Barcelona Spain

Figure 3. King fisher and inspiration for the nose cone of high speed train

Implementing this shape to the front end of the train resulted in power savings and enhancement of the train speed (figure 3).

Biomimetics was coined by the American biophysicist and polymath Otto Schmitt during the 1950s.(Vincent, et al, 2006) It was during his doctoral research that he developed the Schmitt trigger by studying the nerves in squid, attempting to engineer a device that replicated the biological system of nerve propagation.(Bar-Cohen, 2006) He continued to focus on devices that mimic natural systems and by 1957 he had perceived a converse to the standard view of biophysics at that time, a view he would come to call biomimetics. 3(16)

A similar term, Bionics was coined by Jack Steele in 1960 at Wright-Patterson Air Force Base in Dayton, Ohio where Otto Schmitt also worked. Steele defined bionics as "the science of systems which have some function copied from nature, or which represent characteristics of natural systems or their analogues"(Vincent, 2009) During a later meeting in 1963 Schmitt stated,

In 1969 the term biomimetics was used by Schmitt to title one of his papers (Schmitt, 1969) and by 1974 it had found its way into Webster's Dictionary, bionics entered the same dictionary earlier in 1960 as "a science concerned with the application of data about the functioning of biological systems to the solution of engineering problems".

In the 1970s, biomimetics research came along in the fields of chemistry, as "Biomimetic Chemistry," aiming at molecular-level emulation of enzymes and bio-membranes. Around the same period, it became apparent that the X-ray structural analysis had the capability to provide reaction site information on enzymes (i.e., biocatalysts), enabling molecular-level elucidation of biological reactions using techniques of organic chemistry. What was essential in unraveling biological events from chemical standpoints in view of engineering applications was the biology-chemistry collaboration. The artificial photosynthesis research that became active in the 1980s laid the foundation for dye-sensitive solar cells, and the actuator research, using gel, brought about such inventions as synthetic muscles. However, the evolution of molecular biology subsequently turned the mainstream of biology toward the elucidation of life phenomena, where the gene plays a central role. The mainstream of biomimetic chemistry research—an area best described as "biomimetics of molecular systems"—began to show tendencies, accompanied by the rise of molecular electronics research in the latter half of the 1980s, to distance itself from biology, and headed toward the chemistry of molecular assemblies and supramolecules. Entering the 1990s, in spite of the prevalence of the idea "Learn from Living Things," the opportunities for linkage with biology, in effect, almost disappeared. Even the term "Biomimetic Chemistry," which represented an academic field, became almost extinct in the wake of the shift of focus toward molecular nano-technology and nano-biology. The idea "bioinspired —taking cues from living things and surpassing them—

Table 1. Evolution of terms and definitions of the Bio-inspiration process

Year	Definition	Source
1958	"It (bionics) explores systems whose functions are modeled on natural systems, or whose properties resemble those of natural systems, or are analogous to them."	
1958-1960	"(the) science of systems that work like or in the same manner as or in a similar manner to living systems"	
1962	"Bionics is thus the science that investigates biological processes and methods with the goal of applying the results to the improvement of older and the creation of newer machines and systems. One could also say that it is the science of systems demonstrating features similar to those of living organisms."	
1976	"With respect to the present state of development, bionics can be viewed as one of the applied disciplines in the biological sciences with a tendency to integration induced by its objectives, which has as its content the systematic study of life forms for the solution of technical, technological, and architectonic problems; whereby structures and processes serve in their functional relationship in the systems of organisms as a stimulus and pattern, particularly as models for constructions and processes in the various branches of industry and engineering."	
1976	"Bionics: scientific field of integration, with a technically driven problem focus of heterogeneous scientific disciplines. Their scientific matter is characterized by findings that are acquired from biological objects, that embody principles superior to previous technology, and that can lead to a technical utilization; therefore it brings together various disciplines for the solution of specific technical tasks of a varying nature and changing priorities and taps into new types of technical problem-solving approaches."	
1976	"The task of bionics is to investigate biological objects with the goal of modernizing present technical systems or creating new and more accomplished ones and using the results."	
1987	"In general, bionics can be described by three different groups of definition: (1) It is a science for the planning and constructing of systems whose functions emulate those of biological systems. (2) It is a science for the planning and constructing of systems exhibiting characteristic features of biological systems. (3) It is a science for the planning and constructing of organizational structures that emulate the interrelations of patterns of biological organization."	
1993	"Bionics as a scientific discipline looks systematically at the technical conversion and application of constructions, processes, and principles of development in biological systems."	
2002	The definition from VDI-TZ/1993 with the following addition: "Bionics also includes aspects of the interplay of animate and inanimate parts and systems as well as the scientific technical employment of biological organization criteria."	
2005	Bionics = learning from nature to improve technology	
2006	"Biomimetics (which we here mean to be synonymous with biomimesis,' 'biomimicry,' 'bionics,' 'biognosis,' 'biologically inspired design,' and similar words and phrases implying copying or adaptation or derivation from biology) is thus a relatively young study embracing the practical use of mechanisms and functions of biological science in engineering, design, chemistry, electronics, and so on."	
2006	"Bionics as the term for the field of study involving copying, imitating, and learning from biology ... Biomimetics … the term itself is derived from bios, meaning life, and mimesis, meaning to imitate. This new science represents the study and imitation of nature's methods, designs, and processes. While some of its basic configurations and designs can be copied, many ideas from nature are best adapted when they serve as inspiration for human-made capabilities."	

Figure 4. Schematic time-line evolution of Bio-Mimetics

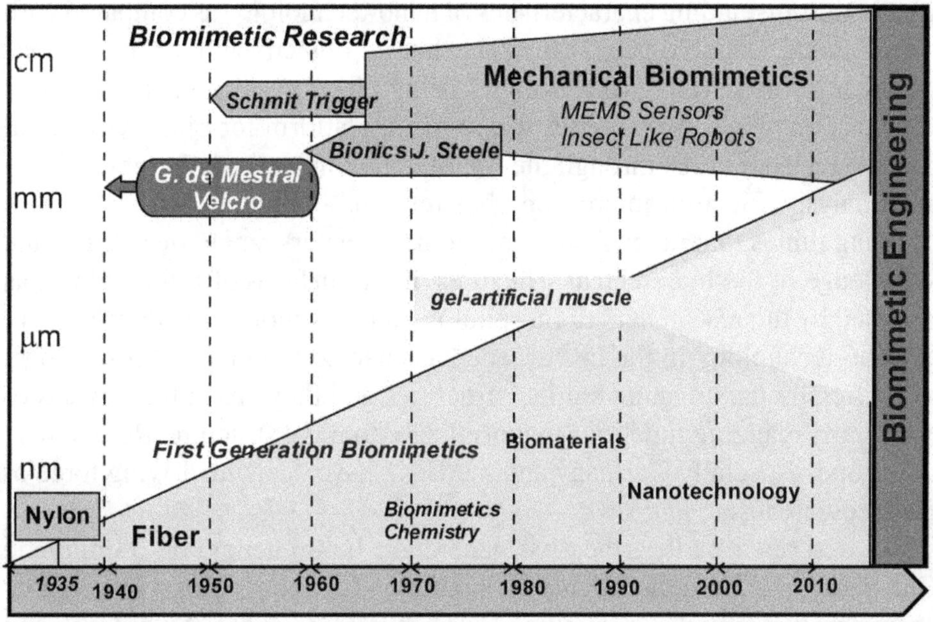

became mainstream, marking the decline of the first generation biomimetics, or biomimetic chemistry.

The Development of Mechanical Biomimetics in the 1970s, biomimetics research also blossomed in such fields as mechanical engineering and fluid dynamics; developments in these fields included robots that mimicked movement of insects and fish, and the sonars and radars that mimicked echolocation capabilities of bats and sensory hairs of insects. Mechanical biomimetics research has continued without decline up to now mainly in such areas as military industries, railways and ships, and aeronautics industries. They have also had an impact on such cutting-edge fields as micromachines and MEMS. In present-day Japan, the term “biomimetics” seems to have more of a connotation synonymous with robotics research.

The Rise of Material Biomimetics, exemplified by the proposal made by the National Academies (U.S.) in 2008, i.e., “Next- Generation Bioinspired Materials,”. With the coming of a new century, new trends of biomimetics have been gaining focus in material research centered in Europe and the United States. A movement is a foot to bring the research results into practical use. In many cases, the surfaces of living things are characterized by hierarchical structures in dimensions ranging from nanometers to micrometers. These ranges of dimension are the targets of nanotechnology. One of the outstanding characteristics of nano-technology, as compared with conventional technologies, is the fact that its objectives have dimensions that require observation and analysis using electron microscopy, and this very fact embraces possibilities to provide a platform for biology-material science collaborations through the use of common methods for observation and analysis. Electron microscopy has revealed the hierarchical structures of living things that range from nano- to miro-meters range. Inspired by the knowledge of the hierarchical structures in the surfaces of living things, as revealed by the researchers of morphology and taxonomy, the development of nano-technology in the last decade has enabled the material researchers to artificially manufacture similar structures and they are in the process of artificially realizing functions that originate from the structure. Researchers in Europe, especially Germany and the U.K., have been the driving force of these researches.

As described earlier, the first generation biomimetics, or Biomimetic Chemistry, was born thorough a collaboration between biology and organic chemistry that aimed at mimicking living things on the molecular level, and X-ray structural analysis built a momentum for this trend. In contrast, the material research, or new generation biomimetics, are considered to have been created through collaborations among natural history, biology, and material nano-technology, where electron microscopy and microfabrication technology provided the common platform. The characteristics of nano-technology in Europe are symbolized by the term "Nano meets Bio." That is, the objective is the fusion and linkage between dissimilar fields. For example, in the research carried out in German universities, linkage between dissimilar fields has become the prerequisite for gaining research funding. The reason that signs of forthcoming activities in new generation biomimetic material research appeared first in Europe can be ascribed to a cultural background that values cross-fertilization as well as to the European science and technology policies that proactively look to fusing together dissimilar fields.

The so-called Da Vinici Index for the U. S., tracks development in bio-inspiration overtime. It considers number of patents linked to bio-inspiration, published peer-reviewed articles, and the number of grants issued by the National Science Foundation (NSF) and the National Institute of Health (NIH), in the United States. This index showed 5 ½ fold increase between the year 2000 and 2012. The increase represented a compound annual growth rate of 15.3%. The index anticipates that Biomimetics, and Bio-inspiration, could account for a staggering $ 425 billion of U. S. gross domestic product (and another $ 65 billion in terms of reduced resource depletion and pollution) by 2025.

TRIBO-ENGINEERED SURFACES IN THE BIOLOGICAL DOMAIN

In the biological world creatures depend on their surface for a multitude of synchronized tasks that include, locomotory, chemical, mechanical, and protective functions among others. Moreover, animals occupy a large niche in the planet. Their respective habitats manifest a wide spectrum of constraints. Applied to tribology, these constraints represent almost all possible tribological challenges faced by a technical surface. Often biology meets the imposed constraints through intriguing surface solutions. This renders biological surfaces a vast resource of specialized surface designs that stand to potentially inspire future paradigms in surface design.

Scale and Variety

Natural systems entertain friction at a wide spectrum of scales. In biological systems, tribological behavior is identifiable at scales as small as few nanometers (figure 5) in case of a DNA helix (Mannion, et al., 2006). Friction at a higher scale, 50 nm for flagellar bacterial motors, provide design for a hostile tribological environment manifested in sustaining rotary speeds of up to 100,000 r.p.m (Berg, 2000). Human blood cells, human hair does also manifest optimized tribological systems at the nano-micro scale (Liskiewicz, et al., 2008). The examples shown in figure 5 are true indicators of the vast reservoir of tribological systems in nature and the associated localized technical solutions that evolved in response to functional constraints.

In what follows we detail few well studied bio-tribological systems where surface design is either the key ingredient in functional optimization, or one of the dominant contributors to optimized function. These biological systems, in most cases, have served as inspiration of some fledgling technological surfaces that are starting to emerge in open literature. Following each of the bio-examples we present the resulting human built system for comparison.

Efficient animal locomotion through liquids depends on drag reduction. Since drag acts at the boundary of the animal and water, skin micro-geometry would play an important role in optimizing its negative effects. One well documented example of drag optimization is that of the fast swimmers: sharks and Dolphins (Amundin, 1974, Au and Weihs 1980, Babenko,1979, Fish, 1998, Gray, 1936, Carpenter,1998). In sharks, the scales over most of the body are small (order of 0.2–0.5 mm). They have very fine, regularly spaced (30–100 µm) longitudinal ridges (riblets). These minute scales can vary in size and shape with placement on the body of the shark as well as with shark species as shown in figure 6 (Carpenter et al.,Pavlov, 2003, Krieger,2004, Lang, et al.,2008, Lee and Choi. 2008). In contrast to the scales over most of the body, those on the leading edges (rostrum) and fins are more firmly embedded in the skin and are completely smooth (Bechert et al.,1986, Bechert et al.,2000 a, b). Previous studies have observed that the body scales of some fast-swimming sharks are moveable to the touch to fairly large angles (Reif, 1982).

Sharks have the ability to alter the flow pattern around their fins. The mechanism responsible is rooted in the material make-up of the stratum compactum. This layer comprises layers of helically wound collagen fibers that form a whole body exo-skeleton. These fibers, to which the base of each scale is attached, form an angle to the longitudinal axis of the shark of approximately 50–60o. This arrangement permits bending flexibility, and results in elastic energy storage during swimming (Reif, 1985a, Reif 1985b, Reif, 1987, Wainwright 1978, Motta 1977, Hebrank, 1980). The erection of the scales takes place through increased overall skin tension at higher swimming speeds. The change in pressure changes the helical orientation of the collagen bundles, to which the scales are attached, thereby erecting the scales. The special configuration of the denticles in swimming actively controls the turbulent boundary layer and in contrast to dolphins the riblets prevent lateral movement across the flow direction. Thus, again, the interaction between morphology, geometry and topology plays a major role in optimization of performance.

SAMPLE EFFICIENT BIOLOGICAL SURFACES

Drag Reduction

Propelling an object through a liquid encounter resistance known as drag. Drag manifests the influence of the viscosity of a liquid. The mechanism through which viscous effects influence motion is similar to that of solid-

Figure 5. Examples from nature illustrate various orders of magnitude of frictional systems from104 meter down to 10-7 meter

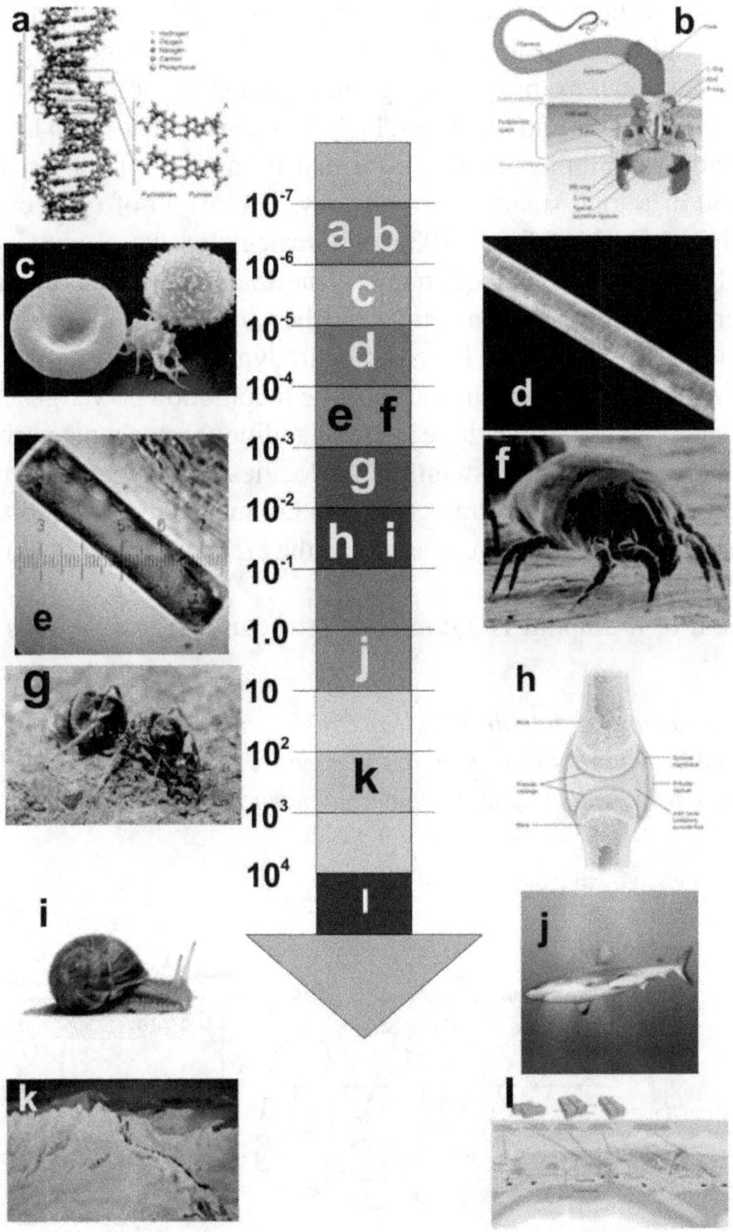

solid interaction in that it is mainly an interfacial phenomenon. For efficient animal locomotion through water, drag reduction is a must. Since drag acts at the boundary of the animal and water, one would expect that the skin micro-geometry would play a role in optimization of the drag effect.

One well known example of drag optimization in biology is that of the fast swimming sharks and Dolphins. Dolphins can attain a speed higher than 11 m/s while swimming. Their performance has been a subject of many studies and of a long-standing controversy. The origin of that controversy is the original work of Gray (1936) who calculated the propulsive power generated by the muscles of the animal and the drag force that the animal would encounter in swimming. Upon comparing the two quantities, Gray reported a discrepancy. He reported that based on hydrodynamic theory alone, the drag force acting to resist swimming exceeds the locomotion power generated by the muscles. He asserted that there has to be additional mechanisms that allow the animal to achieve higher swimming velocities by dissipating or reducing the theoretically calculated drag. These "other mechanisms", as discovered later, are the role of surface geometry and the effect of that geometry on the drag force upon swimming.

The skin of a dolphin is a composite structure (figure 6). It is marked

Figure 6. Structure of dolphin skin. (a) Cross-section; (b) Cut through dermal papillae at AA'; (c) Front view. Key: a, cutaneous ridges or microscales; b, dermal papillae; c, dermal ridge; d, upper epidermal layer; e, fatty tissue.

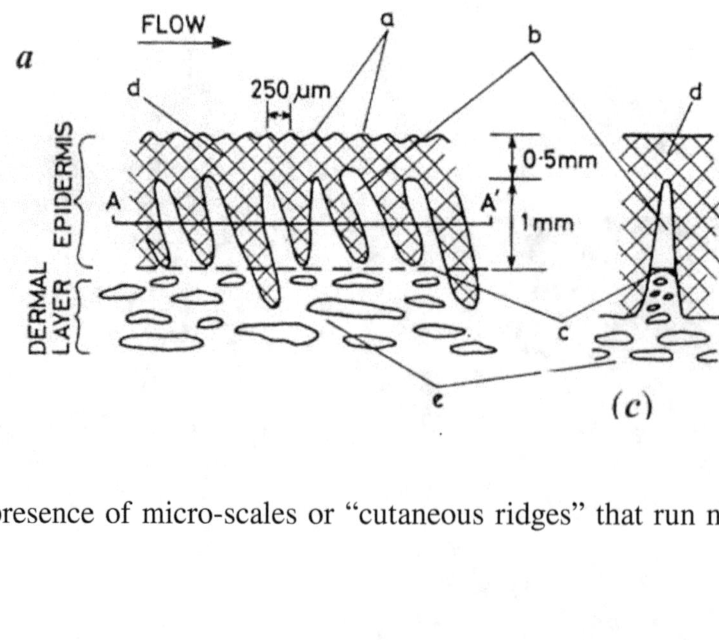

by the presence of micro-scales or "cutaneous ridges" that run normal to the direction of the flow. The upper epidermal layer constitutes a relatively dense elastic membrane capable of transmitting pressure fluctuations to the underlying layers.

These layers and the dermal papillae are made of looser, more hydrated tissue that includes fat cells. The angle the dermal papillae make with the vertical varies on the body from 10-80 degress (Lee and Choi. 2008). The papillae extend upwards from the dermal ridges which run roughly in the stream-wise direction. The dermal papillae are disconnected and are off no ordered pattern to their distribution on the fin of the animal, but are of variable density. They have an elliptical cross section, on the fin planes, and their height decreases gradually from the leading edge to the trailing edge where they almost disappear and merge with dermal ridges (Carpenter et.al., 2000).

Modern views of dolphin hydrodynamics propose a complicated adaptation process by the animal to drag forces of which an interaction between surface geometry and material compliance is at the core. This interaction orchestrates the hydrodynamic behavior (such as unsteady velocity and pressure gradients), and skin (the fluid-solid boundary) response (tension, damping, and compliance).

Pavolv (Pavlov, 2003) studied the role of skin in modulating drag for fast swimming by modeling the 3-D structure of skin from different locations on a harbor porpoise (Phocoena phocoena). He further studied the simulated flow patterns around virtual models of the fins. Pavlov concluded that the distribution of skin morphology parameters correlates with the local flow parameters on the fin surface. Moreover, he pointed out that the structure of the skin allows the flow-skin interface to act similar to a so called "anisotropic complaint wall". The mechanism suggested by Pavlov entails the structure of the compliant skin allowing the displacement of the upper layer of the skin in the regions of favorable and adverse pressure gradients at the edges of the fin. The controlled displacements suppress the instability growth in the boundary layer. The scenario described by Pavlov has at it score the skin morphology parameters (angle of the dermal ridges with the flow, ratio of thickness of skin layers and spatial density of the dermal ridges) actively control skin displacements in the critical regions. That is the interaction between morphology and topology of the skin structure are the main control agents of the boundary layers to accommodate fast swimming.

Sharks, on the other hand, deal with fluid friction in a manner that differs from dolphins (although the same concept of skin morphology-topology still applies). Texturing features in this case manifest the control mechanism that regulates drag.

The skin of sharks is covered by minute dermal bony scales (denticles), which are embedded in the deeper collagenous layer of the skin (stratum compactum). The crown of these denticles is exposed to the water. The scales over most of the body are small (order of 0.2–0.5 mm). They have very fine, regularly spaced (30–100 µm) longitudinal ridges (riblets). The riblets are discretely separated over the whole body, and are regularly aligned along the axis of the body. These minute scales can vary in size and shape with placement on the body of the shark as well as with shark species as shown in figure 2 (Reif, 1985a, Reif 1985b, Reif, 1987, Wainwright 1978, Motta 1977, Hebrank, 1980). In contrast to the scales over most of the body, those on the leading edges of the body (rostrum) and fins are more firmly embedded in the skin and completely smooth (6-8). Previous studies have observed that the body scales of some fast-swimming sharks are moveable to the touch to fairly

Figure 7. Surface morphology of shark skin scales (30, a: Dermal denticles of a lemon shark viewed through a scanning electron microscope (file is licensed under the Creative Commons Attribution-Share Alike 3.0 Unported license. b: Cross section through shark skin (Squalus acanthias) showing the insertion of the placoid scale (http://biodidac.bio.uottawa.ca/thumbnails/filedet.htm/File_name/chon032b/File_type/gif BIODIDAC material is licensed under a Creative Commons Attribution-NonCommercial 4.0 International License).

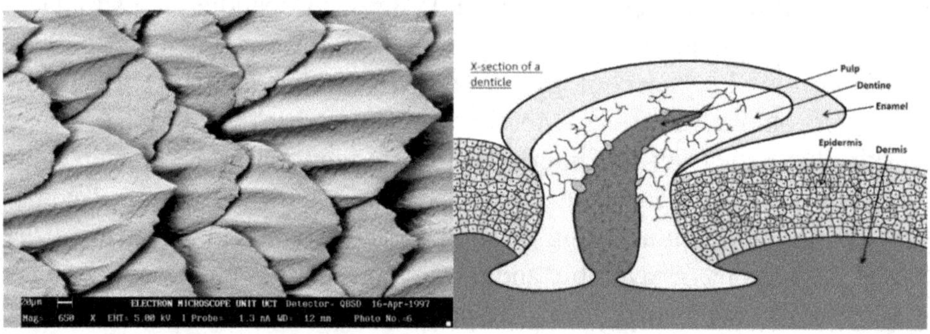

large angles on dead specimens. Such pliable, 'hand-like' scales have been identified over almost the entire body on the short fin mako Isurus oxyrinchus and smooth hammerhead Sphyrna zygaena (Abdel-Aal and El Mansori, 2013).

Sharks may have the ability to bristle their scales at higher swimming speeds, or the scales might erect by passive means during high-speed swimming and thereby alter the flow pattern around their fins. The mechanism responsible for that action is rooted in the nature of the material make-up of the stratum compactum of sharks. This layer is composed of multiple layers of helically wound collagen fibers that form a whole-body exoskeleton. These fibers, to which the base of each scale is attached, form an angle to the longitudinal axis of the shark of approximately 50–60o. This arrangement results in a balancing of longitudinal and hoop stresses in pressurized cylinders, permits bending flexibility, and results in elastic energy storage during swimming. The erection of the scales takes place through increased overall skin tension at higher swimming speeds (typically from 7-14 Pa in a resting lemon shark Negaprion brevirostris to 200 kPa during fast swimming. The change in pressure changes the helical orientation of the collagen bundles to which the scales are attached, thereby erecting the scales.

The special configuration of the denticles in swimming actively controls the turbulent boundary layer and in contrast to dolphins the riblets within the scales prevent lateral movement across the flow direction. Thus, again, the interaction between morphology, geometry and topology plays a major role in optimization of performance.

Hierarchical Structures for Maximum Adhesion

Adhesion plays a fundamental role in the biological world. It ensures locomotion on many tribologicaly challenging surfaces, it aids in survival functions such as prey capturing and defense. In the past decade or so, the topic of adhesion within biological systems received considerable attention. To date, however, the exact physical mechanisms underlying the precise function of many biological attachments are still unsettled. This is because many species, insects, gekos, spiders, and snakes, have the ability to engage a complex array of structural, material, and surface topology features in synchronized function.

Traditionally, the analysis of contact mechanics of biological attachments borrowed from the mechanics of solids. In particular, the statistical provisions made to simplify the analysis of manufactured rough surfaces have affected the analysis of biological attachments. One example is the assumption of hemispherical capped attachments, which mimics the classical assumption of asperities in a solid surface in the Greenwood Williamson model. Adopting a generalized assumption about the geometry of the attachment tips helped mask the mechanistic variations among different species. Several microscopic studies reveal the existence of a vast array of contact shapes (attachment tips) (see figure 8). Most of the attachment terminal ends differ considerably from spherical. Adhesion in many biological systems is a function of the areal density, size and shape of the terminal contact features (with shape exerting

Figure 8. Examples of fibrillar arrays fabricated by different methods (SEM micrographs). Direct nanofabrication without the use of a template by e-beam lithography (a) or chemical vapor deposition (b) Replication of templates by means of molding, using either porous membranes (c),micromachined (d,e),etched (f),or photolithographic templates (h–l).

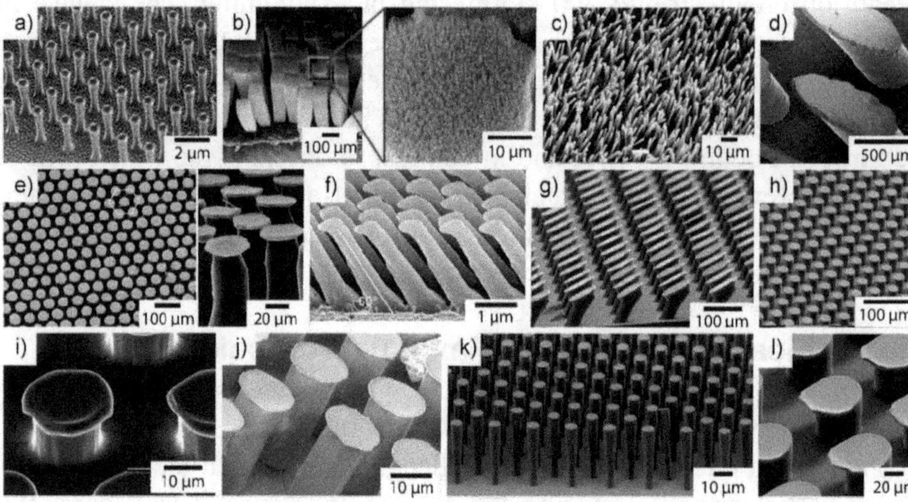

the strongest effect than size). Del campo et al (42) analyzed a wide array of contact shapes and found that mushroom and spatular terminals attain values 30 times higher than those with a flat end.

Arzt et al. (Del Campo et al.,2007) pointed out that the pull-off force in case of a spherical terminal scales with a geometrical length, which gives rise to a size effect. To that effect replacing one large spherical contact of radius Ro say by N smaller ones of radius R=RoN-1/2 with the same total (Hertzian) contact area the pull-off force for simultaneous detachment of these N spheres will then be enhanced by a factor of N1/2

Shape of the terminal ends affect adhesion as well. Spolenak at al (Spolenak et al.,2005) analyzed contact splitting effect for a number of shapes. They assumed negligible friction at the interface and a modest degree of adhesion and deduced that the pull off-force, therefore, varies in proportion to N1/4 for N smaller fibrils with the same area of the fibrillar surface. Thus, for the case of the flat punch, the contact splitting power is less than the case of a spherical terminal shape. Similar considerations for hemispheres suggest that little penalty is sustained as far as the pull-off force is concerned when peeling occurs through the forest of fibrils, rather than simultaneous detachment. One interesting finding of Spolenaks' study is devising a single "master" curve that relates mass and density of the terminal pads given by

$$\log N_A(m^{-2}) = 13.8 + 0.699 \log m(kg) .$$

A consequence of this equation is that increased mass requires finer

Figure 9. Terminal elements (circles) in animals with hairy design of attachment pads. Note that heavier animals exhibit finer adhesion structures
Arzt, E.; Gorb, S.; Spolenak, R., From micro to nano contacts in biological attachment devices", PNAS, 100, 19, 10603-10609, "Copyright (2003) National Academy of Sciences, U.S.A.", reprinted with permission.

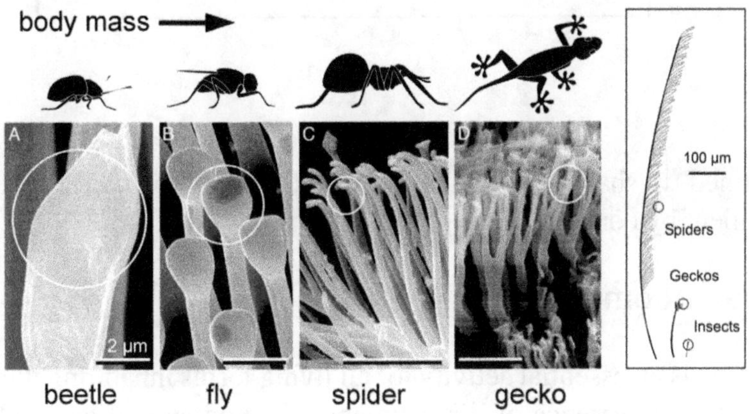

adhesion structures (see figure 9). The master equation is plotted in figure 9 for several groups.

Research data for fibrillar arrays collected over the last decade reveal some general trends summarized as follows. The adhesion force and strength of fibrillar surfaces increase significantly with decreasing fibril radii. Fibrils of varying lengths at constant fibril radius show an increase in adhesion force with increasing aspect ratio. This is attributed to the higher elastic energy dissipated at pull-off. Most of the stored elastic energy dissipated upon detachment, for low aspect ratio fibrils, comes from the deformation of the backing layer. For high aspect ratio fibrils, however, it comes from bending of the fibrils. Mushroom shaped tips have the highest adhesion more so than

Figure 10. Dependence of the terminal element density (NA) of the attachment pads on the body mass (m) in hairy-pad systems of diverse animal groups (logNA(m-2)=13.8 +0.699 logm(kg), R= 0.919).

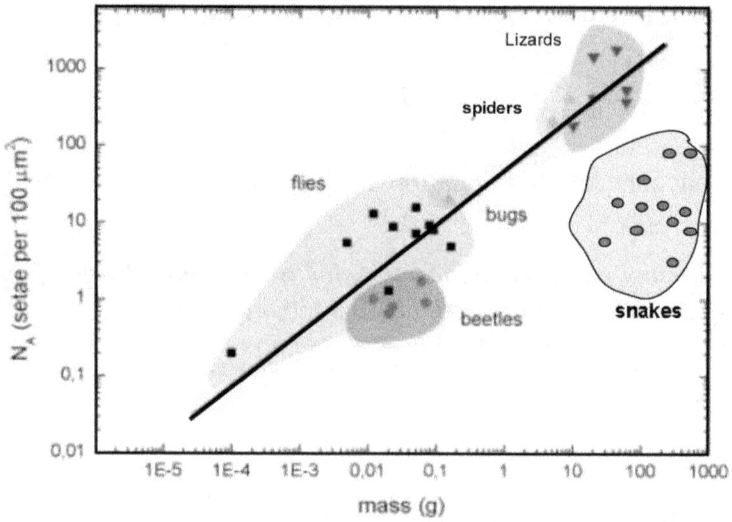

all examined tip shapes. They also show the highest potential for adhesion enhancement by contact splitting.

Legless Locomotors

Locomotion is an essential activity for all living forms inhabiting the planet. Movement is the essence of life. Humans relied on animal locomotion for many centuries for transportation, travel, and as a power source for many applications. It is fundamental therefore, that study of animal locomotion was a subject of interest for many researchers.

The concept of gaits normally facilitates the investigation of locomotion. Originally, such a concept was developed for the study of human locomotion. However, many researchers found it useful in the study of modes of locomotion in land, sea, and air. Gaits refer to patterns of motion. There are as many motion patterns as there are living forms. Each family of animals manifests a general pattern of motion that is governed by speed, surrounding medium, and body build. A generalized pattern does not indicate unified details. Rather, even within the same family of animals there will be considerable differences in the pattern of motion or gaits. Typically, the pattern of motion will differ in specific features such as force, speed, and time of contact of limbs (on the ground for example). Differences also will arise because of the resistance to motion by the surrounding medium (e.g. air drag acting on the body of a bird or water drag on a fish). Friction also could be one of the sources of differences in gait when it affects motion. Such a case arises in limbless locomotion.

Sliding on land is a trait of limbless animals. These are relatively rare among the species on the planet. Typically, legless locomotors (worms, snails, snakes, and few lizards) account for less than two percent of the 1.8 million species known to exist (figure 11). Limbless animals engage in a wide range of activities pertaining to locomotion. Such activities may include slithering, burrowing, tunneling, and several other complex combinations. Most of the activities take place within compositionally complex media (stone, gravel, mud, grass, etc.). The animal engages a complicated assortment of muscles, sensors, and alteration of body shape to propel its mass. Confining the presentation to snakes, the frequency, shape, and use of the locomotry systems also change instantaneously to the extent that a holistic understanding of the locomotion process is still lacking. Never the less, there have been numerous attempts to determine the origin of forces acting on the body of limbless animals and those that are generated by the body for propulsion. Often, however, the analysis is complicated because of the dependence of motion on balancing two frictional influences: the friction needed to prevent the body of the reptile from sliding without, and the frictional resistance encountered due to the nature of the substratum upon which locomotion takes place. The mechanism responsible for the delicate balance between frictional influences is a combination of synchronized muscular activity to manipulate body shape and generate propulsion, sensory, and strategic engagement of the micro-textural features of body scales to control contact (and thereby friction). In other words, the regulation of friction during limbless locomotion entails an intriguing, however complex, manipulation of body form and surface design.

Snakes have a wide distribution all over the planet. They are found almost everywhere on earth except where factors that limit their survivability are present (e.g., higher altitudes where very cold temperatures are predominant year-round). Their ecological diversity, and thereby their diverse habitat, presents a broad range of tribological environments many of which are hostile in terms of sliding and contact conditions. Such a situation requires specific tribological response that manifests itself in functional practices and surface design features. Frictional performance of snakes was a subject of several investigations that confirmed the unique features of the ventral skin of snakes and their optimized tribological response with respect to energy losses and

Figure 11. Comparison of the number of species of limbless locomotors compared to other species

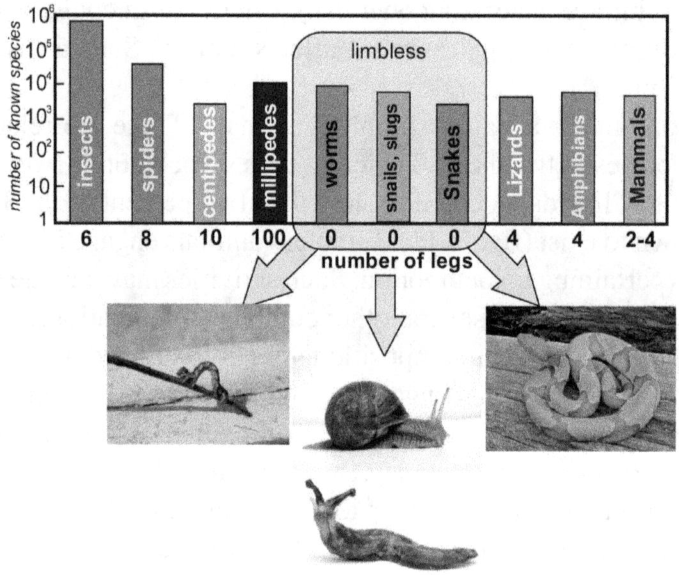

resistance to abrasion and wear (Gray, 1946., Abdel-Aal, 2013, Berthé, et al.,2009, Benz, et al.,2012, Klein et al., 2010, Klein, Gorb, 2012, Abdel-aal, 2015, Abdel-Aal and El Mansori, 2013).

Other results (Klein et al., 2010, Klein, Gorb, 2012, Abdel-aal, 2015) attribute optimization of tribological function to the geometrical patterns and metrology of the ventral micro-texture. Ventral scales in snakes contain micron-sized structures termed as fibrils or denticulations. The fibrils are not unique in shape. Different species manifest different fibril shapes. Figure 12 shows SEM images of the fibril shapes found on the ventral scales of several species. In addition to their variable shapes, fibrils vary in density and proportion and dimensions within the same reptile. That is, within the same species the number of fibrils, their geometrical dimensions, and consequently their density vary by location within the same species.

Textured surfaces and the ventral side of a snake have common features. The snake's ventral side comprises micron-sized fibrils distributed over the skin area in a specialized pattern. Spacing, length, orientation, and shape within the array, in general, are specific to the functional habitat of a particular family of snakes. Textured surfaces, on the other hand, comprise an individual textural building block (cone, dimple, chevron etc.,) distributed on the surface area in an orthogonal pattern. Abdel-aal, and coworkers (Abdel-aal, 2015,

Figure 12. SEM images depicting micro-structural details of the ventral scales of several snake species, scale bar is 5 µm, magnification X=5000.

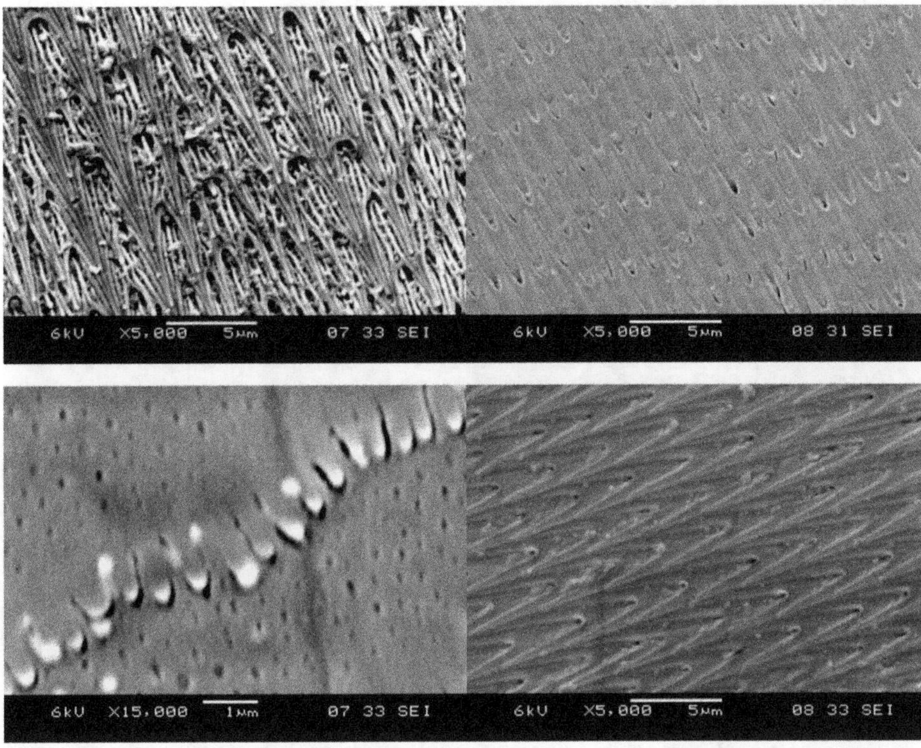

Abdel-Aal and El Mansori, 2013) recognized this fact and compared the geometrical ratios of snake textured surfaces to recommended values for man-texturized surfaces.

It is typical in texturized surfaces practice to evaluate performance in terms of primitive geometrical ratios that describe the deterministic topology of texturization on the surface. These ratios pertain to the aspect ratio of the textural feature, it's relative location on the surface, and the ratio of area occupied by texturization to the nominal area. Figure 13 plots the values of the geometrical ratios for fifteen snake species superposed on the values of man-texturized surfaces recommended by several research groups (see Chapter 5 and table 2). An interesting feature of the comparison is consistency of the ratios for snake surfaces with those recommended values after about 30 years of research in the field. More remarkable is that such compatibility is between two surfaces that evolved along completely separate paths. A snake, however, has to be self-sustaining over a wide spectrum of sliding terrains and conditions, and for survivability; its surface cannot afford a running-in period. Thus, unlike man engineered surfaces, the ventral texturing, has to be functional over the spectrum of contact conditions that the species experiences

Figure 13. Summary comparative plot depicting agreement between performance parameters for reptilian surfaces (Rhombic-symbols) and recommended values resulting from researching man-engineered surfaces. Shaded rectangles represent the full range of recommended values

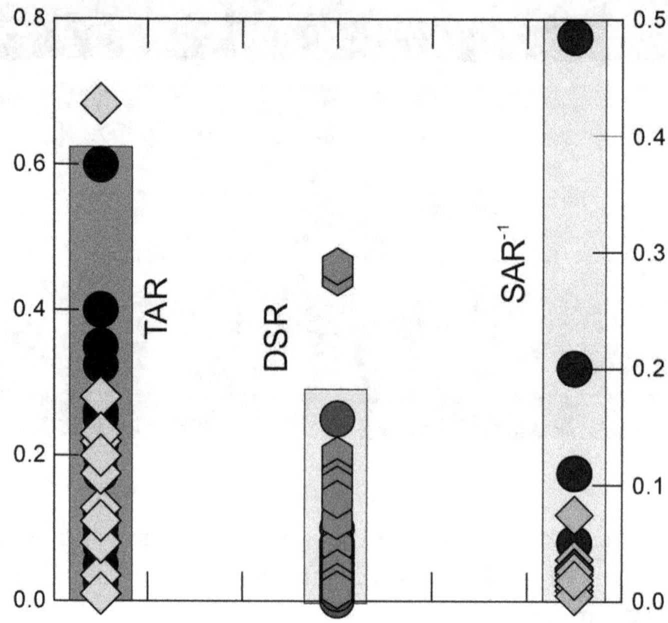

at inception. Such a feature offers an advantage to the natural surface and potentially can inspire creative deterministic solutions for many performance problems faced in technological surfaces.

Self-Cleaning and The Lotus Leaf

Lotus (Nelumbo nucifera) is a semiaquatic plant and develops peltate leaves up to 30 cm in diameter with remarkable water repellency. As an adaptation to the aquatic environment – some of the leaves float occasionally on the water surface – the stomata are located in the upper epidermis. The lower

Table 2. Summary of performance indicators for the species listed in Table 7

Species	Texture Primitive Ratio						
	TAR	SAR min	SAR max	SAR av	DSR max	DSR min	DSR AV
Bittis Gabonica	0.04	13.47	14.52	14.00	0.45	0.30	0.37
Cotton Mouth	0.28	110.45	93.37	101.91	0.03	0.04	0.03
Python Regius	0.18	72.75	66.14	69.44	0.08	0.15	0.11
Taylori Agikstron	0.20	54.10	36.22	45.16	0.04	0.22	0.13
Cerroph Godmani	0.13	28.06	27.00	27.53	0.14	0.17	0.16
Echis Carinatus	0.54	467.18	467.18	467.18	0.01	0.01	0.01
Pseudechis Austarlis	0.18	34.46	30.63	32.54	0.11	0.19	0.15
Agkist Contortrix	0.68	77.50	59.74	68.62	0.03	0.05	0.04
Pituophis Melanoleucus	0.22	201.72	173.70	187.71	0.02	0.03	0.02
Liasis Fuscus	0.20	36.42	39.73	38.08	0.06	0.30	0.18
L Mackloti	0.08	32.83	36.69	34.76	0.15	0.26	0.20
Morelia Amethistina	0.23	42.28	45.71	44.00	0.05	0.29	0.17
Morelia Virdis	0.20	52.50	56.00	54.25	0.04	0.29	0.16
Python Brongersmai	0.01	37.64	40.85	39.24	0.62	0.31	0.47
Python Sebae	0.11	52.98	57.39	55.18	0.06	0.23	0.14

epidermis consists of convex cells covered with wax tubules and contains only few stomata. The upper epidermis features the distinctive hierarchical structure consisting of papillae with a dense coating of agglomerated wax tubules, which is the basis for the famous super-hydrophobicity (figure 14).

Lotus exhibits outstanding water repellency on the upper side of its leaves this behavior however is not unique to Lotus. There are many plants that share the same behavior with some even exceeding the efficiency of lotus. The remarkable super-hydrophobic traits of the lotus leaf originate from a combination of factors, such as optimized surface topography features, robustness and the unique properties of an epicuticular wax. At the core of superior hydrophobicity is the hierarchical surface structure.

Compared to other species that share hierarchical surface structure with the same composition (papillae and wax crystals), the lotus leaf shows specialization of some of its features. The morphology of the papillae, particularly the small tip radius, minimizes the contact area to water drops but also the area where erosion and damaging of the waxes occurs. The robustness of the papillae provides protection of the wax crystals between them. These unique properties render lotus leaves unrivaled when super-hydrophobicity

Figure 14. The lotus plant. (a) Lotus leaves, which exhibit extraordinary water repellency on their upper side. (b) Scanning electron microscopy (SEM) image of the upper leaf side prepared by 'glycerol substitution' shows the hierarchical surface structure consisting of papillae, wax clusters and wax tubules. (c) Wax tubules on the upper leaf side. (d) Upper leaf side after critical-point (CP) drying. The wax tubules are dissolved, thus the stomata are more visible. Tilt angle 15°. (e) Leaf underside (CP dried) shows convex cells without stomata.

and self-cleaning properties are of concern. However, the combination of hierarchical surface structure and minimized contact area may also serve as inspiration to tribological surfaces especially under lubricated contacts where less viscous lubricants can be may be used.

Air Lubrication Lubrication Mechanisms

Penguins cannot fly yet they have to jump out of water onto sea ice sometimes to evade predators. Their physical squat body structure with stubby legs which make it difficult for them to climb ashore especially onto rocky shorelines and

Figure 15. An emperor penguin ascending to break the surface of water with notable air bubble clouds emerging from various regions on a penguin body (51) Aptenodytes forsteri. Images of near-vertically ascending emperor penguins. Note that these drawings, drawn from sequential close-up fields of 2 different penguins, demonstrate that the bubble clouds envelop most of the body and obscure the tail and hind limbs. Note also that the identified anterior edges of the bubble clouds correspond to areas where bright bubbles can be seen against the penguin surface.

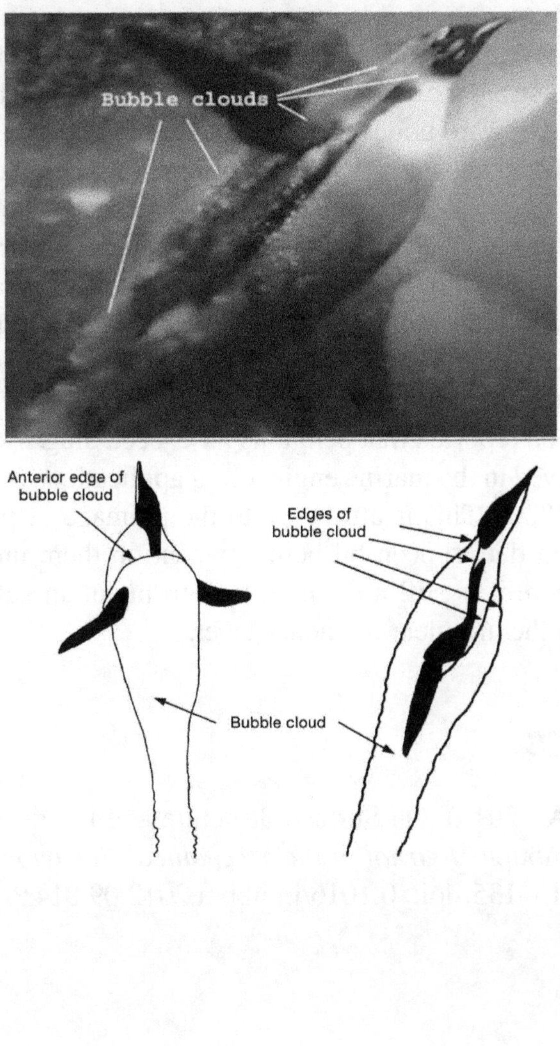

mostly impossible for it to climb onto sea ice. Some species like the emperor penguins can fly as high as 9 feet in air at nearly 5.1 ms-1 and land on ice safely. Davenport et al. (2011) investigated these sudden bursts in emperor penguins. The authors hypothesized that this observed behavior originates in 'Air Lubrication.' That is, injecting air into the boundary layers (a technique that has been used in high speed propulsion of ships and torpedoes through sea water).

Sato and co-workers (2005, 2002) recorded detailed physical the movements of emperor penguins including their speed and distances before (underwater) and after a burst of speed (underwater and in air) of jumping onto sea ice. They found that the velocities required to overcome the effect of gravity correlated with the exit speeds achieved by the penguins. They rationed that the stopping of flipper action some distance below the jump allows for better buoyancy observed in an earlier research with a different species of penguin (Sato et al.,2005), not applicable in case of emperor penguins.

Davenport (Davenport et al.,2011) (see figure 15) analyzed air lubrication phenomenon which aid the emperor penguins to achieve high bursts of speed. He measured the upward underwater ascent speeds of these penguins to be nearly 5.3 ms−1 and showed it to be the maximum speed they can achieve. This high-speed during ascent is attributed to the penguins emitting air bubble clouds into the turbulent boundary layer over most of their body surface. Emission of air bubbles increases the speed as the penguin approaches the surface of the water. The air bubble cloud performs a drag-reducing function whereby it reduces the friction between the solid skin surface of penguin and the water around it.

It has also been reported that penguins far exceed drag/friction reduction than those achieved in the marine engineering applications (Sato et al.,2002, Fukuda et al., 2000). This is attributed to the plumage of penguins which is water repellent due to peen oil being present on them unlike in marine applications where it is still a difficult task to maintain sufficient bubble coverage within the turbulent boundary layer.

REFERENCES

Abdel-Aal, H. A. (2013). On Surface Structure and Friction Regulation in Reptilian Locomotion. *Journal of the Mechanical Behavior of Biomedical Materials*, *22*, 115–135. doi:10.1016/j.jmbbm.2012.09.014 PMID:23582565

Abdel-aal, H. A. (2015). *The structure of Ventral Scale Textures in Snakes in comparison to Texturing of Deterministic Tribological Surfaces.* In J. P. Davim (Ed.), *Processing Techniques and Tribological Behavior of Composite Materials, Rajnesh Tyagi* (pp. 241–320). IGI Global Publication.

Abdel-Aal, H. A., & El Mansori, M. (2013). Tribological analysis of ventral scale structure in a python regius in relation to laser textured surfaces IOP-J. *Surface Topography: Metrology and Properties*, *1*(1), 015001. doi:10.1088/2051-672X/1/1/015001

Alexander, R. McN. (1991). Elastic mechanisms in primate locomotion. *Zeitschrift fur Morphologie und Anthropologie*, *78*, 315–320. PMID:1887661

Amundin, M. (1974). Functional analysis of the surfacing behavior in the harbour porpoise Phocoena phocoena. *Zeit. Saugetier.*, 313–8.

Au, D., & Weihs, D. (1980). At high speeds dolphins save energy by leaping. *Nature*, *284*(5756), 548–550. doi:10.1038/284548a0

Babenko, V. V. (1979). Investigating the skin elasticity of live dolphins [in Russian]. *Bionika (Kiev)*, *13*, 43–52.

Bar-Cohen, Y. (2006): Introduction to Biomimetics: The Wealth of Inventions in Nature as an Inspiration for Human Innovation. In Biomimetics: Biologically Inspired Technologies. Taylor & Francis.

Bechert, D. W., Bartenwerfer, M., & Hoppe, G. (1986). Drag reduction mechanisms derived from the shark skin. *15th ICAS Congress*, 1044–68.

Bechert, D. W., Bruse, M., & Hage, W. (2000a). Experiments with three-dimensional riblets as an idealized model of the shark skin. *Experiments in Fluids*, *28*(5), 403–412. doi:10.1007003480050400

Bechert, D. W., Bruse, M., Hage, W., & Meyer, R. (2000b). Fluid mechanics of biological surfaces and their technological application. *Naturwissenschaften*, *80*(4), 157–171. doi:10.1007001140050696 PMID:10840802

Benz, M. J., Kovalev, A. E., & Gorb, S. N. (2012). Anisotropic frictional properties in snakes. *Proc. of SPIE*, 8339. 10.1117/12.916972

Berg, H. C. (2000). Constraints on models for the flagellar rotary motor. *Philosophical Transactions of the Royal Society of London. Series B, Biological Sciences*, *355*(1396), 491–501. doi:10.1098/rstb.2000.0590 PMID:10836502

Berthé, R. A., Westhoff, G., Bleckmann, H., & Gorb, S. N. (2009). Surface structure and frictional properties of the skin of the Amazon tree boa Corallus hortulanus (Squamata Boidae). *J. Comp. Phys. A*, *195*(3), 311–318. doi:10.100700359-008-0408-1 PMID:19137315

Canjani, C., Andrade, D. V., Cruz-Neto, A. P., & Abe, A. S. (2003). Aerobic metabolism during predation by a boid snake. *Comp. Biochem. Phys. A*, *133*(3), 487–498. doi:10.1016/S1095-6433(02)00150-2 PMID:12443908

Cant, J. G. H. (1987). Positional behaviour of female Bornean orangutans (Pongo pygmaeus). *American Journal of Primatology*, *12*(1), 71–90. doi:10.1002/ajp.1350120104 PMID:31973515

Cant, J. G. H. (1994). Positional behaviour of arboreal primates and habitat compliance. In Current primatology.: Vol. 1. *Ecology and evolution Strasbourg* (pp. 187–193). Universite Louis Pasteur.

Carpenter, P. W. (1998). Recent advances in the use of compliant walls for drag reduction. *Proc. Int. Symp. on Seawater Drag Reduction*, 185–8.

Carpenter, P. W., Davies, C., & Lucey, A. D. (2000). Hydrodynamics and compliant walls: Does the dolphin have a secret? *Current Science*, *79*, 758–765.

Davenport, J., Hughes, R. N., Shorten, M., & Larsen, P. S. (2011). Drag reduction by air release promotes fast ascent in jumping emperor penguins—A novel hypothesis. *Marine Ecology Progress Series*, *430*, 171–182. doi:10.3354/meps08868

Del Campo, A., Greiner, C., & Arzt, E. (2007). Contact shape controls adhesion of bioinspired fi brillar surfaces. *Langmuir*, *23*(20), 10235–10243. doi:10.1021/la7010502 PMID:17722937

Demes, B., Jungers, W. L., Gross, T. S., & Fleagle, J. G. (1995). Kinetics of leaping primates: Influence of substrate orientation and compliance. *American Journal of Physical Anthropology*, *96*(4), 419–429. doi:10.1002/ajpa.1330960407 PMID:7604894

Fejes Tóth, L. (1964). What the bees know and what they do not know. *Bulletin of the American Mathematical Society*, *70*(4), 468–481. doi:10.1090/S0002-9904-1964-11155-1

Fish, F. E. (1998). Comparative kinematics and hydrodynamics of odontocete cetaceans: Morphological and ecological with swimming performance. *The Journal of Experimental Biology*, *201*, 2867–2877.

Fukuda, K., Tokunaga, J., Nobunaga, T., Nakatani, T., Iwasaki, T., & Kunitake, Y. (2000). Frictional drag reduction with air lubricant over a super-water-repellent surface. *Journal of Marine Science and Technology*, *5*(3), 123–130. doi:10.1007007730070009

Gray, J. (1936). Studies in animal locomotion: VI. The propulsive powers of the dolphin. *The Journal of Experimental Biology*, *13*, 192–199.

Gray, J. (1946). The Mechanism of Locomotion in Snakes. *The Journal of Experimental Biology*, *23*, 101–120. PMID:20281580

Hales, T. C. (2001). The Honeycomb Conjecture. *Discrete & Computational Geometry*, *25*(1), 1–22. doi:10.1007004540010071

Hebrank, M. R. (1980). Mechanical properties and locomotor functions of eel skin. *The Biological Bulletin*, *158*(1), 58–68. doi:10.2307/1540758

Klein, M.-C. G., Deuschle, J., & Gorb, S. N. (2010). Material properties of the skin of the Kenyan sand boa Gongylophis colubrinus (Squamata, Boidae). *J. of Comp. Phys. A: Neuroethology, Sensory, Neural, and Behav. Phys.*, *196*, 659–668.

Klein, M-C. G., & Gorb, S. N. (2012). Epidermis architecture and material properties of the skin of four snake species. *J. R. Soc. Interface*. doi:10.1098/rsif.2012.0479

Krieger, K. (2004). Do pool sharks really swim faster? *Science*, *305*(5684), 636–637. doi:10.1126cience.305.5684.636 PMID:15286360

Lang, A. W., Motta, P., Hidalgo, P., & Westcott, M. (2008). Bristled shark skin: A microgeometry for boundary layer control? *Bioinspiration & Biomimetics*, *3*(4), 1–9. doi:10.1088/1748-3182/3/4/046005 PMID:18838758

Lee, S. J., & Choi, Y. S. (2008). Decrement of spanwise vortices by a drag-reducing riblet surface. *Journal of Turbulence*, *9*, 1–15. doi:10.1080/14685240802251517

Liskiewicz, T., Morina, A., & Neville, A. (2008). Friction in nature. *WIT Transactions on Ecology and the Environment: Design and Nature IV*, *114*, 263-272.

Mannion, J. T., Reccius, C. H., Cross, J. D., & Craighead, H. G. (2006). Conformational analysis of single DNA molecules undergoing entropically induced motion in nanochannels. *Biophysical Journal*, *90*(12), 4538–4545. doi:10.1529/biophysj.105.074732 PMID:16732056

Markham, R., & Groves, C. P. (1990). Weights of wild orangutans. *American Journal of Physical Anthropology*, *81*, 1–3. doi:10.1002/ajpa.1330810102 PMID:2405688

Mehta, R. S., & Burghardt, G. M. (2008). Contextual flexibility: Reassessing the effects of prey size and status on prey restraint behaviour of macrostomate snakes. *Ethology*, *114*(2), 133–145. doi:10.1111/j.1439-0310.2007.01437.x

Motta, P. (1977). Anatomy and functional morphology of dermal collagen fibers in sharks. *Copeia*, *1977*(3), 454–464. doi:10.2307/1443263

Murphy, J. C., & Henderson, R. W. (1997). *Tales of giant snakes: a historical natural history of anacondas and pythons*. Malabar, FL: Krieger Publishing Company.

Pavlov, V. V. (2003). Wing design and morphology of the harbor porpoise dorsal fin. *Journal of Morphology*, *258*(3), 284–295. doi:10.1002/jmor.10135 PMID:14584030

Reif, W. E. (1982). Hydrodynamics of the squamation in fast swimming sharks N. *Jb. Geol. Palaont.*, *164*, 184–187.

Reif, W. E. (1985a). Morphology and hydrodynamic effects of the scales of fast swimming sharks. *Fortschritte der Zoologie*, *30*, 483–485.

Reif, W. E. (1985b). Squamation and ecology of sharks. Courier Forschungsinstitut Senckenberg, 78.

Reif, W. E. (1987). Squamation of hammerhead sharks: Variation in the polarity field Zool. *Journal of Anatomy*, *115*, 153–161.

Sato, K., Naito, Y., Kato, A., & Niizuma, Y. (2002). Buoyancy and maximal diving depth in penguins: Do they control inhaling air volume? *The Journal of Experimental Biology*, *205*, 1189–1197. PMID:11948196

Sato, K., Ponganis, P. J., Habara, Y., & Naito, Y. (2005). Emperor penguins adjust swim speed according to the above-water height of ice holes through which they exit. *The Journal of Experimental Biology*, *208*(13), 2549–2554. doi:10.1242/jeb.01665 PMID:15961741

Schmitt, O. (1969). *Third Int. Biophysics Congress. Some interesting and useful biomimetic transforms.* Academic Press.

Boback, Hall, McCann, Hayes, Forrester, & Snake. (2012). Modulates constriction in response to prey's heartbeat. *Biol. Lett. J, 8*(3), 473-476.

Spolenak, R., Gorb, S., & Artz, E. (2005). Adhesion design maps for bio-inspired attachment systems. *Acta Biomaterialia, 1*(1), 5–13. doi:10.1016/j.actbio.2004.08.004 PMID:16701776

Thorpe, S. K. S., & Crompton, R. H. (2005). Locomotor ecology of wild orang-utans (Pongo pygmaeus abelii) in the Gunung Leuser ecosystem, Sumatra, Indonesia: A multivariate analysis using log-linear modelling. *American Journal of Physical Anthropology, 127*(1), 58–78. doi:10.1002/ajpa.20151 PMID:15386279

Thorpe, S. K. S., & Crompton, R. H. (2006). Orangutan positional behavior and the nature of arboreal locomotion in Hominoidea. *American Journal of Physical Anthropology, 131*(3), 384–401. doi:10.1002/ajpa.20422 PMID:16617429

Vincent, J. F. V. (2009). Biomimetics a review. *Journal of Engineering in Medicine. Proceedings of the Institution of Mechanical Engineers. Part H, 223*(8), 919–939. 10.1243/09544119JEIM561

Vincent, J. F. V., Bogatyreva, O. A., Bogatyrev, N. R., Bowyer, A., & Pahl, A.-K. (2006). Biomimetics: Its practice and theory. *Journal of the Royal Society, Interface, 3*(9), 471–482. doi:10.1098/rsif.2006.0127 PMID:16849244

Wainwright, S., Vosburgh, F., & Hebrank, J. H. (1978). Shark skin: A function in locomotion. *Science, 202*(4369), 747–749. doi:10.1126cience.202.4369.747 PMID:17807247

ENDNOTES

1 Varro actually gave two possible reasons for the six-sided honeycombs: (1) "Does not the chamber in the comb have six angles, the same number as the bee has feet?", (2) "The geometricians prove that this hexagon inscribed in a circular figure encloses the greatest amount of space." The second, though premature, has two thousand years later proved accurate. Varro's comments were not always so accurate, as when he wrote that the bees ". . . follow their own king wherever he goes . . ." Apparently

it was not discovered until the seventeenth century that the “king” was a queen.

2 An account of his endeavour was given by the Moroccan historian Ahmed Mohammed al-Maqqari (d. 1632), “Among other very curious experiments which he made, one is his trying to fly. He covered himself with feathers for the purpose, attached a couple of wings to his body, and, getting on an eminence, flung himself down into the air, when according to the testimony of several trustworthy writers who witnessed the performance, he flew a considerable distance, as if he had been a bird, but, in alighting again on the place whence he had started, his back was very much hurt, for not knowing that birds when they alight come down upon their tails, he forgot to provide himself with one.”

3 Biophysics is not so much a subject matter as it is a point of view. It is an approach to problems of biological science utilizing the theory and technology of the physical sciences. Conversely, biophysics is also a biologist’s approach to problems of physical science and engineering, although this aspect has largely been neglected. —Otto Herbert Schmitt, In Appreciation, A Lifetime of Connections: Otto Herbert Schmitt, 1913 – 1998.

Chapter 7
Filtering Texture From Biological Surfaces to Technological Surfaces:
Case Study

ABSTRACT

This chapter discusses the impact of absence of a holistic surface-design methodology in the technological realm. The authors show that manifesting designs that merge function, form, and topography to achieve lean performance is currently a bottleneck in the field of tribology. The presented material shows that merging function and topography, while not matured within the realm of manmade surfaces, is advanced in natural designs, especially within the scaled reptiles (Squamata). This prompts many engineers to scour biological analogues for design alternatives. However, the problem of evaluating the viability of a natural surface and judging its suitability for a technological application is frequently encountered. The chapter adresses this problem through a detailed case study that involves extracting metrological information for snakes that engage in rectilinear locomotion.

INTRODUCTION

Theoretical analysis identified several dimensional groups that influence the tribological performance of a textured surface (Ryk, 2002, Ronen et. al.,2001, Golloch, et. al., 2004, Dumitru, et. al, 2004, Hiroki et. al., 2011, Hu and Ding, 2012, Borghi, et al., 2008, Qiu and Khonsari, 2011, Brizmer and Kligerman,

DOI: 10.4018/978-1-7998-1647-8.ch007

2012, Eichstät, et al. 2011, Kovalchenko et al., 2005, Liu, et al. 2010). To date, however, there is no agreement on the optimal values to be implemented given a particular surface. More importantly, a well-defined methodology for the generation of textures for optimized surface designs is virtually non-existent. Such a process is still viewed as a "black art" mainly because of the absence of universal standards that rate design and manufacturability from a targeted-performance perspective. That is, the absence of a holistic surface-design methodology that merges function, form and topography to achieve lean performance within a wide range of contact conditions. Such an approach, while in essence, has not matured as of yet within the realm of human surface engineering is advanced in natural designs especially within the scaled reptiles (squamata). Squamate Reptiles present diverse examples where surface structure, texturing, and modifications through submicron and nano-scale features, achieve frictional regulation manifested in: reduction of adhesion (Arzt, et al. 2003), abrasion resistance (Rechenberg, 2003), and frictional anisotropy (Hazel, et al.1999).

Squamata comprises two large clades: Iguania and Scleroglossa. The later comprises 6,000 known species, 3100 of which are referred to as "lizards," and the remaining 2,900 species as "snakes" (Vitt, et al.,2003). They are found almost everywhere on earth. Their diverse habitat presents a broad range of tribological environments. This requires customized response that manifests itself in functional practices and surface design features. This, potentially, can inspire deterministic solutions for many technical problems. Many authors studied appearance and structure of snakeskin in relation to friction (Jayne, 1988-a, Rivera, et al., 2005, Berthé, et al., 2009, Saito, et al. 2000, Shafiei, and Alpas, 2008, Shafiei and Alpas, 2009, Abdel-Aal, et al.,2010, Abdel-Aal and Mansori, 2011, Abdel-Aal, 2012, Gray, 1946, Benz, et al., 2012, Klein, and Gorb, 2012, Schmidt and Gorb, 2012). Results pointed at the relation of surface topography in snakes to tribological performance.

Locomotion of snakes takes place through different modes: rectilinear, concertina, side winding, lateral undulations, and slide pushing. These modes entail the generation of cyclical contraction-relaxation waves of particular muscle groups. The skin of the reptile is the medium that simultaneously generates and accommodates the tractions necessary for motion. Dissipation and transformation of the kinetic energy associated with motion also takes place within the skin. Topology of the skin and its' interaction with the surrounding environment, is, therefore, crucial for control and optimization of the tribological response of during motion.

The motion of a snake is a delicate balance between the propulsive forces generated by the muscles and the friction tractions due to contact with the substratum. In some cases, the snake makes use of friction to generate thrust. However, for economy of effort, the COF needs to be minimized (especially in rectilinear locomotion) since friction opposes motion. As such, a self-regulating mechanism to control frictional tractions should exist in the snake. The texture of the surface (i.e., the micron sized fibril structures or denticulations) are a major component of such a mechanism. The geometry and topology of the fibril structures allow the snake to modify the frictional profiles in response to changes in contact situations. The presence of the fibrils contributes to the dynamic control of the real area of contact between the skin and the substratum upon sliding (Hoppermann and Kordt, 2002). The function of the denticulations (micron-sized fibrils) in this sense is similar to that of the deterministic textures used to regulate friction. Such a similarity raises a curious question. Namely, if we consider the denticulations present on the ventral side of a snakeskin as deterministic textures, would their descriptive parameters coincide with, or at least fall within, the range of values recommended by researchers for technological surfaces based on laser texturing? Furthermore, in case of the validity of such a preposition, can the texturing of a snake inspire a new paradigm in deterministic surface texturing?

SIMILARITIES OF NATURAL AND MANMADE SURFACES

There are several similarities between a LTS and the ventral side of a snake, both in metrology and topology. The main similarity is the dependence of the metrological features on the scale of observation. That is both of the snakeskin and the LTS belong to the so-called multi-scale surfaces (Jiang, and Whitehouse, 2012). In addition, the basic building block in each case (snakeskin and LTS) is a textural element that is repeated in an array distribution. The snake's ventral side is composed of identical micron-sized fibrils (denticulations) distributed over the skin area in a particular pattern. Spacing, length, orientation and shape of denticulation are in general common to a particular family of snakes. LTS, on the other hand, by their very definition comprise an individual textural building block (cone, dimple, chevron etc.,) that is distributed on the surface area in array form. Therefore, both types of surfaces share common constructal features. A snake, however, has to be self-sustaining over a wide spectrum of sliding terrains and conditions.

The ventral texturing, therefore, has to be efficient over the spectrum of contact conditions that the species experiences. Such a feature offers an advantage to the natural surface over an engineered LTS where the dimples satisfy efficient performance over a narrow domain within possible contact conditions. The question, therefore, becomes how to optimize LTS in order to extend efficient performance, similar to that of a snake, over a wider range of contact conditions. In other words, how to capture the essence of texturing within the ventral side of a snakeskin and then incorporate it into a manmade surface to enhance tribological performance.

The answer to this question is rather complex due to the many factors involved. A simple answer is to mimic, or replicate, the textural designs present on the reptile in a technological surface. However, the success of this approach depends on the existence of a one-to-one match between the function and perhaps, the material, of the target surface and that of a reptile. This match will limit the replication process to polymeric surfaces (because the skin is elastomeric). However, bio-inspired surface texturing has a broader goal than merely replication. In essence, it seeks to extend the potential tribological benefits of reptilian surfaces to the domain of man-engineered surfaces. Transfer of design benefits requires developing multi-aspect compatibility metrics in order to evaluate the feasibility of the inspiration process itself (i.e., basing the design of a technical surface on that of a biological analogue). Two features are important in this context: the quality of frictional performance of the snake and matching of the deterministic surface parameters of the biological target to those recommended for the technological surface.

Frictional performance of snakes was a subject of several investigations that confirmed the unique features of the ventral skin of snakes and their optimized tribological response with respect to energy losses and resistance to abrasion and wear (Klein, and Gorb, 2012, Schmidt and Gorb, 2012). Other results (Shafiei, and Alpas, 2008, Shafiei and Alpas, 2009, Abdel-Aal, et al.,2010, Abdel-Aal and Mansori, 2011, Abdel-Aal, 2012) attribute optimization of tribological function to the geometrical patterns and metrology of the ventral micro-texture. These results, while promising, need to be complimented by assessing the deterministic features of the ventral textures along the same guidelines followed in examining LTS (and manmade surfaces in general). In this chapter, therefore, we present a comparative study of the deterministic surface parameters of the ventral texturing in a snake (P. regius) and those parameters recommended from LTS.

CLASSIFICATION OF SNAKES

Snakes comprise eighteen families (see table 1). Each family divides into genera, and these, further sub-divide into species. The number of species per genera depends on biological similarities between candidates. The family Colubridae is the largest family with around 1850 species. Families are arranged in order of evolution (from the most primitive to the recently evolved). The rigidity of the skull distinguishes three snake families Leptotyphlopidae, Anomalepidae, and Typholopidae-differ as the most primitive among snakes. These families also share other features such as very small eyes and being covered with scales. The most advanced families include all venomous snakes. These are the Acrochordidae, Clubridae, Attactaspididae, Elapidae, and Viperidae. Snake families may be divided into subfamilies as well. One example is the Boidae which is divided into Boinae and the Erycinae.

Table 1. Major families of snake species

Family Name	Common Name	No.	Distribution
Acrochordidae	Wart snakes	2	Australia, E Indies, SE Asia
Boidae	Pythons, boas, woodsnakes	60	Tropical regions
Colubridae	Terrestrial, arboreal and aquatic snakes	>1850	Worldwide
Elapidae	Cobras, mambas, coral snakes	170	Asia, Africa, N and S America
Hydrophiidae	Sea snakes	50	Indian and Pacific oceans
Letotyphlopidae	Slender blind snakes	40	N and S America, SW Asia, Africa
Typhlopidae	Blind snakes, worm snakes	200	Tropical regions
Uropeltidae	Shieldtail snakes	50	S Asia
Viperidae	Vipers, rattlesnakes, moccasins	180	Asia, Europe, Africa, Americas
Xenopeltidae	Sunbeam snake	1	India

Ventral scales in snakes have constant number for the particular species. This number is often used to distinguish the species. The average areas of the ventral scales vary also by species. Table 2 shows a sample of the variation in the number and average areas of ventral scales for some species.

The data show that the number of ventral scales, and the average area, are not generally proportional to weight or length.

Ventral scales in snakes contain micron-sized structures termed as fibrils or denticulations. The fibrils are not unique in shape. Different species manifest different fibril shapes. Figure 1 shows SEM images of the fibril shapes found on the ventral scales of several species. In addition to their variable shapes, fibrils vary in density and proportion and dimensions within the same reptile. That is, within the same species the number of fibrils, their geometrical dimensions, and consequently their density vary by location within the same species. Figure 2 shows an example of such variation. The figure depicts the details of the fibril distribution for a P. regius at three positions on the body. The geometry of the micron-sized fibrils within the three regions, chosen on the body, is similar. However, the spacing and density of the fibrils (and thereby the ratio of area occupied by fibrils to the rest of the area of a scale) differ in each region.

Table 2. Sample length, weight, and ventral scale counts in some snake species

Species		Length (cm)	Weight (kg)	Ventral Scales	Max Area/ Scale (mm^2)
Bittis Gabonica	true-viper	150	8	132	377.87
Agkistrodon Piscivorus	pit-viper	89	2.624	138	235.27
Python Regius	non-viper	150	1.3	208	102.93
Taylori_Agikstron	pit-viper	87	0.2526	132	112.69
Cerrophidion_Godmani	pit-viper	84	0.227	139	165.85
Pseudechis_Austarlis	pit-viper	152	1.36	205	123.5
Agkistrodon Contortrix	non-viper	135	0.95	148	168.12
Pituophis Melanoleucus	non-venom	154	1.414	232	155.26

Analogy Between LTS and Ventral Snake Skin

AFM imaging of the fibrils, shown in figure 3 (a and b), reveals the elevation of the tips above the overall ventral cell plateau. Fibril tips protrude above the general plateau of the ventral scale. In that sense, fibril tips resemble the protrusions of a textured surface (similar to the micro-features created in a laser texturing process). A distinct difference, however, is arrangement of texturing in each case. For the shed-skin, the fibril tips on the ventral scale manifest a wave arrangement (as opposed to the orthogonal array arrangement observed in manmade surfaces). The function of texturing is to modify the frictional behaviour of the surface upon sliding. In this sense, both manmade and snake surface share the devices of tribological response modifiers: texturing. This observation provides the essence of the comparison between the topological features of the skin of a Python and an LTS.

Figure 1. SEM images depicting structural details of the ventral scales at three locations

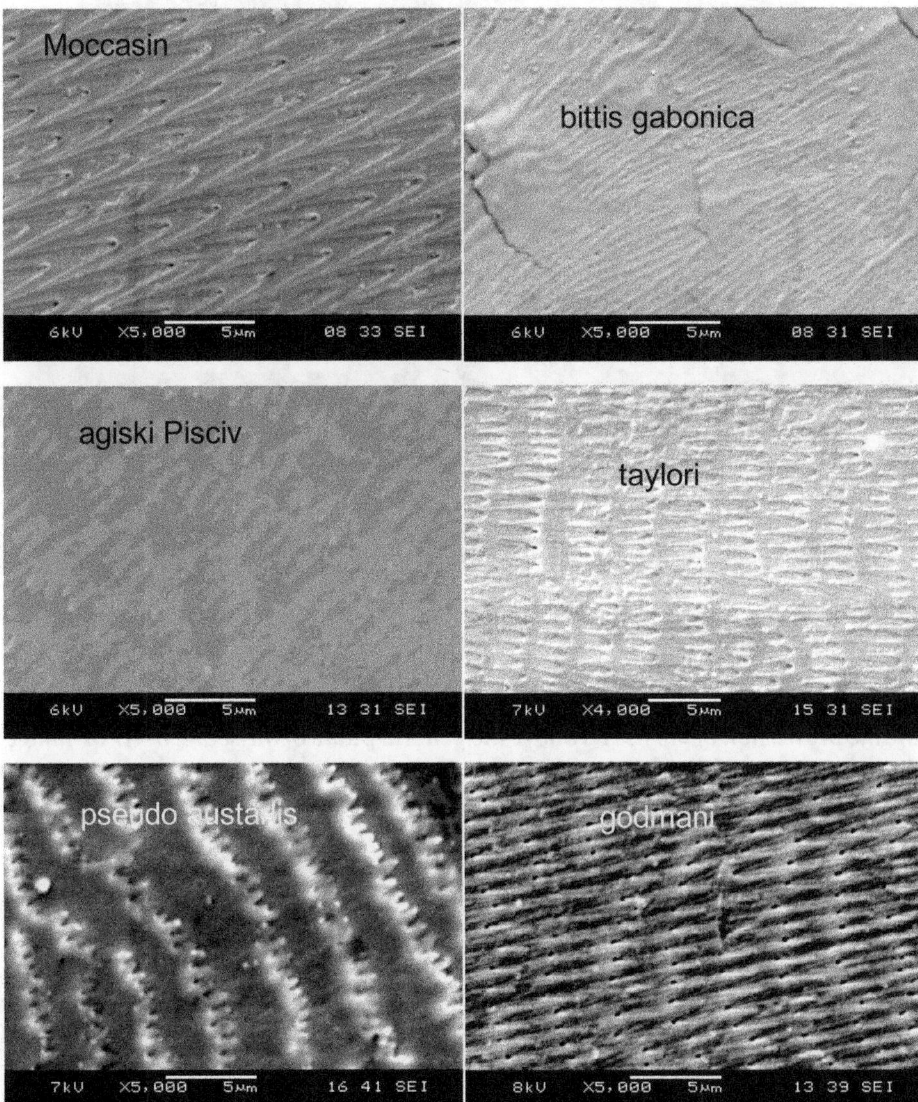

Figure 2. SEM images depicting structural details of the ventral scales at three locations, TDP-S, MP-S, and BDP-S respectively. Images on the left hand side are taken at a magnification of X=250 (length marker 100 μ m) and those on the right hand side are taken at a magnification X=10,000 (length marker 1μm). Note the arrangement of the fibrils in wave like rows, and the non-uniform distribution of the fibril dimensions.

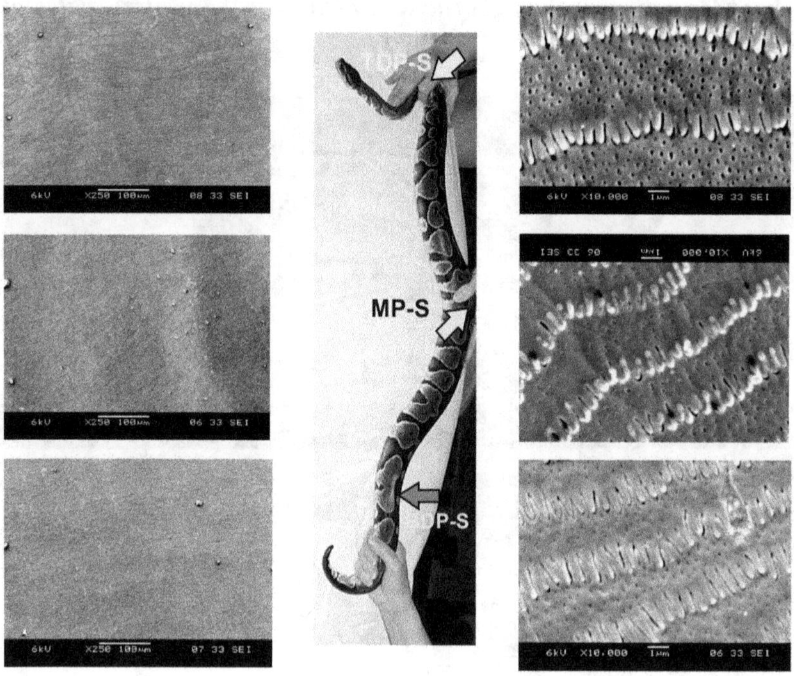

Figure 3. Three dimensional AFM-images of the fibril structures within the ventral scales of the Python, a- general scan of a 45 μm x 45 μm area of a ventral scale. Note the protrusion of the fibrils above the general plane of the scale, b- AFM-image of a 5 μm x 5 μm area of the ventral scale. Note the orientation of the fibrils along the AE-PE Axis

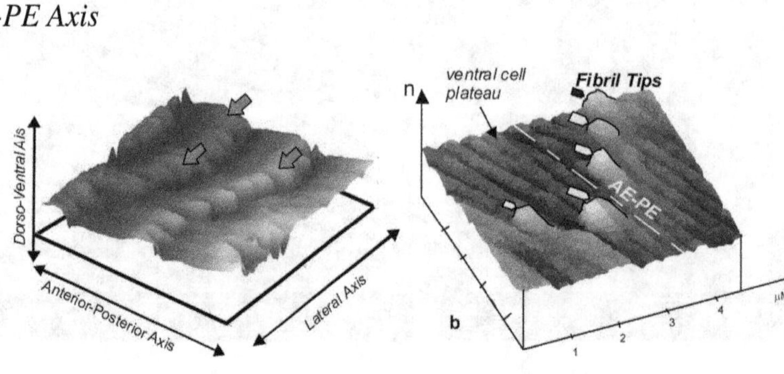

To compare the texture of the snakeskin to that of the LTS, there is a need to define texture descriptors for reptilian skin along the same lines used for LTS. The direct approach is to redefine the LTS descriptors, introduced in chapter V, in terms of the surface features present on the snakeskin. To this end, we use the subscript (s) to denote surface texture features present on the snakeskin, where as we use the subscript (p) to denote descriptor of LTS. Therefore, for the snakeskin the DSR is defined as the ratio of the fibril height (hs) to the base diameter of the fibril tip (Φs). The SAR meanwhile is redefined as the distance between two consecutive fibril waves (λs) averaged over the entire body and the protrusion distance. Finally, we redefine the TAR for the snake as the total area of the fibrils to the total surface area. Table 3 presents a summary comparison of the formulas used in computing the additional surface parameters for an LTS and a snake.

Table 3. Summary of formulas used to calculate parameters used in comparing LTS and Python Skin

Parameter	LTS	Python Skin
	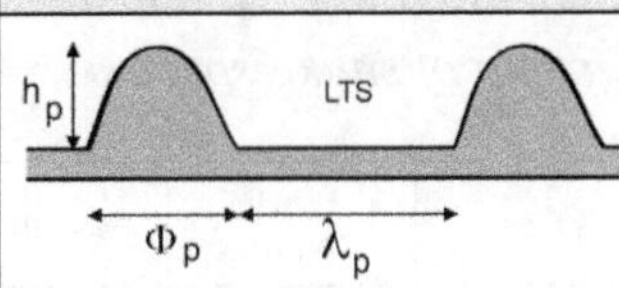	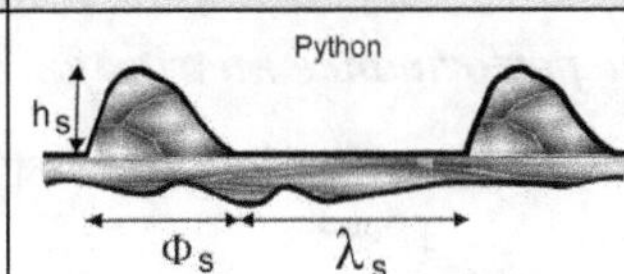
Total Area Ratio (TAR)	$\frac{\textit{Total area of dimples}}{\textit{total surface area}}$	$\frac{\textit{Total area of fibril tips}}{\textit{total surface area}}$
Dimple Slenderness Ratio (DSR)	$DSR_p = \frac{h_p}{\Phi_p}$	$DSR_s = \frac{h_s}{\Phi_s} \approx \frac{R_a}{\Phi_s}$
Surface Aspect Ratio (SAR)	$SAR_p = \frac{\lambda_p + \Phi_p}{h_p}$	$SAR_s = \frac{\lambda_s + \Phi_s}{h_s}$

The total fibril height of the snake, hs, manifests the protrusion of the fibril tip over the cell plateau. This is approximately equivalent to the average roughness of the surface, Ra, along the orientation of the fibrils (i.e., along the AE-PE-axis). Accordingly, to compare the performance indicators of the shed skin to those of dimpled LTS, one first needs to calculate the four parametric ratios listed in chapter V for each surface. Note that the descriptors of an LTS depend on the primitive geometry of the surface, for a shed skin, however, these are functions of metrological parameters (dimensional and textural). As such, for the shed skin, characterization of the essential metrological parameters should precede the calculation of the performance indicator ratios. Once the necessary parameters are determined, calculation of the four ratios of table 3 may take place.

Table 4 presents a summary of the primitive metrological dimensions needed for computing the performance metrics for a group of snake species. The table includes two data sets for the distance between the textural features ($\lambda s+\phi s$). The first represents the distance along the AE-PE-axis, whereas the second with a subscript l, represents the same distance in the lateral direction. Again, the data does not reflect considerable directional anisotropy. The data also indicate that there is no particular dominance of directional values between species (i.e., there is no trend for the values along the AE-PE-axis to be consistently greater, or smaller, than those values in the lateral direction according to species).

Table 4. Summary of the primitive metrological dimensions needed for computing the performance metrics for a group of snake species

Species	Lf(Fs)	Fibril Width	λs+Φ s	λs+Φ sl	Dist rws
Bittis Gabonica	0.75	1.10	4.50	4.85	1.60
Cotton Mouth	1.75	1.00	4.85	4.10	0.33
Taylori_Agikstron	2.50	0.50	6.05	4.05	0.66
Cerrophidion_Godmani	1.00	0.85	4.00	3.85	0.74
Echis_Carinatus	5.50	5.50	20.50	20.50	0.26
Pseudechis_Austarlis	1.30	0.75	4.95	4.40	0.76
Agkistrodon Contortrix	3.90	2.12	7.75	5.97	4.40
Pituophis Melanoleucus	1.75	1.00	5.40	4.65	0.18
Pythonides					
Liasis Fuscus	1.50	0.30	3.30	3.60	1.80
L Mackloti	0.70	0.40	3.40	3.80	2.70
Morelia Amethistina	1.80	0.30	3.70	4.00	1.90
Morelia Virdis	2.00	0.30	4.50	4.80	2.50
Python Brongersmai	0.20	0.40	4.70	5.10	4.50
Python Sebae	1.50	0.40	4.80	5.20	3.30
Python Regius	1.00	0.50	5.50	5.00	

Table 5 presents a summary of the performance indicators for the species of table 4. The table entries represent values computed using the Ra value along the AE-PE-axis. This value, presented in the first column in the table, is an average of 32 equally-spaced readings along the AE-PE-axis of the particular reptile. Consequently, the table also reports the aximum, minimum, and the average performance metrics directly affected by the Ra value (namely the SAR and the DSR).

Table 5. Summary of the performance indicators for the species of table 4

Species	Texture Primitive Ratio							
	Ra	TAR	SARmin	SARmax	SARav	DSR max	DSRmin	DSRAV
Bittis Gabonica	0.33	0.04	13.47	14.52	14.00	0.45	0.30	0.37
Cotton Mouth	0.04	0.28	110.45	93.37	101.91	0.03	0.04	0.03
Python Regius	0.08	0.18	72.75	66.14	69.44	0.08	0.15	0.11
Taylori Agikstron	0.11	0.20	54.10	36.22	45.16	0.04	0.22	0.13
Cerroph Godmani	0.14	0.13	28.06	27.00	27.53	0.14	0.17	0.16
Echis Carinatus	0.04	0.54	467.18	467.18	467.18	0.01	0.01	0.01
Pseudechis Austarlis	0.14	0.18	34.46	30.63	32.54	0.11	0.19	0.15
Agkist Contortrix	0.10	0.68	77.50	59.74	68.62	0.03	0.05	0.04
Pituophis Melanoleucus	0.03	0.22	201.72	173.70	187.71	0.02	0.03	0.02
Pythonides								
Liasis Fuscus	0.09	0.20	36.42	39.73	38.08	0.06	0.30	0.18
L Mackloti	0.10	0.08	32.83	36.69	34.76	0.15	0.26	0.20
Morelia Amethistina	0.09	0.23	42.28	45.71	44.00	0.05	0.29	0.17
Morelia Virdis	0.09	0.20	52.50	56.00	54.25	0.04	0.29	0.16
Python Brongersmai	0.12	0.01	37.64	40.85	39.24	0.62	0.31	0.47
Python Sebae	0.09	0.11	52.98	57.39	55.18	0.06	0.23	0.14

Comparison to Laser Textured Surfaces

In this section, we draw a "performance prediction" comparison between the surface of the ventral side of the python and LTS. The scope of the comparison at this stage of the investigation is confined to comparing the numerical ranges of the performance indicators as computed from the geometry of the skin to those reported in tribology literature for LTS (whether based on computational or experimental studies).

The strategy of the comparison process entails constructing performance indicator maps (whenever possible) from data reported in open literature. Thereafter, the computed quantities for the skin are superimposed on these maps. The performance indicators for the skin of the snake, as mentioned earlier, are functions of metrological parameters (both dimensional and topographical). The metrological parameters, on the other hand, differ by position and direction of profile extraction (refer to table 5). As such, indirectly performance indicators for the snakeskin will depend on the evaluated region on the body of the reptile. To generalize the comparison, we compute four variations of the performance indicators for the skin (and thereby we superimpose these on the constructed maps). As such, we compute the performance indicators for each of the pre-identified regions on the skin (TDP-S, MP-S, and BDP-S) using local data. In addition, we compute an average value for each of the comparison ratios for the skin based on averaged metrological values. This procedure helps identify which region, if any, is more suitable for in depth analysis for studying feasibility of surface-mimicry (i.e., simulations and experimental friction studies for surface replicas). The starting point for this procedure is to compare the shed skin to the experimental parameters of Kovalechenko et al (Kovalchenko et al, 2011).

Kovalechenko and coworkers performed a series of experiments to rate the tribological performance of a set of LTS. Experiments employed a pin-on-disk apparatus. The authors used a speed range of 0.15-0.75 m/s and nominal contact pressures that ranged from 0.16 to 1.6 MPa. They also used two lubrication oils with different viscosities (54.8 and 124.7 cSt at 40 o C). Table 6 presents a summary of the parameters describing the surfaces used in the experiments of Kovalchenko et al. and the equivalent parameters for the snakeskin.

Table 6a. Surface designation used in the experiments of Kovalchenko et al

Surface	Description
S-3	Standard LTS (SLTS)
S-4	Higher Dimple Density (HDD)
S-5	Standard unlapped (SU)
S-6	Lower dimple density (LDD)
P-AV	Python skin based on averaged metrological quantities

(Kovalchenko et al, 2011)

Table 6b. Parameters used in the experiments of Kovalchenko et al

Parameter	S-3 SLTS	S-4 HDD	S-5 SU	S-6 LDD	Python (average)
Depth of dimples h (μm)	5.5	5	6.5	4	0.076
Surface roughness between dimples Ra (μm)	0.03	0.06	0.07	0.09	0.05
Diameter of dimples Φ (μm)	78	58	80	58	1.0
Distance between dimples λ (μm)	200	80-100	200	200	4.5
Dimple area density	0.12	0.15	0.12	0.7	0.1

(Kovalchenko et al, 2011)

Figure 4 presents a summary of the results obtained by Kovalchenko and co-workers. The figure depicts evolution of the coefficient of friction (COF) of the LTS used in the experiments. Data pertain to lubricated sliding in the presence of high viscosity motor oil. From the figure, we can distinguish the general trend of the COF. Depending on the sliding speed, two regimes take place upon sliding. The first is characterized by high friction associated with low load carrying capacity. The second regime meanwhile takes place at higher sliding speeds. Here the fluid film builds up to reach a thickness capable of establishing a hydrodynamic lubrication regime. Upon increasing the sliding speed, the thickness of the fluid film increases and the load carrying capacity increases consequently. This causes the friction to drop. The value of the COF and the rate of transition to a full hydrodynamic lubrication regime constituted the basis of the criterion used to rate the tested surfaces. Accordingly, a better performing surface, thus, would exhibit low COF, and will transit to a full hydrodynamic regime at a sliding speed lower than that of a poor performing surface. Based on such criteria, and as shown in figure 4, the best performing surface is surface S-3. This surface exhibited the lowest COF of all the tested surfaces. Surface number S-6 exhibited lower COF than the rest of the surfaces (except surface S-3). As such, it was rated the second-best performing surface and so on.

Ranking of the surfaces originates from their performance. Performance, in turn, is a function of the topographical features of the individual surfaces. Therefore, in principle ranking the surfaces stems from the ability of the lubricating oil to establish a film of a thickness capable of supporting the contact loads and separating the contacting surfaces. Ability of the lubricant to build a suitable film depends on the efficiency of the laser texturing of the surface. The efficiency meanwhile is manifested in optimal values of surface parameters. Naturally, not all parameters are equally influential on lubrication quality and friction reduction

Figure 4. Friction behavior of the surfaces used by Kovalchenko et al (2011)

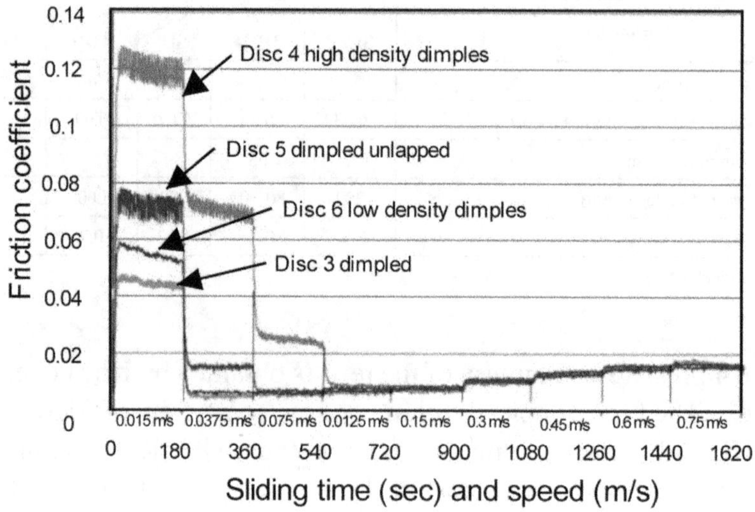

Surface geometrical parameters may be classified in two categories: primary and auxiliary. Primary parameters directly influence the quality of lubrication; meanwhile auxiliary parameters play a role only when two surfaces share the same value of a primary parameter (or primary parameter set). For example, in table 6 surfaces S-3 and S-6 share the same distance between dimples, yet the COF exhibited by surface S-3 is less than that exhibited by surface S-6. This indicates that the better performance of surface S-3 relates to the difference in other parameters (dimple area density and dimple diameter for example). Similarly, surfaces S-3 and surface S-5 share the same dimple area density (0.12). Yet, again, the performance of surface S-3 is better than that of surface S-5. So that the different values of the roughness parameter, Ra, and the diameter between the two surfaces for example should be the source of the contrast in performance. As such, if complete mapping of the performance of tested surfaces in relation to their respective individual surface parameters (both primary and auxiliary) takes place, rating a surface of unknown performance should be possible.

Ranking the performance of an unknown surface may be achieved by locating the position of the test surface on a comparison map that includes surfaces of known ranked performance. For example, suppose that we construct a Ra-area-density map for which the respective parameters of the surfaces given in table 6 constitute the entries. Suppose, further, that we plot the Ra and area density values of a surface of unknown performance on the same map. Then the ranking of the unknown surface will depend on the location of the unknown surface with respect to surfaces S-3 and S-6. If the new surface is located between surfaces S-3 and S-6 it is likely to perform better than surfaces S-4 and S-5. The overall ranking of the new surface will also depend on how closely it is located to surface S-3 on the Ra-area density map. Repeating such a process with maps constructed by plotting the combination of parameter pairs of the surfaces in table 3, allows ranking of the unknown surface. Abdel-Aal and El Mansori (Abdel-Aal and Mansori, 2011) applied this approach to project the tribological performance of Python-inspired surfaces. Herein, we extend the same procedure to project the tribological performance of the species listed in table 4 using the average performance metrics, where applicable, to develop the respective performance maps. Figure VII.5 shows the resulting maps. The figure comprises eight plots, labeled a through h, that compare the family pythonidae to other species. The reason for such a choice stems from the allometry of the snake families, presented in figure 1, which shows that pythons are among the heaviest of the species. As such, in effect, figure 5 compares natural surfaces that sustain relative high load to other species. To this end, the figure includes the control surfaces labeled "S". These are the test surfaces used by Kovalechenko et al (2011). Note the scale of presentation in the figure. In particular, for the Pythonidae family, the domain of the figure is almost one third of the domain of the other species. This indicates that the performance indicator metrics for Pythons are closer to those of the control (manmade) surfaces than metrics of the other species. This conclusion may be verified for example by comparing figures 5 (e and f). The figures compare the TAR ratio versus Ra for the snake species examined to those of the family pythonidae. Due to the scale of presentation, figure 5-e deceivingly imply that the vipers Agkistrodon Taylori (AT), and cerroph Godmani (CG) are closer to the values of the best performing surface S-3. However, observing figure f, and noting the difference in scaling, we deduce that almost all members of the family Pythonidae fall closer to the best performing surface S-3. Naturally, there are exceptions to this general trend. For example, figure 5-c, which maps the SAR vs DSR, shows that the Agkistrodon Taylori falls closer to the control surface S-3 more than the Python Sebae. Similarly, within figure 5-e the Colubride Pituophis Melanoleucus seems to fall closer to S-3 than any of the tested Pythons. The overall picture that emerges from the preliminary analysis is that members of the family Pythonidae are good candidates for further investigation. Particularly the

Python regius species emerged as a strong candidate for further study especially with the low computed COF in lubricated conditions. Therefore, members of the family Pythonidae are candidates for further mining to search feasible biological surfaces for possible mimicry.

The criterion of selection, so far, was based on geometrical features (i.e., construction of the surface). This poses questions regarding adequacy of geometrical features as a sole metric for performance prediction and thereby for deciding the feasibility of a bio-surface for mimicry or for conceptual purposes. The straightforward answer is that a single criterion is not adequate. However, the process of selecting a bio-analogue for mimicry is more complex.

In principal, setting feasibility criteria for a surface should stem from in-depth analysis of the perceived function of the target surface, its loading and its operating environment. The requirements of surfaces that sustain higher loads, but have to maintain high wear resistance in dry conditions, differ from demands on surfaces operating in lubricated contacts. There are many constraints on the structure of the later class of surfaces. The designer in this case should decide on the type of lubrication to be maintained, what type of lubricant, permissible COF, and at which distance from the entrance the lubricant should establish maximum separation. These demands are functions of both surface texture, flow conditions, and lubricant physical properties. In all the process is not simple and a tree of decisions may result in any given situation. When considering deterministic, or textured, surfaces therefore, deciding on a feasible bio-analogue is not as simple. In addition, the lack of cumulative design experience with deterministic surface features adds another layer of difficulty to the design process. To this end, there is always a need for additional criteria to rank candidate bio-surfaces or to explore the feasibility of strong candidate bio-inspired textures. The primary purpose of this additional criteria is to rank the predicted performance of the snakeskin-inspired surfaces under lubricated contact conditions. For this purpose, we adopt the model introduced in chapter V. The COF in case of lubricated sliding of textures in this model has the form,

$$\mu_f = \left[\frac{\nu U}{W}\right]\left[\frac{\left[R_a^s\left(1+\left(TAR\right)^2\right)\right]+1}{\left(\eta+1\right)}\right]\cdots\cdots\eta \geq 0 \tag{1}$$

Figure 5. Mapping the predicted tribological performance of candidate analogue species based on the criteria utilized in man-made surfaces.

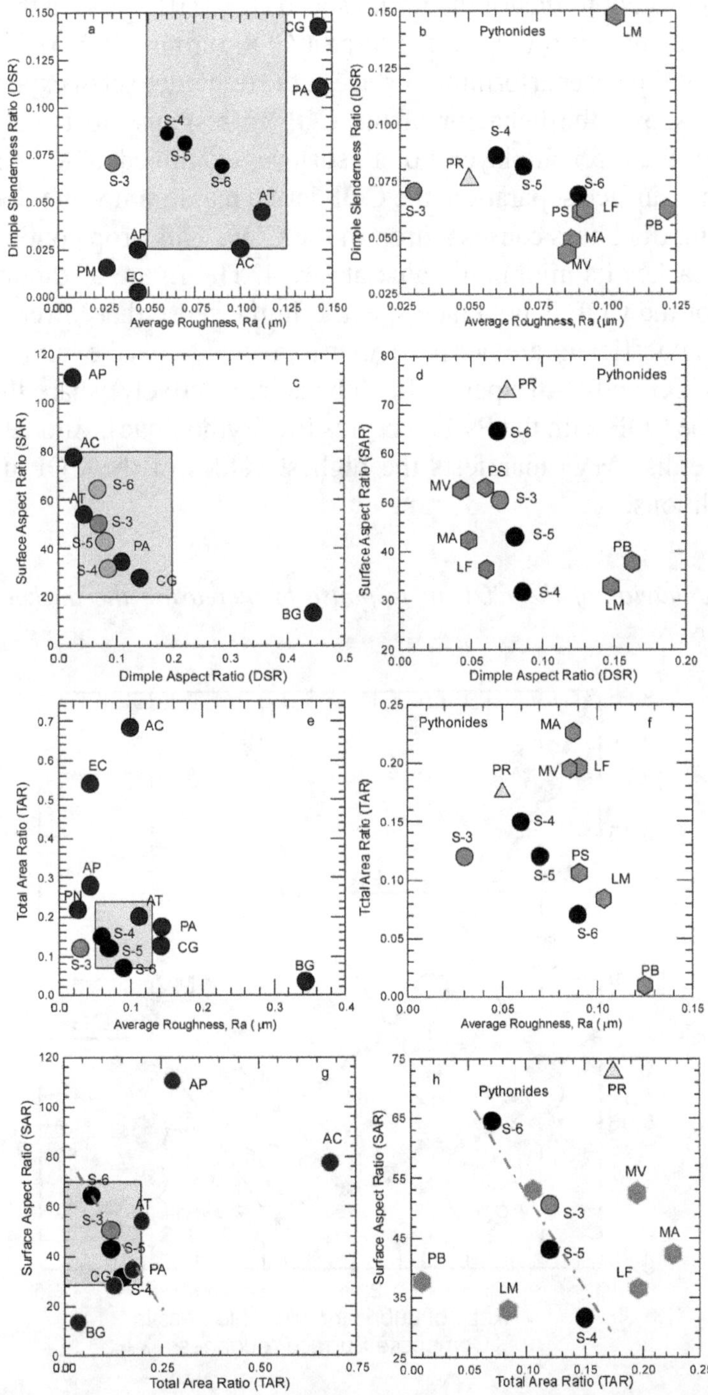

Figure 6 plots the results of applying equation 7 to selected snake surfaces. The plots were generated for a bearing number of (νU/W) =10-5. This value is equivalent to that of the initial speed (U= 0.015 m/s) in the work of Kovalchenko since it produced the largest COF (around 0.055) for surface S-3 (which is the best performing surface in the reference set of experiments). The figure shows the behavior of the COF in response to increasing the thickness of the lubricant layer. For all surfaces examined, at low thickness, and minimal surface separation, the COF is at a maximum value. Increasing the lubricant thickness causes a drop in the COF. This drop continues until the COF reaches its minimum value at $\lambda \approx 4$. The figure also implies that the value of the COF is not clade specific in the sense that lower values of the friction coefficient are not particular to a specific snake family. In fact, the snakes CG and PA (viper and colubride respectively) share the lowest values of the COF with the Python regius PR (Pythonidae). Another Python Morelia Verdis (MV) manifests the highest values of the COF under the same conditions.

Figure 6. Behavior of the COF in response to increasing the thickness of the lubricant layer

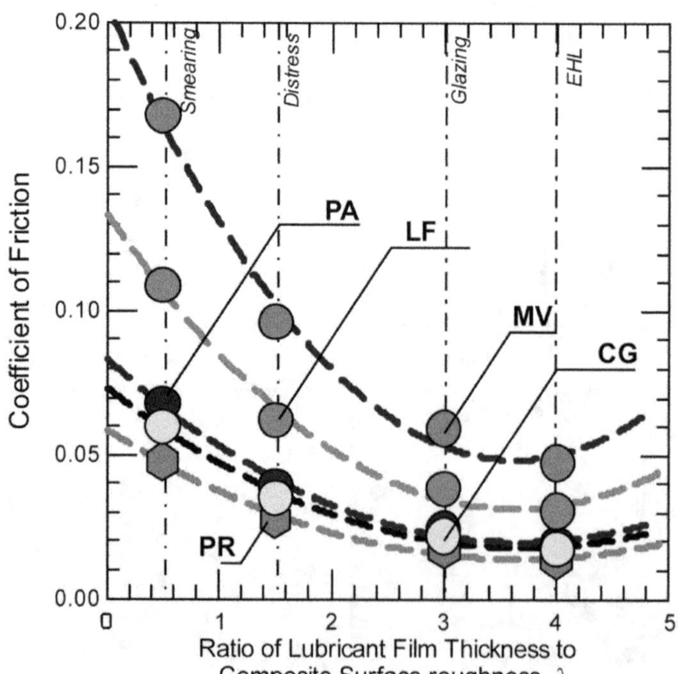

Table 7 lists the COF values obtained for all species examined in this work computed at the specified critical values of the lubricant thickness ratio. The table lists the mass of the individual snakes as well. The numbers manifest a curious trend. For the family Pythonidae, the COF values increase with the increase of mass, whereas for the rest of the species the COF decreases with the increase of mass. This is paradoxical as one would expect that the COF for heavier snakes will be inversely proportional to the mass in order to reduce the frictional losses while moving.

Table 7. COF values obtained for all species examined in this work computed at the specified critical values of the lubricant thickness ratio.

Species	Mass	Smearing	Distress	Gazing	EHL
bittis gabonica	8	0.034	0.021	0.013	0.010
cotton mouth	2.62	0.257	0.140	0.086	0.068
taylori_agikstron	0.2526	0.159	0.091	0.056	0.045
Cerrophidion_Godmani	0.227	0.060	0.035	0.022	0.017
Echis_carinatus	0.28	2.611	1.182	0.678	0.531
Pseudechis_austarlis	1.36	0.068	0.039	0.024	0.019
Agkistrodon contortrix	0.95	0.270	0.110	0.060	0.047
Pituophis melanoleucus	1.41	0.452	0.256	0.157	0.125
Pythonides					
Liasis Fuscus	3.076	0.109	0.062	0.038	0.031
L Mackloti	3.625	0.070	0.042	0.026	0.021
Morelia amethistina	18.79	0.141	0.079	0.049	0.039
Morelia Virdis	3.076	0.168	0.096	0.059	0.047
Python Brongersmai	2.336	0.050	0.030	0.019	0.015
Python Sebae	30.24	0.154	0.091	0.057	0.045
python regius	2.012	0.048	0.028	0.017	0.014

Given the variation in the results obtained, especially with the close COF values for at least three species, there is a need to determine which of the examined skin surfaces is the most feasible for mimicry, at best, or for in-depth study at least. One simple criterion for further filtering of surfaces is to consider the comparison of the COF values obtained in figure 15 and attempting to normalize the individual differences in geometry and metrology. It follows that the best surface would be that which yields the lowest COF value for all transitional limits given equal constraints of relevant geometrical and metrological parameters. To carry this comparative scheme there is a need to predefine a measure for comparison along the same lines discussed in the previous section. This condition is satisfied, in this work, by adopting the COF values for surfaces S-6, as an upper bound, at the slowest speed used in the work of Kovalchenko et. al. For mapping surfaces however, there is a need to neutralize the effect of Ra variation between the surfaces used in figure 10 and the extra species examined. For this purpose, we plot the COF, resulting from application of equation 7, against the geometrical ratio Lc/ Φ. This quantity represents the ratio between the so-called characteristic-length of the surface feature to its' base diameter. The characteristic length defines the side of an imaginary square that encloses the surface feature. This is given by (Huang, et al., 2012),

$$\frac{L_c}{\Phi} = \frac{1}{2}\sqrt{\frac{\pi}{(TAR)}} \tag{8}$$

The resulting plot represents a comprehensive map of the predicted COF of all examined skins. It follows that values falling with the bounds the parameters of surface S-6 will represent the snakeskins, and thereby the species, that are feasible to consider for further analysis.

Figure 7 (a-d) depicts the plots resulting from following the aforementioned procedure. Four plots are displayed in the figure, each represents the variation of the COF for the particular skin at the particular ratio Lc/ Φ. To determine the domain of acceptable COF values for the skin, we calculated the ratio Lc/Φ for surface S-3 (around 1.25). As such, the domain of acceptable COF values for feasibility of mimicry is bound by:

$$0 \leq COF \leq 0.055, 0 \leq \frac{L_c}{\Phi} \leq 1.25 \tag{6}$$

The shaded rectangle in each plot depicts this domain, as such, for a given snake to be considered feasible to mimic, the predicted COF has to fall within the shaded rectangular area shown in each figure. Comparison of the figures reveals that three skins satisfy this condition (i.e, there predicted COF falls within the pre-designated limits). These are the Python regius (PR), the Pseudoechis Australis (PA), and the Cerrophidion Goodmani (CG). Two other species, belonging to the Agkistrodon family (Agkistrodon Contortix (AC) and the Agkistrodon Taylori (AT), meet the COF cut off criterion at higher film thicknesses.

The skin of the Python regius (red circle in all plots), however, consistently yields the lowest COF at all transitional limits. It is interesting to note that the three snakes that meet the COF cut off do not belong to the same snake family. The CG is a member of the Viperidae family, a pit viper, PA is an Elapidae, whereas the PR is a member of the Pythonidae family. The masses of these snakes are diverse as well (the PR is the heaviest (see table 7)). Despite that, the control snakes MV and LF are members of the Pythonidae family their predicted COF was greater than that of the PR (almost three times the value).

Figure 7. Variation of the COF for the particular skin at a required value of the ratio Lc/ Φ

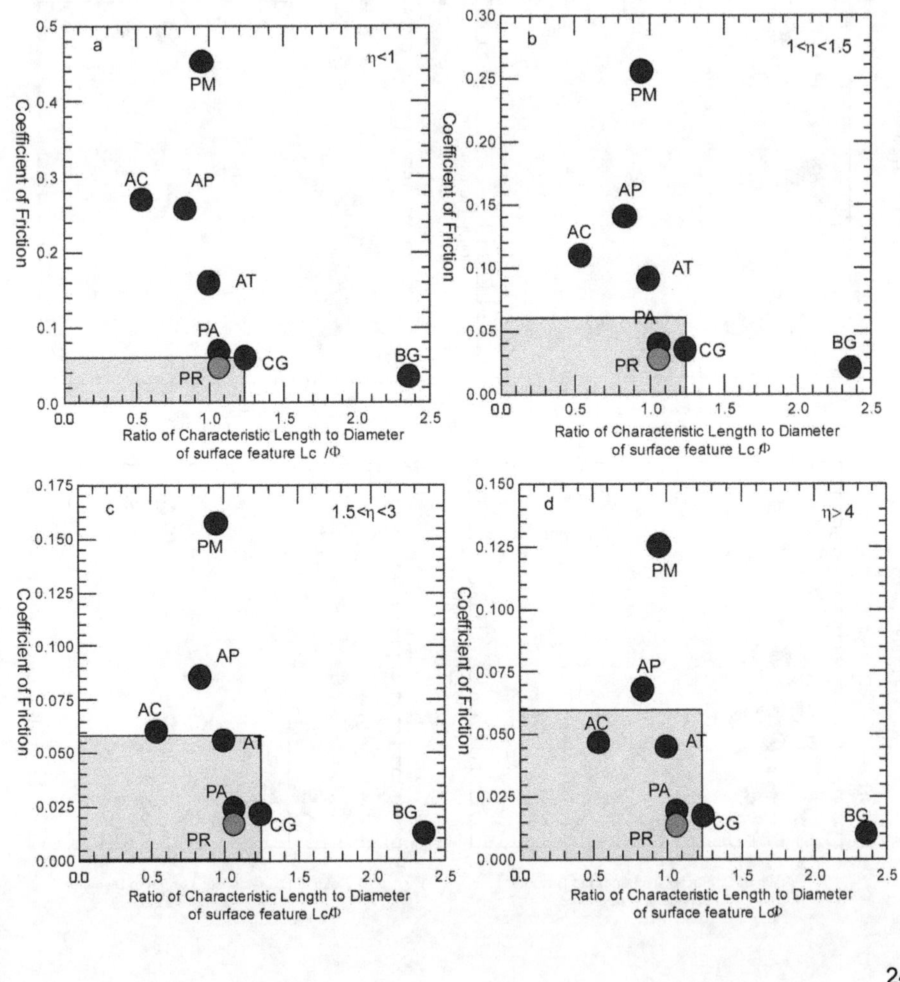

It is noted that other members of the python family fall within the cut-off limits as well especially for the glazing and EHL critical limits ($1.5 < \eta < 3$, and $\eta < 4$). Naturally, primary focus would be on a surface that falls within the cut-off limits for all transition values.

To verify the rank of the Python regius surface, we re-plotted figure 8 (a-d) while superposing the data for all the examined Pythons. The new data is shown as red circles in all plots. The data point of the Python regius is shown as a yellow circle in all plots. Adding the COF values for the pythons confirms that the python regius surface is the most feasible surface for further study for mimicry or design inspiration.

Figure 8. Comparison of the variation of the COF for the skin of Pythonides and other species at a pre-determined value of the ratio Lc/Φ

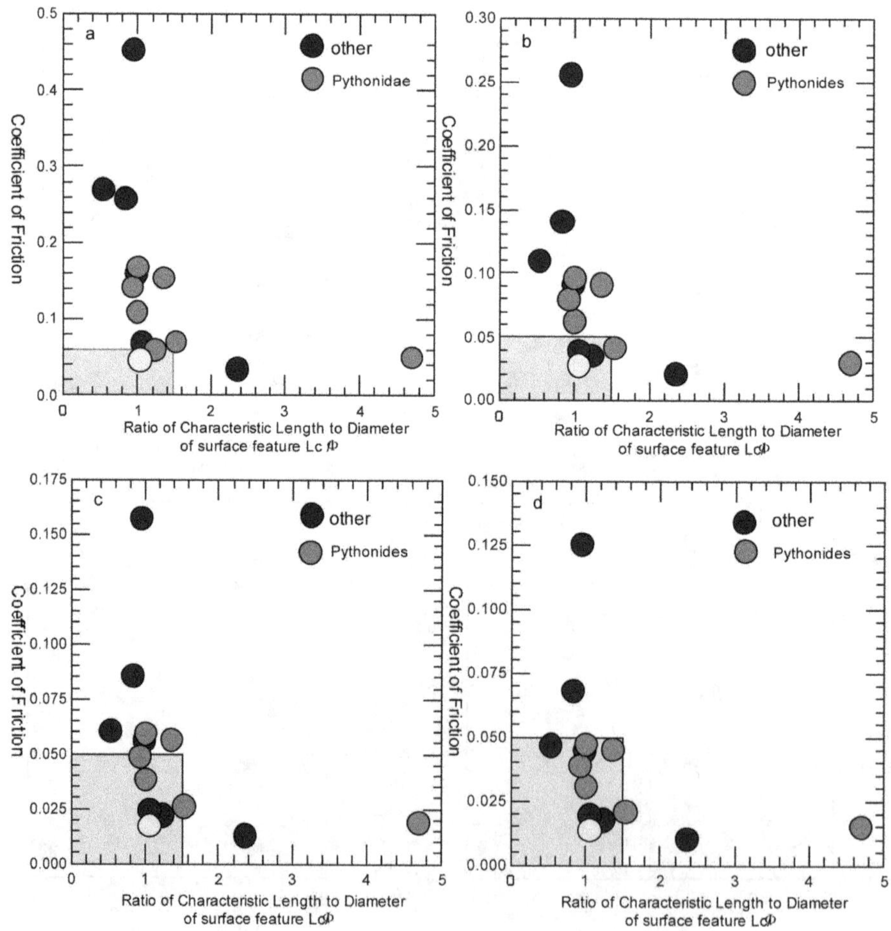

Comparison to Recommended Values

Mimicry of reptilian surfaces does not suffice by itself for constructing a surface of optimal tribological quality. Rather it is the deduction of design rules, combined to observing reptilian dimensional proportionality that contributes to successful texturing. Observing the layout of dimples on many engineering surfaces reveals an array arrangement in which dimple arrangement is equally-spaced and is in patterns of parallel rows and columns. Equal spacing is hardly observed, if at all existent, in snake ventral scales (which are the principal frictional frontier in snakes). Rather, a wave arrangement is present in all sliding directions. An additional observation is the orientation of the fibrils along the body.

The orientation of the fibrils along the AE-PE axis always points toward the posterior end. The fibrils are not necessarily parallel to the A-PE axis. This orientation contrasts the orientation of the hexagonal ventral scales (in which the lateral diagonal is always perpendicular to the AE-PE axis). Combined to the waveform distribution of spacing, the orientation features of the fibrils enhance both the friction and wear performance of the skin. Huang et al (2012) performed exhaustive studies on the effect of textural orientation on friction and wear behavior of surfaces. The authors used several LTSs in lubricated sliding. They also varied the contact pressures and the pressure of lubricant supply. Yuan and coauthors concluded that the orientation of the micro-textural elements has a strong influence on friction performance of sliding surfaces. For relatively low contact pressure small sized grooves (order of few microns), that are oriented perpendicular to the sliding direction, reduce friction by about 40% compared to un-textured surfaces. Under high contact pressure, bigger grooves (order of 20 μm), but parallel to the sliding direction, yield better friction performance and reduced contact stresses. The best performance reported in the work of Yaun pertained to textural elements oriented at an angle to the principal direction of motion (regardless of size). This orientation resulted in reducing the friction by a factor of two.

Shinkarenko and co-investigators presented a theoretical model to evaluate the feasibility of using LST in the form of spherical micro-dimples for soft elasto-hydrodynamic lubrication. They performed an extensive parametric investigation to identify key parameters of the problem. The parameters identified were the aspect ratio of the dimples (equivalent to DSR in our study) and the area density (equivalent to the SAR in our study). The authors also provided optimal ranges for the key parameters.

The aspect Ratio of the dimple (DSR) defines the average slope of the converging film thickness zone of each dimple. This parameter is known to be the main factor affecting load capacity in slider bearings. Shinkarenko deduced the optimal range of this parameter to fall within the range 0.06<DSR< 0.08. The equivalent value for the Python, see table 6, is 0.076 which falls within the suggested optimal range. Further, the Dimple Area Ratio (equivalent to TSAR) was identified as 0.05 <SAR<0.5 with SAR equal to 0.2 being the recommended value. The analogous value for the Python, is 0.1.

Costa and Hutchings (2007) extensively investigated the influence of surface topography on lubricant film thickness. They measured film thickness using a capacitance technique in reciprocating sliding of patterned plane steel surfaces. The study employed circular patterns (pockets), grooves, and chevrons. They also varied the sliding orientation relative to the texture. The results of Costa and Hutchings concluded that a TAR of 0.11 seems to achieve the maximum film thickness when circular pockets are used for texturing. It is interesting to note that the equivalent ratio for the Python (0.1) fits closely to that recommended value. The same authors also found that maximum film thickness, for a sample with a TSAR ratio of 0.06, was achieved when the DSR was about 0.07. Consequently, they rationalized that such a value is optimal for texturing. Again, for the analogous DSR value for the Python (0.076) falls very close to such a recommendation.

Table 8. Summary of recommended performance indicators extracted from open literature

TAR	DSR	SAR
	0.06-0.08	0.05-0.5
0.025-0.1	0.07	0.11
	0.05	
		0.05-0.2
0.25	0.02-0.25	
0.05	0.01-0.02	
0.03-0.12		
0.6		
0.075-0.2	0.03-0.08	
0.085		
0.2-0.4	0.03-0.1	
0.1753	0.076	0.017

Table 8 presents a collective comparison between the performance indicator values computed for the snakeskin and those recommended by several authors. To visualize the extent of agreement between the skin values and the recommended values data in table 8 were graphically represented. Figure 9 presents the resulting plot. The figure depicts plots for three performance indicators: TAR, DSR, and SAR (note that the figure plots the reciprocal of the later quantity for compatibility with literature values). The plot indicates that the values for the python skin are in very god agreement with the recommended values. The SAR values fall at the lower boundary of the recommended values.

Figure 9. Summary plot to compare the performance indicators recommended in literature and those computed for the ventral skin of the Python.

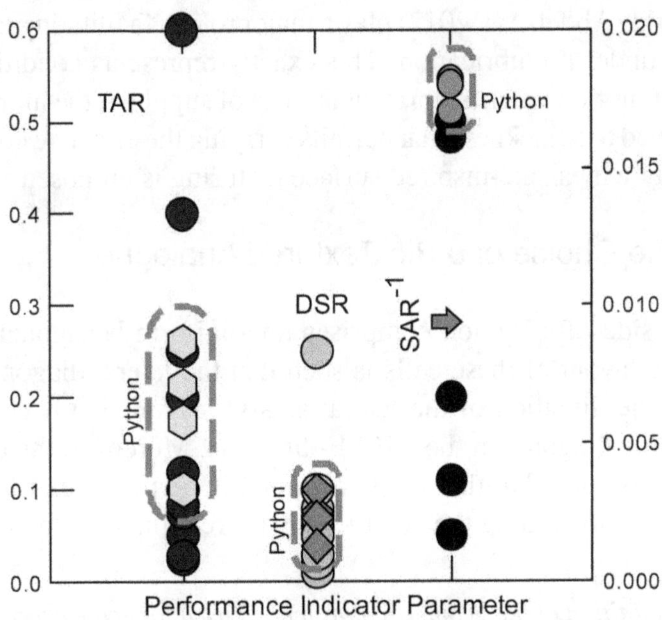

The extracted data were obtained under different conditions (different materials, different shapes of textural elements, different lubricants and lubrication conditions, etc.,). Despite these differences, it is remarkable that most of the recommended values converge to a relatively narrow band of optimal values. More remarkable is the fit of the computed snakeskin values within this narrow band, which points at the optimal construction of the topographical features of the reptile. This finding encourages additional in-depth studies of surface design of other snake species. Results of Pettersson and Jacobson (Pettersson and Jacobson, 2003, 2004), Kovalchenko et al. (2011), and Costa and Hutchings (2007) support the preposition that texturing of surfaces does not always effectively influence friction. They found that, although surface texturing reduces friction under boundary lubrication conditions, friction is likely to be insensitive to texturing when a full film exists between the sliding surfaces surface texturing has little or no effect on friction. Most of the damage sustained in rubbing therefore takes place before building up of the film thickness. That is where texturing is efficient.

Several authors, (Swift, and Smith, 2001, Baumgartner, et al., 2007, Zhang, et al., 2006, 2008), reported the high resistance to wear and abrasion in addition to the considerably low COF in reptilian motion. This remarkable tribological performance result from surface topoology. In particular, the topographical features and the multi scale characteristics of the roughness (Abdel-Aal and Mansori, 2011, Abdel-Aal, 2012) play a major role in facilitating low friction locomotion under no lubrication. This exactly represents conditions in the beginning of motion when a lubricant is short of supply or the lubricant layer has not evolved to a thickness that permits carrying the effective contact load. It is precisely why snake-inspired surface texturing is proposed.

Refining the Choice of a Bio-Textured Analogue

The ventral side of a Python comprises non-uniform hexagonal unit cells (scales). The layout of these cells is such that the lateral diagonal (i.e, the diagonal in the direction of the lateral axis-see figure 10) is considerably longer than the diagonal in the AE-PE-direction. Moreover, the orientation of the scales is such that the lateral diagonal is perpendicular to the main direction of motion (along the AE-PE-axis for rectilinear locomotion).

Figure 10. Definition of the geometric parameters used for scale characterization, (a) definition of the surface area of the ventral scale, Avs, and Ventral Scale Aspect Ratio (VSAR), (b) definition of the fibril aspect ratio.

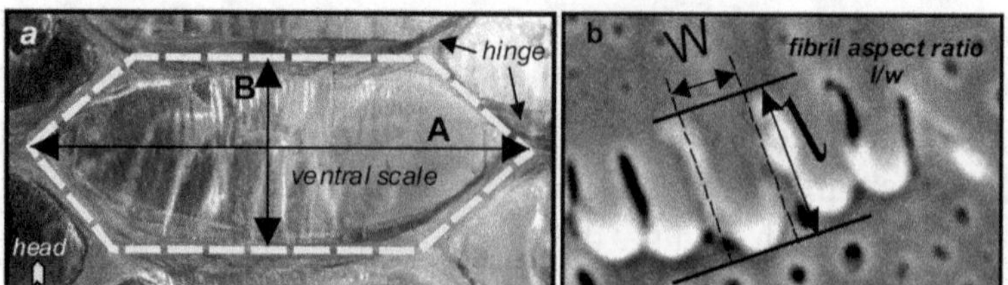

The variation in the length of the lateral and longitudinal diagonals of the ventral scales implies that the aspect ratio of the building block of the ventral side is greater than unity. Moreover, the aspect ratio for the ventral scales is not uniform along the AE-PE-axis. Therefore, as a point of entry to refining the model of texturing there is a need to evaluate metrological parameters of the micron-sized fibrils as a function of body location. Table 6 lists the extracted Ra values in both directions (lateral and AE-PE).

Table 9. Values of the Profile Arithmetic Mean Deviation, Ra (μm), at selected regions within the skin

Region	AE-PE-Axis	SD	Lateral-Axis	SD
TDP-S	0.041	± 0.001	0.126	± 0.01
MP-S	0.075	±0.0072	0.06	±0.008
BDP-S	0.039	± 0.0085	0.14	± 0.006
Average	0.052	± 0.0056	0.1087	± 0.008

The data imply that the roughness in the AE-PE direction differs from that in the transverse direction. It follows that the surface roughness parameters (Ra, Rμ, etc.,) will show directional dependence "i.e. texture anisotropy". It follows also that we have to recompute the surface descriptor-ratios in by location and taking anisotropy in account. Table 10 presents a summary of the computed results. The table includes two data sets. The first pertains to the AE-PE axis. This set resulted from using the Ra values for profiles along the AE-PE axis. The second set meanwhile resulted from using the Ra values of profiles along the lateral axis. It is noted that the TAR value for both directions is equal. This is a result of the definition of the TAR parameter, which in essence is the percentage of the area occupied by the fibrils within the area of the scale. This is not direction dependant. Values of the DSR and the SAR manifest some difference. For each of the computed quantities we also included the average value (entries in the last raw of table 10).

The slight difference between the computed values along the axes is rather deceptive. This is because of the sensitivity of the data and the original size of the surface features involved. To illustrate this point, we introduce figure 11 (a and b). The figure is a plot of the effect of the textural anisotropy on the computed performance parameters.

Table 10. Value of the computed performance indicators along the two main axis AE-PE and the lateral axis

Region	AE-PE-Axis				Lateral Axis			
	Ra	TAR	SAR	DSR	Ra	TAR	SAR	DSR
TDP	0.041	0.21	55	0.1	0.126	0.21	43.65	0.13
MP	0.076	0.26	58.51	0.076	0.06	0.26	91.67	0.06
BDP	0.039	0.1	56.12	0.039	0.14	0.1	39.28571	0.14
Average	0.052	0.19	56.54	0.072	0.109	0.19	58.20	0.109

Values at the abscissa represent the textural anisotropy in Ra (defined as the ratio of Ra, in the lateral direction, to that in the AE-PE direction). The quantity represented in the ordinate differs according to the figure. In figure 11-a, the ordinate depicts data for the SAR anisotropy ratio (defined as SAR in the lateral direction to the SAR value in the AE-PE direction). For figure 11-b, however, the ordinate depicts data for the DSR anisotropy ratio (defined as DSR in the lateral direction to the DSR value in the AE-PE direction). In both plots, the dashed lines represent a quadratic regression function. The symbol labeled "average" in each plot marks the average value of the particular ratio of anisotropy. This quantity was computed based on average values.

The DSR anisotropy ratio manifests a trend that contrasts that of SAR anisotropy ratio. In particular, the DSR ratio increases with textural anisotropy whereas SAR anisotropy drops with textural anisotropy. A question that arises given the difference in directional values is which of the data sets to be used upon comparing the textural construction of the skin to recommended optimal values for LTS? The answer depends on the principle direction of motion of the surface in question.

Pythons are heavy snakes. Due to their weight, they move via rectilinear locomotion where the displacement is mainly along the AE-PE axis of the reptile. As such, the use of data along the AE-PE axis in all comparisons is appropriate.

Filtering Competing Body Zone Within a Single Bio-Analogue

Ranking of the Python skin is performed by the aid of four maps figure 13 (a-d). The plots map the following parameters: Ra versus DSR (figure 13-a), DSR versus SAR (figure 13-b), Ra versus Total Area Ratio TAR (figure 13-c), and SAR versus TAR (figure 13-d).

Figure 11. Variation in the Ratio of Anisotropy for the SAR and DSR with textural anisotropy

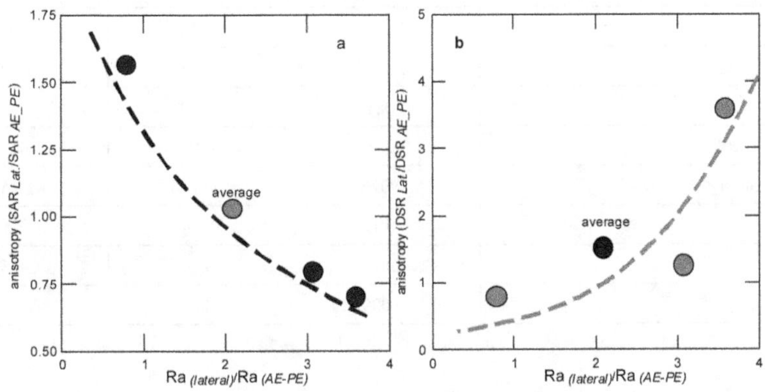

Figure 12. Selective geometrical descriptive ratios for fibril-textural elements of a P. regius

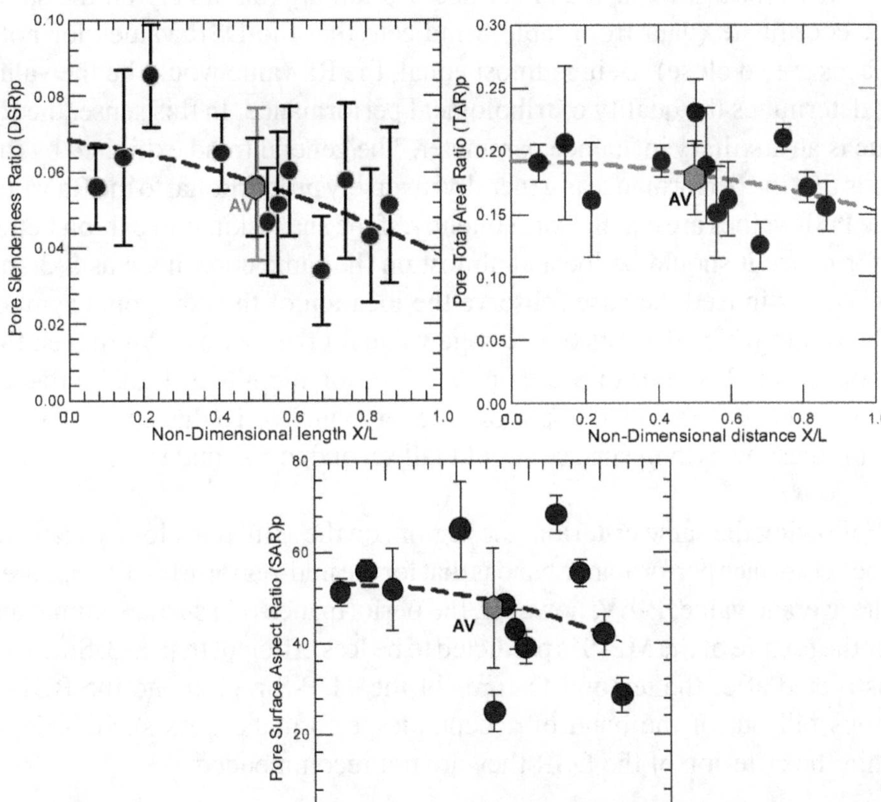

Table 11. Value of parameters used in comparing LTS and Python Skin

Surface	DSR	TAR	SAR
S-3	0.0705	0.12	50.5
S-4	0.0862	0.15	31.6
S-5	0.0812	0.12	43
S-6	0.0689	0.07	64.5
Python (average)	0.076	0.1753	56.54
TDP-S	0.1	0.214	55
MP-S	0.07	0.26	58.51
BDP-S	0.039	0.1	56.122

Figure 13-a shows that the Ra parameter is not a principal performance influence parameter. It is noticed that the DSR value is what primarily distinguishes between the better performing surfaces and those with lower performance. Surfaces S-3 and surface S-6 fall approximately on the same DSR coordinate (data from table 8 indicate that the DSR values for both surfaces are so close). Being almost equal, the Ra value would be the value that determines the quality of tribological performance. In this sense, the Ra value is an auxiliary influence parameter. The general trend exhibited by the data is that the performance is generally inversely proportional to the Ra value if the DSR values are equal. Consequently, if the snakeskin is to exhibit better performance it should be located almost on the same coordinate as S-3 and S-6. This is indeed the case (observe the location of the hexagonal symbol in the figure which denotes the average value for the python). Moreover, the python skin falls closer to S-3 than to S-6 (implying higher Ra than that of S-6) whence it is projected to perform better than S-6. To that end, a surface that mimics the python surface should fall second to S-3 under the same test conditions.

Following the same criterion, the region, on the skin, that closely matches the better surface performance band is that for the mid-region MP-S. Compared to the average value, P-AV, however, the performance of a surface mimicked after the texture of the MP-S is predicted to be less efficient than S-3. Surfaces constructed after the textural features of the TDP-S region and the BDP-S regions fall out of the band of acceptable performance. As such, at least within the criterion of the DSR they are not recommended.

Figure 13. Comparison maps between snake skin and Laser Textured Surfaces, a- Ra versus DSR b-DSR versus SAR, c- Ra versus Total Area Ratio TAR, d-SAR versus TAR.

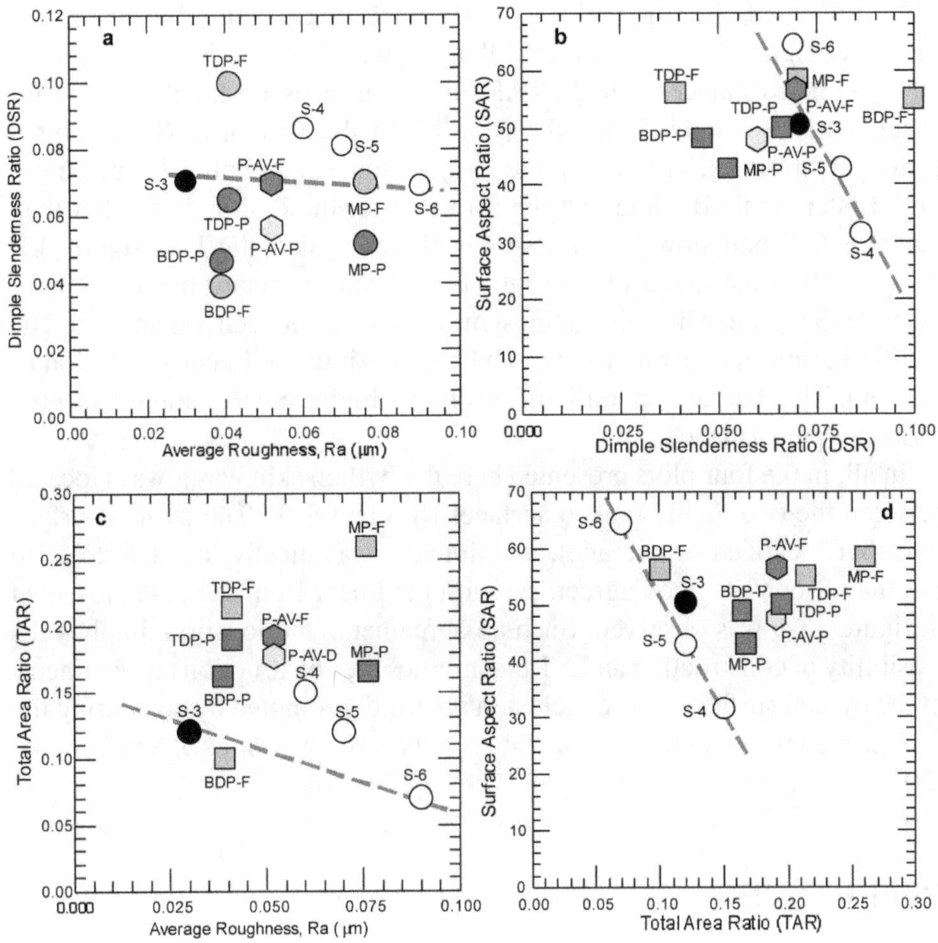

The DSR in itself, however, may act as an initial filter for surfaces when it is mapped against the SAR. This is noticed in figure 13-b. Here the best performing surfaces are located on a narrow band of SAR values. Surfaces S-4 and S-5, for which the tribological performance is of lesser quality, are located at the higher end of the DSR axis. A linear fit of the data considered (the straight line shown in the figure) implies that the quality of performance is directly proportional to the SAR (the higher the SAR the better the performance). Again, it is noticed that if the surfaces are of equal DSR values a lower SAR is better. Interestingly, the snakeskin (pentagon symbol) is directly located closer to S-3 and directly on the linear fit. Similar to the trend reflected in figure 13-a, among all regions within the skin texture within the MP-S region is predicted to offer the best performance.

Upon considering the variation in the Total Area Ratio (TAR) with the Ra value (figure 13-c) it is noticed that a lower Ra value is a precondition for better performance when the TAR for two surfaces are equal. Observe for example that S-3 and S-5 have the same TAR value (0.12), yet S-3 exhibited better performance. The skin of the Python, square symbol, is located close to S-3 (star symbol). Note that the Ra value for the Python is smaller than that of S-6. When, however, the SAR is plotted against the TAR figure (13-d), the most of the snake textures fall out of the performance band except for the BDP-S region, which falls almost mid-distance between S-3 and S-6. The BDP-S region, moreover, seems to be a good fit to the linear regression of the data. The data in general indicates that a higher SAR value is essential for better performance.

In all, in the four plots presented here the python skin was always located between the two highly ranked surfaces (S-3 and S-6). The python surface (whether localized or average), additionally, was mostly located closer to S-3 than S-6. Very good agreement with the linear fit of the data presented in figure 13-b was observed. Such an agreement, and location, implies the feasibility of constructing an LST that mimics the construction and geometry of the python surface. This conclusion is further promoted by comparing the surface parameters of the Python (tables 8 and 9) to the results of Shinkarenko (2011).

REFERENCES

Abdel-Aal, H. A. (2012). On Surface Structure and Friction Regulation in Reptilian Locomotion. *Journal of the Mechanical Behavior of Biomedical Materials*. doi:10.1016/j.jmbbm.2012.09.014 PMID:23582565

Abdel-Aal, H. A., & El Mansori, M. (2011). Python Regius (Ball Python) Shed Skin: Biomimetic Analogue for Function-Targeted Design of Tribo-Surfaces. In D. Bruckner & P. Gruber (Eds.), *Biomimetics - Materials, Structures and Processes. Examples, Ideas and Case Studies*. Springer.

Abdel-Aal, H. A., El Mansori, M., & Mezghani, S. (2010). Multi-Scale Investigation of Surface Topography of Ball Python (Python regius) Shed Skin in Comparison to Human Skin. *Tribology Letters*, *37*(3), 517–528. doi:10.100711249-009-9547-y

Arzt, E., Gorb, S., & Spolenak, R. (2003). From Micro to Nano Contacts in Biological Attachment Devices. *Proceedings of National Academy of Sciences (PNAS), 100.*

Baumgartner, W., Saxe, F., Weth, A., Hajas, D., Sigumonrong, D., Emmerlich, J., ... Schneider, M. (2007). The Sandfish's Skin: Morphology, Chemistry and Reconstruction. *Journal of Bionics Engineering*, *4*(1), 1–9. doi:10.1016/S1672-6529(07)60006-7

Benz, M. J., Kovalev, A. E., & Gorb, S. N. (2012). Anisotropic frictional properties in snakes. *Proceedings of the Society for Photo-Instrumentation Engineers*, *8339*. doi:10.1117/12.916972

Berthé, R. A., Westhoff, G., Bleckmann, H., & Gorb, S. N. (2009). Surface structure and frictional properties of the skin of the Amazon tree boa Corallus hortulanus (Squamata, Boidae). *Journal of Comparative Physiology. A, Neuroethology, Sensory, Neural, and Behavioral Physiology*, *195*(3), 311–318. doi:10.100700359-008-0408-1 PMID:19137315

Borghi, A., Gualtieri, E., Marchetto, D., Moretti, L., & Valeri, S. (2008). Tribological effects of surface texturing on nitriding steel for high-performance engine applications. *Wear*, *265*(7-8), 1046–1051. doi:10.1016/j.wear.2008.02.011

Brizmer, V., & Kligerman, Y. (2012). A laser surface textured journal bearing. *Journal of Tribology*, *134*(3), 031702. doi:10.1115/1.4006511

Christine, V. (2012). *Gorb: Snake Scale Microstructure*. Phylogenetic Significance and Functional Adaptations.

Costa, H. L., & Hutchings, I. M. (2007). Hydrodynamic lubrication of textured steel surfaces under reciprocating sliding conditions Trib. *Int.*, *40*, 1227–1238.

Dumitru, G., Romano, V., Weber, H. P., Gerbig, Y., Haefke, H., Bruneau, S., ... Sentis, M. (2004). Femtosecond Laser Ablation of Cemented Carbides: Properties and Tribological Applications. *Applied Physics. A, Materials Science & Processing*, *79*(3), 629–632. doi:10.100700339-004-2675-1

Eichstät, J., & Römer, G. R. B. E. (2011). Towards friction control using laser-induced periodic Surface Structures. *Physics Procedia*, *12*, 7–15. doi:10.1016/j.phpro.2011.03.099

Golloch, R., Merker, G. P., Kessen, U., & Brinkmann, S. (2004). Benefits of Laser-Structured Cylinder Liners for Internal Combustion Engines. *Proceedings of the 14th Int. Colloquium Tribology*, 321–328.

Gray, J. (1946). The mechanics of locomotion in snakes. *The Journal of Experimental Biology*, *23*, 101–120. PMID:20281580

Hazel, J., Stone, M., Grace, M. S., & Tsukruk, V. V. (1999). Nanoscale design of snake skin for reptation locomotions via friction anisotropy. *Journal of Biomechanics*, *32*(5), 477–484. doi:10.1016/S0021-9290(99)00013-5 PMID:10327001

Hoppermann, A., & Kordt, M. (2002). *Tribological Optimisation Using Laser-Structured Contact Surfaces, Oelhydraulik und Pneumatik*. Vereinigte Fachverlage Mainz.

Hu, N., & Ding, Q. (2012). Combination of laser surface texturing and solid lubricant film. *Surface and Coatings Technology*, *206*(24), 5060–5066. doi:10.1016/j.surfcoat.2012.06.014

Huang, Z., Liu, H., Shen, Z., Li, P., Hu, Y., Liu, H., ... Wang, X. (2012). Process parameters analysis on surface texturing under laser shock peening. *Chinese Journal of Lasers*, *39*, 107–113.

Jayne, B. C. (1988, January). Mechanical behavior of snake skin. *Journal of Zoology*, *214*(1), 125–140. doi:10.1111/j.1469-7998.1988.tb04991.x

Jiang, X.J., & Whitehouse, D.J. (2012). Technological shifts in surface metrology. *CIRP Annals - Manufacturing Technology, 61*(2), 815-836.

Klein, M.-C. G., & Gorb, S. N. (2012). Epidermis architecture and material properties of the skin of four snake species. *Journal of the Royal Society, Interface*, *9*(76), 3140–3155. doi:10.1098/rsif.2012.0479 PMID:22896567

Koch, K., Bhushan, B., & Barthlott, W. (2009). Multifunctional surface structures of plants: An inspiration for biomimetics Prog. *Materials Science*, *54*, 137–178.

Kovalchenko, A., Ajayi, O., Erdemir, A., & Fenske, G. (2011). Friction and wear behavior of laser textured surface under lubricated initial point contact. *Wear*, *271*(9-10), 1719–1725. doi:10.1016/j.wear.2010.12.049

Kovalchenko, A., Oyelayo, A., Erdemir, A., Fenske, G., & Etsion, A. (2005). The effect of laser surface texturing on transitions in lubrication regimes during unidirectional sliding contact. *Tribology International*, *38*(3), 219–225. doi:10.1016/j.triboint.2004.08.004

Liu, H., Han, H., Xue, Y., & Li, J. (2010). Influence of laser surface texturing distribution patterns on the hydrodynamic lubrication. *Advanced Materials Research*, *97-101*, 1429–1432. doi:10.4028/www.scientific.net/AMR.97-101.1429

Pettersson, U., & Jacobson, S. (2003). Influence of surface texture on boundary lubricated sliding contacts. *Tribology International*, *36*(11), 857–864. doi:10.1016/S0301-679X(03)00104-X

Pettersson, U., & Jacobson, S. (2004). Friction and wear properties of micro textured DLC coated surfaces in boundary lubricated sliding. *Tribology Letters*, *17*(3), 553–559. doi:10.1023/B:TRIL.0000044504.76164.4e

Qiu, Y., & Khonsari, M. M. (2011). Experimental investigation of tribological performance of laser textured stainless steel rings. *Tribology International*, *44*(5), 635–644. doi:10.1016/j.triboint.2011.01.003

Rechenberg, I. (2003). *Tribological Characteristics of Sandfish*. In Nature as Engineer and Teacher: Learning for Technology from Biological Systems, Shanghai, China.

Rivera, G., Savitzky, A. H., & Hinkley, J. A. (2005). Mechanical properties of the integument of the common gartersnake, Thamnophis sirtalis (Serpentes: Colubridae). *The Journal of Experimental Biology*, *208*(15), 2913–2922. doi:10.1242/jeb.01715 PMID:16043596

Ronen, A., Etsion, I., & Kligerman, Y. (2001). Friction-reducing surface texturing in reciprocating automotive components. *Tribology Transactions*, *44*(3), 359–366. doi:10.1080/10402000108982468

Ryk, G., Kligerman, Y., & Etsion, I. (2002). Experimental investigation of laser surface texturing for reciprocating automotive components. *Tribology Transactions*, *45*(4), 444–449. doi:10.1080/10402000208982572

Saito, M., Fukaya, M., & Iwasaki, T. (2000). Serpentine locomotion with robotics snakes. *IEEE Control Systems Magazine*, *20*, 64–81.

Shafiei, M., & Alpas, A. T. (2008). Fabrication of biotextured nanocrystalline nickel films for the reduction and control of friction. *Materials Science and Engineering C*, *28*(8), 1340–1346. doi:10.1016/j.msec.2008.02.006

Shafiei, M., & Alpas, A. T. (2009). Nanocrystalline nickel films with lotus leaf texture for superhydrophobic and low friction surfaces. *Applied Surface Science*, *256*(3), 710–719. doi:10.1016/j.apsusc.2009.08.047

Swift, J. A., & Smith, J. R. (2001). Microscopical investigations on the epicuticle of mammalian keratin fibres. *Journal of Microscopy*, *204*(3), 203–211. doi:10.1046/j.1365-2818.2001.00957.x PMID:11903797

Vitt, L. J., Pianka, E. R., Cooper, W. E. Jr, & Schwenk, K. (2003). History and the global ecology of squamate reptiles. *American Naturalist*, *162*(1), 44–60. doi:10.1086/375172 PMID:12856236

Yamakiri, H., Sasaki, S., Kurita, T., & Kasashima, N. (2011). Effects of laser surface texturing on friction behavior of silicon nitride under lubrication with water. *Tribology International*, *44*(5), 579–584. doi:10.1016/j.triboint.2010.11.002

Zhang, X., Shi, F., Niu, J., Jiang, Y., & Wang, Z. (2008). Super hydrophobic surfaces: From structural control to functional application. *Journal ofMaterials Chemistry*, *18*(6), 621–633. doi:10.1039/B711226B

Zhang, Z. L. (2006) Ultrastructure and reducing-resistance mechanism of ventral scales of Trimeresurus stejnegeri. J Jiangsu University (Natural Sc. Ed), 27, 471-474.

Chapter 8
Examples of Implemented Technological Bio-Inspired Surfaces

ABSTRACT

This chapter examines the multi-scale nature of biological materials. It is shown that this characteristic motivated several design attempts within the field of tribological surfaces. These designs were not easy to implement because of a lack of technological means. Until the push for nanoscale material manipulation, many designs, although conceived and conceptually verified, were not technologically possible. The leap in technologies that matured within the past decade resurrected efforts to manufacture many discarded designs on a commercial scale. The material within this chapter presents samples of existing bio-inspired tribological surfaces. The examples are either a direct replica of the bio-analogue or represent a modification of the surface through a combination of chemical and geometrical changes.

INTRODUCTION

Intense probing of surface phenomena has accelerated towards the 1980s. As a result, our current understanding of surface interactions and functional requirements and different applications has improved dramatically. Currently, our level of comprehension of the functional surfaces, and their interactions with their environment, permits accurate determination of technical needs

DOI: 10.4018/978-1-7998-1647-8.ch008

in a given application. Such an advanced knowledge has encouraged efforts for customization of surface features i.e. the so-called surfaces-by-design. There are many customization lists for surfaces in optics, energy, informatics, and of course tribology. The list of enabling technologies, have also grown larger, and many techniques that enable manipulation of matter at almost any scale are currently available.

Advances in technical understanding and availability of material manipulation techniques induced rush for miniaturization. Miniaturization, many would argue, is a necessity condition for materializing efficient functionalization of surfaces. That is, implementing the so-called surfaces-by-design doctrine. However, despite the availability of techniques and advanced understanding of nature of surface function, many technical problems still exist. These problems, as detailed in earlier chapters, mainly pertain to developing suitable design paradigms. As mentioned before, also, the major obstacle is conceiving surface design paradigms capable of producing through multi-functional surfaces. That is materializing of the surface construct of which performance is an integral part of its enclosed system, and in the same time caters to the multiple needs of that system. Such a requirement was expressed in chapter 1, in terms of ability of the surface to change, and respond to, the various information with its surroundings. Within the realm of tribology, these requirements add to the level of complexity encountered. The also represent a design constraint. As detailed in chapter 7, a possible remedy for such a situation is to learn from natural surfaces since, by default, natural surfaces specially in biology are multi-functional.

Biology offers a wide spectrum of multi-scale and multi-functional materials. Variation in material scale (from nano to micro to macro) allows for adopting many topographical features and textural topologies. As a result, biological surfaces, despite facing the same physical constraints as man-made surfaces, manifest better performance and efficiency. The multi-scale nature of the building blocks contributes to the high level of endurance, structural integrity, and functional adaptation.

Scouring the multi-scale nature of biological materials ensued many decades ago. Synonymous with the intense work on surface phenomena, parent work on biological materials took place. The emerging picture showed that natural materials, and thereby natural surfaces, are in essence multi-functional and not mono-functional. This multi-functionality, it was found, has its roots in the multi-scale nature of the material building blocks of the surface. Examples of some materials are shown in table 1.

Table 1. Typical biological materials with function integration.

Biological Materials	Functions	Ref.
Butterfly wing	Super-hydrophobicity, directional adhesion, structural color, self-cleaning, chemical sensing capability, fluorescence emission functions	(Zheng Y, et al., 2007, Potyrailo et al, 2007, Vukusic and Hooper, 2005, Sato O, et al, 2008., Aizenberg, et al., 2001)
Brittlestar	Mechanical and optical functions	(Sun, et al.,2005)
Cicada wing	Anti-reflection, super hydrophobicity	(Liu, et al., 2009)
Fish scale	Drag reduction, superoleophobicity in air, superoleophobicity in water	(Hansen, et al.,2005)
Gecko foot	Reversible adhesive, super hydrophobicity, self-cleaning	(Feng, et al.,2002)
Lotus leaf	Super hydrophobicity, low adhesion, self-cleaning	(Gao et al.,2007)
Mosquito compound eye	Super hydrophobicity, anti-reflection, anti-fogging	(Mayer et al.,2005)
Nacre	Mechanical property, structural color	(Tan et al., 2004; Zi et al.,2003)
Peacock feather	Structural color, super hydrophobicity	(Stegmaier et al., 2009)
Polar bear fur	Optical property, thermal insulation	(Feng et al., 2008)
Rice leaf	Super hydrophobicity, anisotropic wettability	(Gao et al.,2007)
Rose petal	Super hydrophobicity, structural color, high adhesion	(Feng et al., 2010, Bhushan and Her 2010, Ball 1999)
Shark skin	Drag reduction, anti-biofouling	(Sundar et al.,2003)
Spicule	Mechanical and fibre-optical properties	(Aizenberg et al.,2004, Aizenberg et al.,2005, Zheng et al.,2010)
Spider capture silk	Water collection ability, mechanical property, elasticity, stickiness	(Becker et al., 2003, Swanson 2007, Vollrath and Knight 2001)
Spider dragline silk	Mechanical property, super contraction, torsional shape memory	(Work 1985, Shao and Vollrath 1999, Shao et al., 1999, Pérez-Rigueiro et al., 2003, Liu et al., 2005, Bell et al., 2002, Emile et al., 2006, Gao and Jiang 2004)
Water strider leg	Durable and robust super hydrophobicity	(Lu et al.,2014)

The multi-scale nature of biological materials motivated several design attempts within the field of tribological surfaces. However up until the push for nanoscale material manipulation, many designs, although conceived, were not technologically possible. The leap in technologies that matured within the past decade or so, have resurrected efforts to manufacture commercially viable surfaces in the field of tribology. This chapter presents samples of existing bio-inspired tribological surfaces that are either a direct replica of the bio-analogue or represent a modification of the surface through a combination of chemical and geometrical changes.

SHARK INSPIRED SURFACES

Although shark skin is noted for drag reduction, recently it served as an inspiration for improving lubrication quality. Lu et al (2014) recognized the effect of the arrangement of textural grooves on the quality of friction behavior. The authors were hindered by previous studied reports that reflected the disagreement on the optimal value for spacing between textural features. In seeking a solution for the textural arrangement, they sought to mimic the shark effect by producing grooved surfaces of rhombic textural features (figures 1,2). The design exploited two functions performed by the placoid scales present on the skin of a shark. Namely, regulation of the boundary conditions of the flow and accommodating the space necessary for the flow of the fluid within the riblets present on the scales. In that sense, the surface design of Lu and his co-workers is similar to a honing pattern on a cylinder liner, except that the scale of the unit feature is far smaller than a typical honing cell. One effect that Lu attempted to capture is the ability of the micro-riblets on the placoid of the shark to reduce shear stress due to fluid flow. To validate the concept of design Lu used a femtosecond laser micro-machining system to pattern 2CrBsteel rings (8 mm thick, outer diameter 52 mm and inner diameter 30 mm) with micro scale rhombus grids (figure 2). Pin-on-ring lubricated experiments combined to a theoretical computational model verified the effectiveness of the rhombic cells in building up the fluid film to provide an efficient pressure build-up capable of fully lifting the flat moving counter surface. The build-up fluid film and the build-up of the pressure improved the tribological performance by separation of the rubbing surfaces. The experiments highlighted the dependence of the quality of lubrication on the depression width ratio. When this ratio is greater than 0.45 the lowest COF is obtained. Accordingly, the recommendation was to use a large depression ratio combined to a small rhombus apex angle.

Figure 1. (a) SEM photography of shark's skin and (b) bionic texture surface (35)

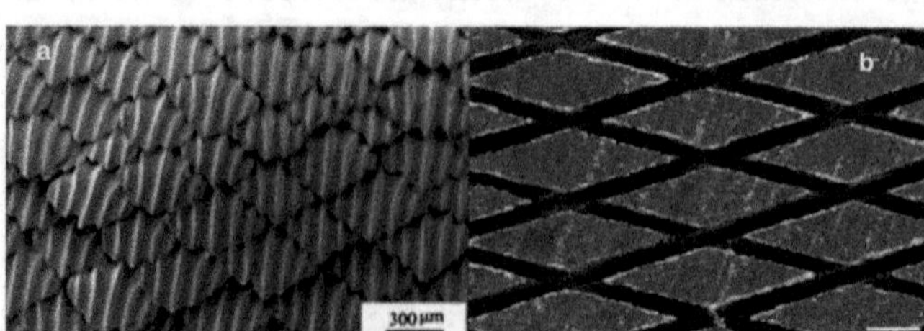

Figure 2. OSP profiles of the originally machined microtexturingin the work of Lu surface: (a) a = 100 μm, b = 50 μm, α = 90 deg, d = 50 μm; (b) a = 450 μm, b = 180 μm, α = 45 deg, d = 800 μm; (c) a = 900 μm, b = 180 μm, α = 30 deg, d = 800 μm; and (d) a = 900 μm, b = 180 μm, α = 60 deg, d = 800 μm
(Lu et al.,2014)

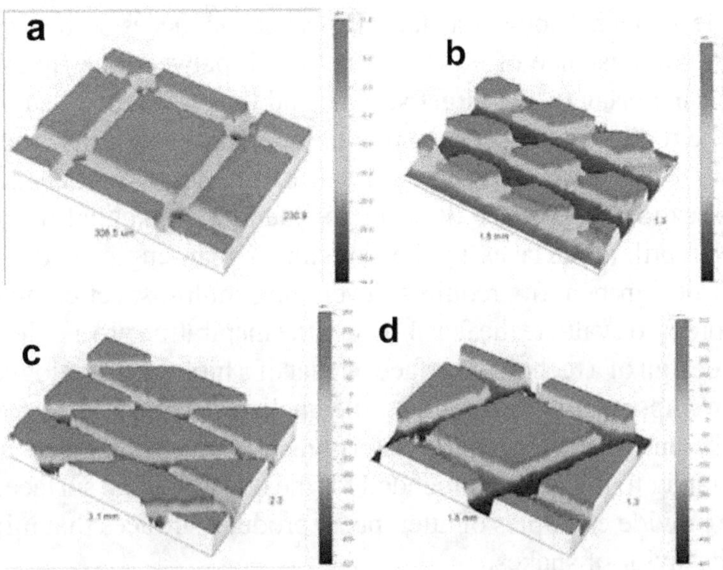

SNAKE INSPIRED SURFACES

There are several similarities between a Texturized Technological Surfaces and the ventral side of a snake, both in metrology and topology. The main similarity is the dependence of the metrological features on the scale of observation. That is both of the snakeskin and the TTS are multi-scale entities. In addition, the basic building block in each case (snakeskin and TTS) is a textural element that is repeated in an array distribution. The snake's ventral side is composed of identical micron-sized fibrils (denticulations) distributed over the skin area in a particular pattern. Spacing, length, orientation and shape of denticulation is in general common to a particular family of snakes. TTS, on the other hand, by their very definition comprise an individual textural building block (cone, dimple, chevron etc.,) that also is distributed on the surface area in array form. Therefore, both types of surfaces share common traits. A snake, however, has to be self-sustaining over a wide spectrum of sliding terrains and conditions. The ventral texturing, therefore, has to be efficient over the spectrum of contact conditions that the species experiences.

Such a feature offers an advantage to the natural surface over an engineered LTS where the dimples satisfy efficient performance over a narrow domain within possible contact conditions.

The answer to this question is rather complex due to the many factors involved. A simple answer is to mimic, or replicate, the textural designs present on the reptile in a technological surface. However, the success of this approach depends on the existence of a one-to-one match between the function and perhaps, the material, of the target surface and that of a reptile. This match will limit the replication process to polymeric surfaces (because the skin is elastomeric). However, bio-inspired surface texturing has a broader goal than merely replication. In essence, it seeks to extend the potential tribological benefits of reptilian surfaces to the domain of man-engineered surfaces. Transfer of design benefits requires developing multi-aspect compatibility metrics in order to evaluate the feasibility of the inspiration process itself (i.e., basing the design of a technical surface on that of a biological analogue). Two features are important in this context: the quality of frictional performance of the snake and matching of the deterministic surface parameters of the biological target to those recommended for the technological surface. In this section we provide examples of attempts to produce surfaces that mimic the frictional behavior of snakes.

In this section we present examples of successful surfaces of which design originated from snakes. We distinguish between attempts of direct replication and what is termed as "snake-inspired" attempts, where observing the metrological proportions of the geometrical features, not direct replication, is the design claim.

Direct Replication of Snakeskin Texture Features

Muhlbergeret al (2015) applied powder injection molding combined to ultra violet light assisted nano-imprint lithography to replicate the delicate features of a Cobra (Najanigrocollis). Combining both techniques allowed direct copying of the biological surface and then imprinting unit textural features that are three times the length of actual features of the snake. A comparison between the two surfaces resulting is given in figure 3. Pin-on-disc experiments in parallel and anti-parallel sliding directions, with respect to the orientation of texturization, verified the benefits of the snake ornamentation.

Figure 3. Biological surface: AFM images (20x20 µm2) of natural cobra (Najanigricollis) skin structure: (a) positive epoxy resin structure after a two-step molding process, (b) PEG diacrylate positive structure after UV-NIL imprint. The head of the snake is towards the bottom of the images, so the moving direction of the snake is towards the bottom of the images. Artificially designed structure: AFM images (50x50 µm2) of (a) PMMA master (negative/recessed structures) with 15 µm long denticles displaying shallow tapers at the bottom and 1.4 µm high pins at the top, (b) PDMS stamp (positive/elevated structures).

Reprinted with permission from: M. Mühlberger, M. Rohn, J. Danzberger, E. Sonntag, A. Rank, L. Schumm, R. Kirchner, C. Forsich, S. Gorb, B. Einwögerer, E. Trappl, D. Heim, H. Schift, I. Bergmair, UV-NIL fabricated bio-inspired inlays for injection molding to influence the friction behavior of ceramic surfaces, Microelectronic Engineering, 141, 2015, 140-144,.

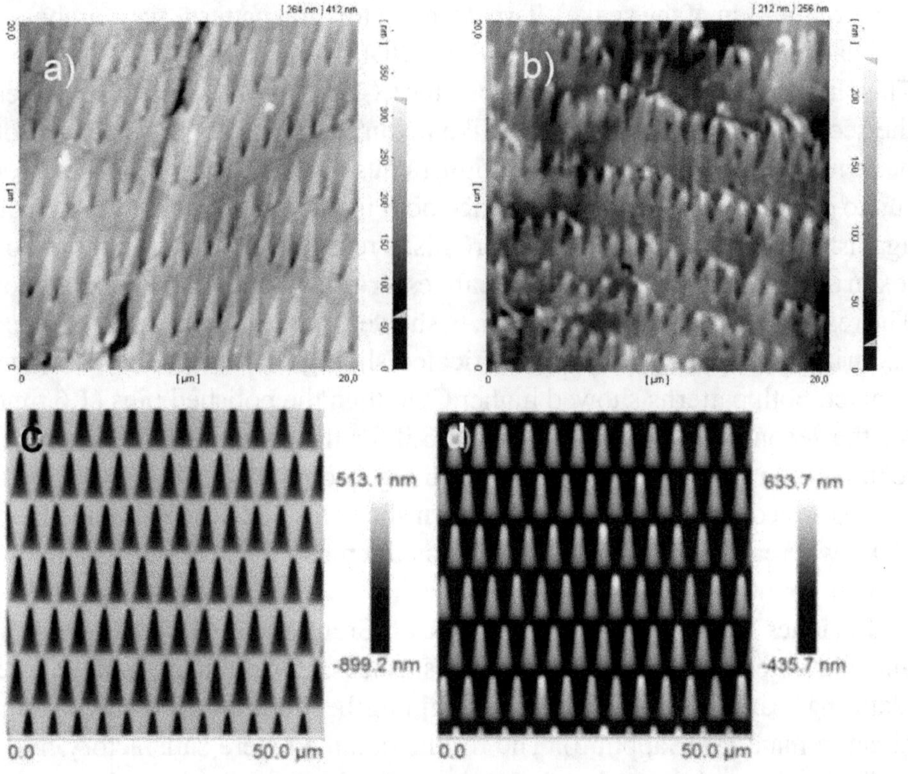

The experiments compared the friction behavior of textured and polished ceramic samples (Commercial Zirconia and Alumina Ceramic powders and a propriety binder). The results favored the textured surfaces. Snake skin textured samples running in the parallel direction gave a COF that is approximately 30% lower than that of the polished non-textured sample, and also lower than samples sliding in anti-parallel direction.

The work of Muhlberger also pointed at the importance of using an "efficient" replication method to produce the required fine size texturization. In his work the samples with features almost of the same length as that of the original skin did not perform well not because of the tribological inefficiency of the geometrical features, but because of the inadequacy of the replication technique used to transfer the texturization to the ceramic material.

Another form of direct replication was endeavored by Greiner and Schäfer (2015). Their work attempted to mimic the overlapping structure of the scales (and not the micro structure within the scales). They applied laser ablation to produce two sets of overlapping textures that mimic the sub-caudal scales of a python regius (again, no mimicry or adaptation of the primitive geometrical ratios of the skin of the snake). Two different textural patterns were embossed on bearing steel surfaces (100Cr6-Fe with 1.5% Cr and 1.0 C %) figure 4. The first pattern overlapped in both the X-and the Y-directions, whereas the second had overlapping scale-like texturization arranged in rows. Both patterns had quasi-round scales of 25μm radius. Pin-on-disk experiments were run to evaluate frictional performance both in dry and lubricated conditions against Sapphire and against 100Cr6 disks respectively. For control, a non-texturized steel pin was tested with all tests running in the parallel direction. The results of Greiner, see figures 5, 6, showed that in general, this particular texturization is not efficient in lubricated sliding. In the case of lubricated contact, both patterns showed higher COF than the polished pins (1.6 times for the second pattern and a factor of 3.0 for the first pattern). Benefits of texturization were realized, however, in dry sliding where, compared to the un-texturized sample, the biaxial pattern showed some 20% reduction in the COF, whereas the row arrangement was the most efficient with some 40% reduction in the COF.

Johannes (2018) extended the work of Greiner and examined both dry and lubricated behavior of the textured surfaces. His initial attempts entailed texturing bearing steel surfaces and examining the tribological behavior against ceramic material (Sapphire). The results obtained were satisfactory in dry sliding and speed dependent in lubricated sliding. Such a behavior negates the perceived value of texturing advocated by many authors. One explanation for this behavior was size effect that is critical to performance which, again, is compatible with reports by other investigators. This size effect impedes performance and mixed lubrication regimes and in transition from static to kinematic friction. Johannes therefore experimented with textured bearing steel against several materials in dry and lubricated sliding the counter face is in his work where metallic, polymeric, and ceramic bodies sliding at different

Figure 4. Scanning electron microscopy and white light profilometry images of laser-textured bearing steel (100Cr6) surfaces with biologically inspired scale-like surface morphologies. (a,b) A linear arrangement in parallel rows as tested under dry and lubricated sliding against metallic, polymeric and ceramic counter bodies. (c,d) Arrangement with an offset. The secondary electron SEM images in (a,c) were taken under a tilt angle of 30°. The sliding direction for the textured pins is indicated by the arrow.

Printed with permission from: Schneider, J.; Djamiykov, V.; Greiner, C. Beilstein J. Nanotechnol. 2018, 9, 2561–2572. subject to the terms of the Creative Commons Attribution Noncommercial

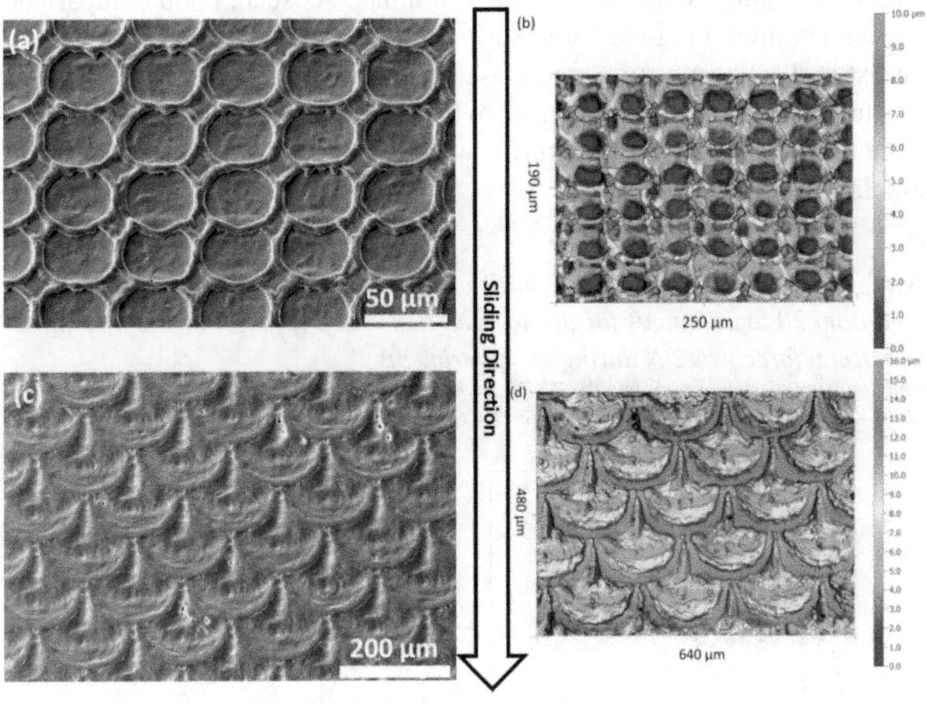

speeds. He further instigated the size effect for dry conditions against Sapphire to compare with the results of Greiner.

Results of the experimental program of Johannes, for dry and lubricated sliding, are shown in a Stribeck-type curve in figure 5-a and b. The dry sliding results, figure 5-a, show the effect of speed on friction. The coefficient of friction is uniform for all speeds in textured surfaces and shows a slight fluctuation for the untextured control surfaces. The value of the friction coefficient for the textured surfaces is considerably lower than that of the untextured surfaces. Thus, for dry sliding textured surfaces are favored and perform as expected. The results for lubricated contacts, however, display serious departure from the dry behavior (figure 5-b). In the lubricated regime, the absolute values of the friction coefficients decreased by almost an order of magnitude with speed. The efficiency of performance, i.e. realization of the benefits of texturing, depends on increasing the speed. Thus, by increasing the speed lubrication regime changes and the thickness of the lubrication film increases until for separation between surfaces is achieved. At this speed considerable reduction in the coefficient of friction takes place (in this case from approximately 0.2 at 10 mm/s to 0.01 at 75 mm/s). Increasing the speed raises the coefficient of friction slightly to about 0.015 at 150 mm/s. As such, upon comparison between textured and untextured surfaces in this case, Johannes concluded that textured surfaces demonstrate classical Steinbeck curve behavior, whereas untextured surfaces are not influenced by any change in sliding speeds. This indicates that the benefits of texturing would be realized only within a certain domain of sliding speeds.

Figure 5. Stribeck curve-like pin-on-disc experiments for a variation in sliding speed from 20 to 170 mm/s for dry (a) and lubricated (b) steel-on-steel contacts. The normal force was 2 N during all experiments.

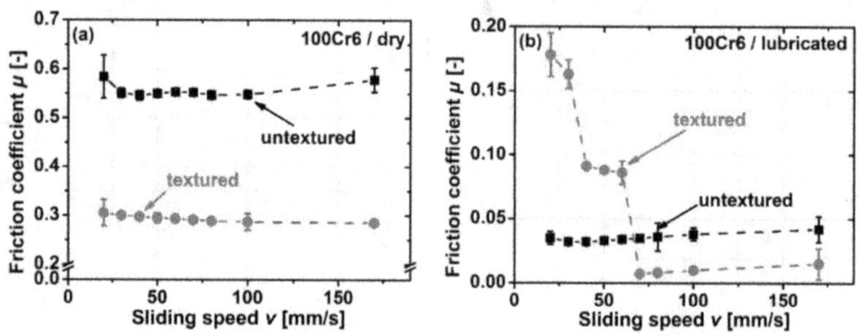

Sliding against the polymeric counter face for lubricated and for dry testing resulted in similar qualitative trend with the sliding speed. That is, an initial small drop in the coefficient of friction followed by an increase in function at higher speeds than a slight cramping in friction with increasing speed. In lubricated conditions, friction decreased compared metal to metal sliding. The textured surfaces displayed reduce friction over the entire range of sliding speeds tested. More important, in this case, is that no transition from beneficial to a detrimental effect, by introducing texturing, was observed for small sliding speeds.

In sliding against ceramics, figure 6 a-b, textured and untextured samples manifested attend similar to metallic samples. In dry mode, the coefficient of friction, for both samples grew was sliding speeds up the critical value (around 80 mm/s) then dropped slightly. At all speeds, friction for textured surfaces was always lower than that for untextured surfaces. In lubricated contacts, however, the behavior was contrasting. The efficiency of texturing dependent on sliding speed. At relatively slow sliding speeds, the friction increased then it dropped with increasing the speed of sliding. The coefficient of friction dropped with increasing speeds up to the critical speed where it became significantly lower than that for the untextured surfaces. Thus, again, similar to metallic sliding pairs, feasibility of texturing depends on the speed of sliding.

Figure 6. Stribeck curve-like pin-on-disc experiments for a variation in sliding speed from 20 to 170 mm/s for dry (a) and lubricated (b) steel-on-ceramic contacts. The normal force was 2 N during all experiments.
Printed with permission from: Schneider, J.; Djamiykov, V.; Greiner, C. Beilstein J. Nanotechnol. 2018, 9, 2561–2572. subject to the terms of the Creative Commons Attribution Noncommercial

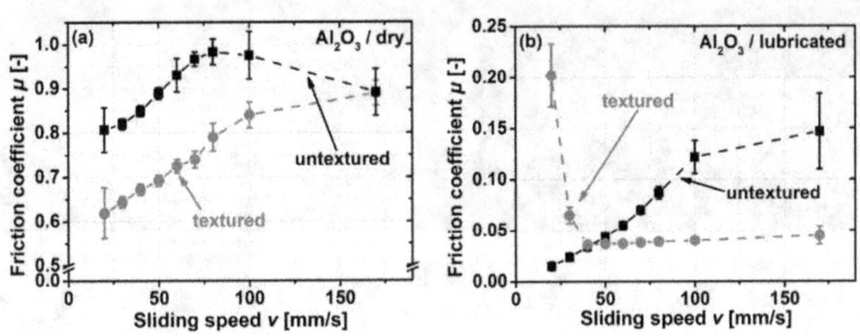

One effect crucial to the work was the effect of the hardness on anisotropic of friction. This effect was not investigated. However, his results pointed to the possibility of an anisotropic frictional effect influencing the results. This effect, which stems basically from the shape of the scale-like texturing, has the potential of changing contact conditions and thereby altering the frictional behavior.

The tribological environment in machining is very hostile. High strain rates in the order of 20,000/s combined to stresses in excess of 2 GPa and several hundred degrees temperature rise may exist. Coupling between thermal and mechanical states intensifies the hostility especially in dry machining. Bio-inspired texturing recently found its way to this field through a P.

regius texturing scheme. Fatima and Mativenga (2015) reported successful application of cutting inserts where bio-inspired texturing ornamented the flank face (figure 7). Textures were implemented using femto-second Sapphire laser. Similar to the work of Greiner, textural elements were replication of the shape of the ventral scale of the Python (i.e, the shape of the element of texture copied the outer shape of a P. regius ventral scale with an aspect ratio of 4-figure 26-b and d).

Figure 7. Details of the Bio-inspired rake face texture of tungsten carbide cutting inserts

Reprinted with permission from Anis Fatima, P. T. Mativenga, On the comparative cutting performance of nature-inspired structured cutting tool in dry cutting of AISI/SAE 4140, Proceedings of the Institution of Mechanical Engineers, Part B: Journal of Engineering Manufacture December 16, doi: 10.1177/0954405415617930

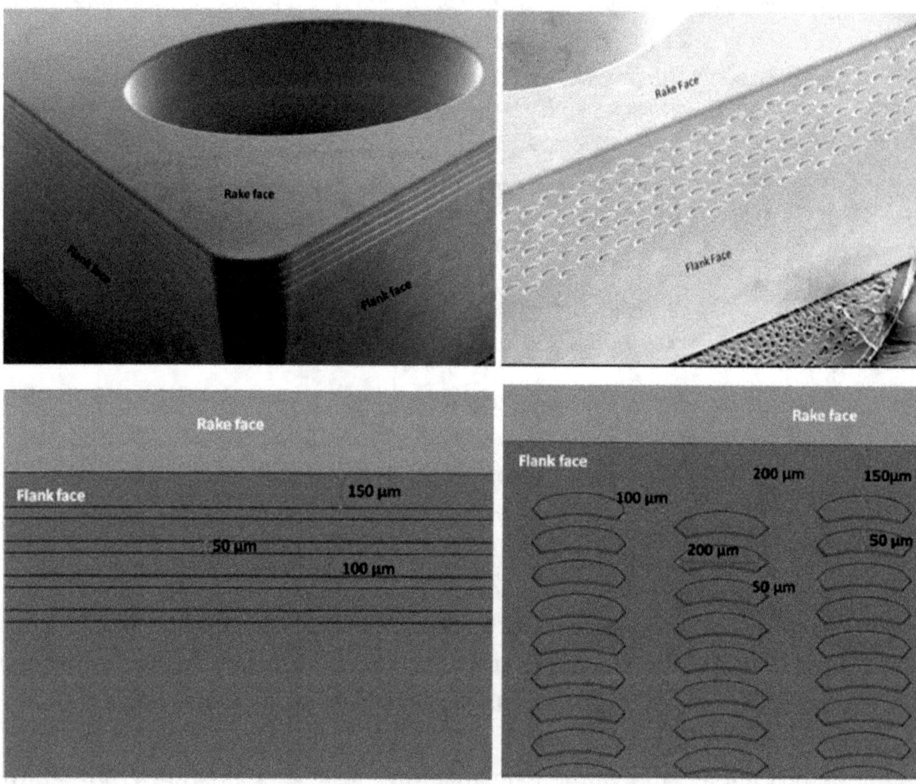

As shown in the figure layout of the features constituted an orthogonal matrix, with no apparent observation of the actual spacing between features on the bio-analogue. The authors compared the performance of the textured insert to a conventionally insert where grooves (figure 6-a and c) are used to enhance tribological performance. The comparison entailed lubricated cutting of AISI/SAE 4140 steel tubes in orthogonal cutting configuration using a lubricant on the flank face of the tool. Two aspects of the results are of interest in this work. The first is the temperature rise during machining, whereas, the second is the feed force which represents the friction force.

Bio-inspired textured inserts consistently outperformed a control insert and a conventionally textured insert at each of the cutting speeds used. Table 2 provides a summary of the results. Bio-inspired inserts showed remarkable reduction in the temperature rise compared to the non-structured control insert (65% for the slow speed and 49% for the high speed used in the experiments).

Table 2. Comparison between the performance of the tool inserts used in the work of Fatima and Mativenga

Speed m/min		283	Percent Change	628	Percent Change
Temperature C	Unstructured	153	100%	170	100%
	Conventional	134.75	88%	185.3	109%
	Bio-Inspired	53.55	35%	86.7	51%
Feed force N	Unstructured	525	100%	512.5	100%
	Conventional	509.375	97%	493.5	96%
	Bio-Inspired	493.75	94%	487.5	95%

Considerable reduction in the feed force, which implies reduced friction force, was registered as well. This reduction ranged between 8% and 16% for the slow and the high velocities respectively. The reduction in temperature and friction force led to saving in the overall power used in cutting and to reduced wear (Fatima and Mativenga reported an average power saving of 38% and 51% for the low and high speeds used compared to the conventional structured tool).

Snakeskin "Inspired" Surfaces

Friction as a phenomenon is multi-faceted. Mitigation of the phenomenon depends on optimizing the interaction and the interference between all contributing factors (geometry, material, chemistry, and mechanics). Thus, an alternative to direct replication of bio-texturing is to design surfaces with variable "controlled" contributions of each of the contributing effective factors. This was attempted by Baum et al (2014) who set to examine the influence of the interaction of micro-structure and geometrical metrology of the surface of snakes (shape, dimension, and orientation) on friction performance. The biological analogue used in her work was the California King Snake (L. Getuta Californae). A deterministic, wave like texturization was embossed on epoxy resin (see figure 8 and table 3). Frictional characterization took place by sliding a glass ball over the epoxy surfaces under Hertzian contact conditions. The results obtained in the work of Baum pointed at the existence of an optimal threshold of interaction between the various contributions to a frictional process. In particular, because of the nature of the polymeric surfaces tested a tradeoff between dimensions and structure on the one hand and the reduction in the real contact area and interlocking between surfaces on the other hand took place. This tradeoff manifests the interference between surface energy represented by the adhesion and the mechanical contribution of friction represented by the true contact area. Thus, a successful surface design is one which manages the complex interference between chemistry and geometry. The results also showed that the most pronounced reduction in the COF depends on the orientation of the micro textural features with respect to the direction of sliding. When the surface roughness (texturization features) are of minimal size and the adhesion is eliminated due to orientation of sliding (parallel to the friction force-see figure 9), maximum reduction in friction is possible.

Table 3. Dimensions of the microstructures of the ventral scales of the snake L. g. californiae and the SIMPS.

	Microstructure Wavelength (μm)	Denticulation Length (μm)	Denticulations Mean Width (μm)
L. g. californiae	1.1 ± 0.1	2.5 ± 0.5	0.6 ± 0.1
SIMPS	1.2 ± 0.1	1.1 ± 0.2	0.7 ± 0.2

Figure 8. From snake skin to SIMPS. a) Photograph of L. g. californiae, the California King Snake; b) SEM micrograph of a ventral scale of L. g. californiae; c) SEM micrograph of the snake-inspired microstructured polymer surface – SIMPS. Scale bars: 10 μm. 3-dimensional image of d) snake skin from L. g. californiae and e) SIMPS based on AFM data.

Figure printed from Baum MJ, Heepe L, Fadeeva E, Gorb SN. Dry friction of microstructured polymer surfaces inspired by snake skin. Beilstein Journal of Nanotechnology. 2014;5:1091-1103. doi:10.3762/bjnano.5.122, subject to the terms of the Creative Commons Attribution Noncommercial License

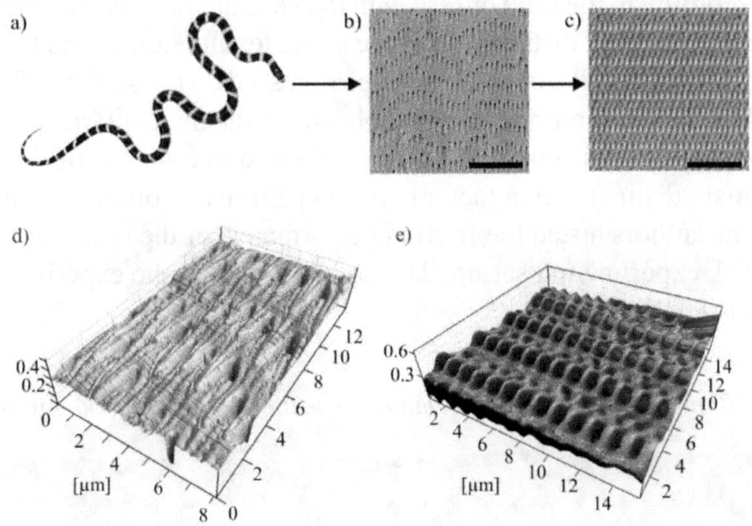

Figure 9. Scheme of the directionality of frictional measurements on polymer surfaces and its geometry. a) Directions of measurement relative to the topography of the SIMPS and angles of the microstructure depending on the measurement directions: α: along, β: against and λ: lateral to the microstructure. b) Diagram of the periodic groove-like profile. MS: microstructure, A: periodicity of the structure, W: width of the structure and D: depth of the grooves.

Figure printed from Baum MJ, Heepe L, Fadeeva E, Gorb SN. Dry friction of microstructured polymer surfaces inspired by snake skin. Beilstein Journal of Nanotechnology. 2014;5:1091-1103. doi:10.3762/bjnano.5.122, subject to the term sof the Creative Commons Attribution Noncommercial License

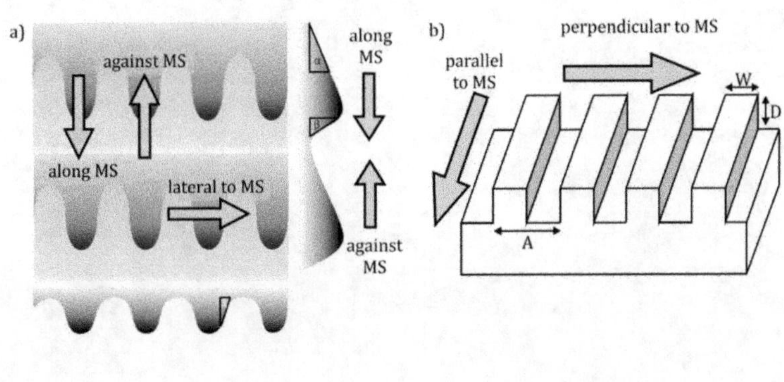

Adaptation of the primitive geometrical ratios influential to performance of texturized surfaces found on the skin of a snake was attempted by Cuervo et al (2018). In this work, the group used Photo-Lithography to texturize the surface of Titanium (Ti6Al4V) pins with elliptical patterns figure 10. Elliptical texture patterns were used before to modify surfaces as dimples. However, the new aspect in the work of Cuervo was not the geometrical shape of the texturing feature. Rather, it is the topology of the features. The patterns were placed according to the ratio observed in the ventral scales of a P. regius thus mimicking the spatial distribution of the snake texture rather than the size or shape of the features. Two patterns were embossed on the surface of the pins (figure 11). Both had elliptical features however the main difference was the height of the features (that is the so-called slenderness ratio of the feature was the main distinguishing factor between patterns). Consistent with other studies, the authors tested the friction performance of the titanium pins in a pin-on-disk experimental set-up. The counter-body in the experiments was a polymer (UHMWPE).

Figure 10. Typical elliptical pattern embossed on titanium using photo Lithography.

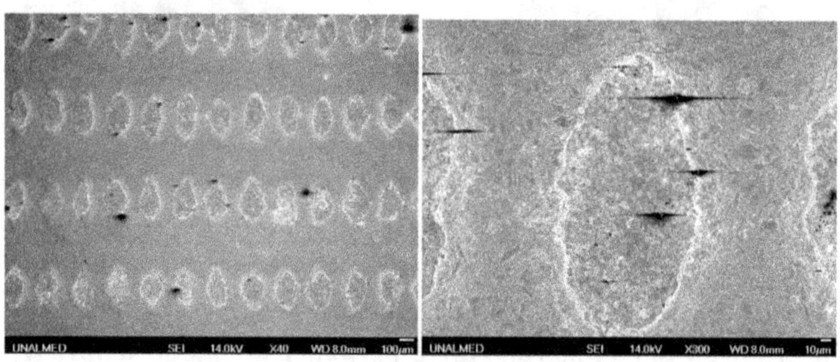

Figure 11. Typical 3D-images of snake-inspired surfaces studied in the work of Cuervo et al (40), a) Texturized, pattern 1, b) Texturized, pattern 2.

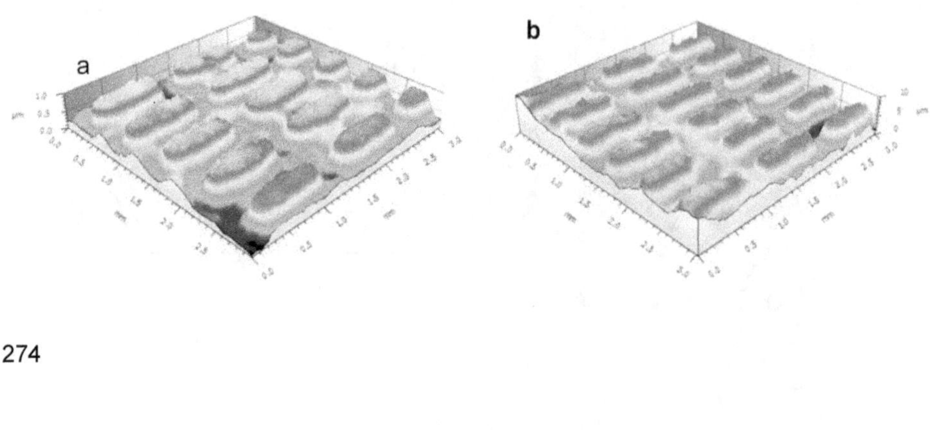

Two control specimens were also tested, a sand-blasted and an as manufactured pin with no texturization patterns). The results, figure 12, confirmed the feasibility of mimicking the spatial ratios found on the snake surface. Consistent with other studies, friction anisotropy with the lateral COF being greater than the COF in parallel sliding was observed. However, the remarkable find of the study was the dependence of the friction on the slenderness ratio of the feature. This variable represents the ratio between the heights of the feature to the length of the axis of the ellipse along the direction of sliding. For features with the same surface footprint, the COF was directly proportional to the feature height. Given the size of the features and the polymeric nature of the counter body in the experiments this raises a question about control of the adhesive forces.

Figure 12. Average COF measured after the running-in period for textural patterns used in the work of Cuervo (2018), P indicated sliding along the major axis of the elliptical

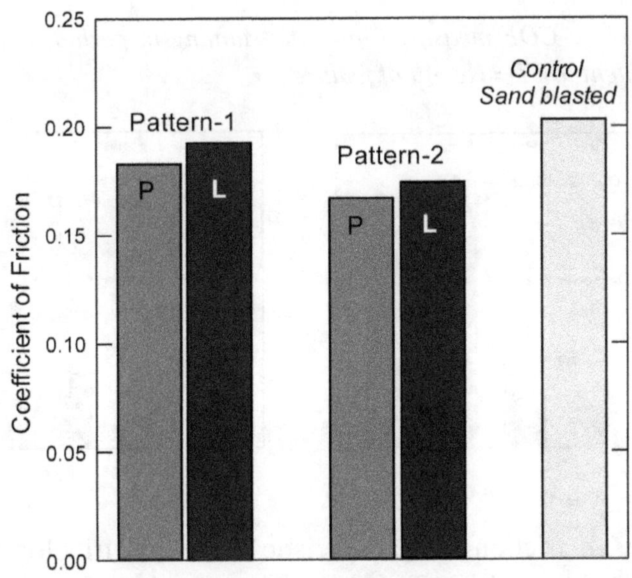

The surface-to-volume ratio of micron-sized features should affect adhesion, the higher the ratio the higher the adhesive contribution. This ratio, however, in the work of Cuervo reduces to the reciprocal of the feature height, which indicates that features with the same footprint but with larger feature height should have smaller adhesion contribution and vice versa. A clear investigation of that point was not attempted, which leaves an open question about the optimized value of the slenderness ratio to minimize friction.

The average COF of the texturized surfaces increase in direct proportion to the increase in feature height. An example is figure 13-a where the average COF for pattern-I, with a feature height of 14 μm, is greater than that measured when the feature height is 5μm. Moreover, in both plots, regardless of the feature aspect ratio (FAR), the lateral coefficient of friction in lateral sliding is slightly larger than the value in parallel sliding. Values of the COF for each pattern suggest the existence of an optimal range of feature heights beyond which the friction of the texturized surface will exceed that of the machined sample (i.e, the lowest friction coefficient for the particular load). An interesting feature of the plots shown in figure 15 is anisotropy of friction. The value of the friction force (and thereby the COF) when sliding in the lateral direction, for the same pattern, is greater than the value measured when sliding in the parallel direction. Frictional anisotropy, however, is not pronounced ($1.05 \leq \mu lat/\mu\ parl \leq 1.15$).

Figure 13. Average COF measured after the running-in period. PH=Protrusion Height, equivalent to FH=Height of feature.

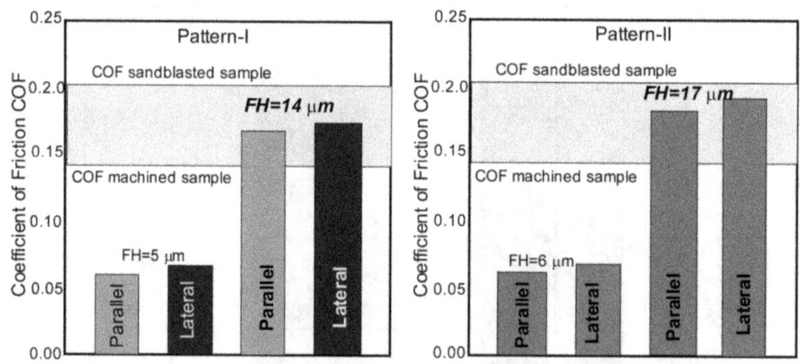

Anisotropy of friction is characteristic of several tribological systems in biology (notably snakes). Two categories of anisotropy are reported in literature. The first concerns the ratio of the friction coefficient in backward motion to that in forward motion (i.e. μ_{back} / μ_{for}), while the second, also reported for man-engineered systems, is associated with the ratio of the COF in lateral direction to that in longitudinal direction (i.e. μ_{lat} / μ_{for}). Accordingly, the anisotropy implied in the current work belongs to the second category.

Anisotropy of the first category originates from the geometry of the fibril structures within the ventral scales in snakes and the orientation of the keratin structures within the skin layers. Anisotropy of the second kind has not been fully investigated. However, within the context of textured surfaces, and periodic surfaces, the phenomenon was linked to the orientation of texture and to the variation in the adhesion length as a function of the sliding orientation. The origins of anisotropy in this case seem to originate from difference in adhesion length.

Because of the geometry and orientation of the texture-feature array, surfaces examined in this work establish contact in a periodic, intermittent, manner. This can be visualized in reference to figure 14 (a and b) along with figure 13. During sliding, the protrusions establish contact first through the leading edge of the particular texturing feature. Continuous sliding leads to the growth of the contact area to and maximum then to gradual decrease in the area of contact until disengagement of the protrusion from the particular spot on the surface. Following contact, and due to protrusion of the features above the plateau of the surface, a small time-period, through which the contact is relieved, will take place. Periodicity of contact (i.e., the scenario of engagement-disengagement of contact) repeats in a frequency that is firstly a function of the sliding speed, and secondly is independent of the orientation of sliding.

LOTUS INSPIRED SURFACES

As mentioned earlier, super-hydrophobicity is well recognized for the leaves of a lotus plant (figure 15). However, one unique advantage of the topographical features on the surface of a lotus is minimization of the area of contact due to the spherical caps of surface protrusions. Yoon et al (2003) attempted to exploit this quality to enhance tribological performance on the micro-scale. At the micro-nano-scale, a large surface to volume ratio results in high surface forces (especially adhesion) and friction. Increased surface forces decrease the performance of MEMS devices for example. A possible route to face this challenge is to minimize the area of true contact between the two rubbing surfaces through controlling the topography of the interface. In a series of investigations Yoon and his group (2003, 2005, 2006), studied the performance of silicon wafer surfaces, of which texturizing was a direct replica of the natural leaves of two lotus species: Nelumbonucifera and Colocasiaesulenta. The replicated surfaces (figure 16) were tested for micro-

Figure 14. Representative excerpts of the time evolution of the COF with time for pattern I texturing elements. The figure depicts variation of the friction in parallel and lateral sliding. Observe the periodic nature of the curves that follow the growth of the contact area of the texturing elements and their respective relief from the surface.

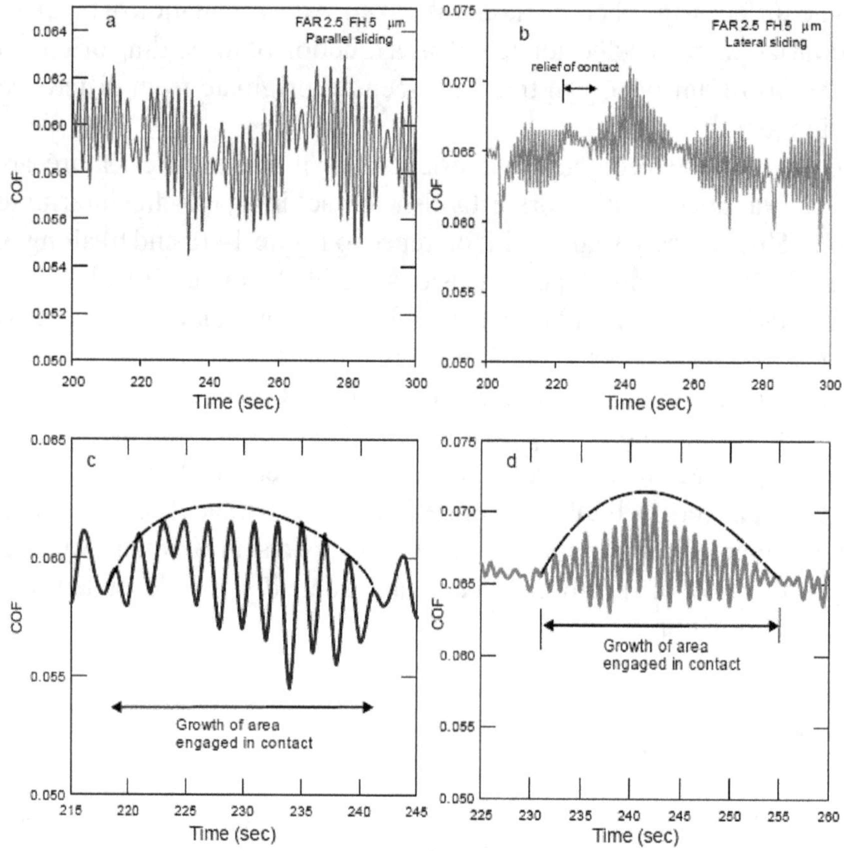

friction against Soda lime glass balls (0.5 mm-radius) under dry reciprocating motion. The results showed that it is possible to reduce the coefficient of friction of texturized Silicon by a factor of five compared to that of PMMA thin films sliding under the same conditions.

Further work attempted to enhance the micro-friction of Si (100) which possesses high interfacial energy and whence unfavorable tribological properties. The route to enhancing frictional performance combined modification of surface geometry by the fabrication of micro-patterns on Si (100) and surface chemistry modification by coating DLC/Z-DOL on the surface. That is mimicking both the Lotus wax effect and the geometry.

Figure 15. The lotus plant. (a) Lotus leaves, which exhibit extraordinary water repellency on their upper side. (b) Scanning electron microscopy (SEM) image of the upper leaf side prepared by 'glycerol substitution' shows the hierarchical surface structure consisting of papillae, wax clusters and wax tubules. (c) Wax tubules on the upper leaf side. (d) Upper leaf side after critical-point (CP) drying. The wax tubules are dissolved, thus the stomata are more visible. Tilt angle 15°. (e) Leaf underside (CP dried) shows convex cells without stomata.

Figure printed from Ensikat HJ, Ditsche-Kuru P, Neinhuis C, Barthlott W. Superhydrophobicity in perfection: the outstanding properties of the lotus leaf. Barthlott W, Koch K, eds. Beilstein Journal of Nanotechnology. 2011;2:152-161. doi:10.3762/bjnano.2.19. subject to the terms of the Creative Commons Attribution Noncommercial License

However, in this work instead of replicating the lotus surface a deterministic pattern that comprised the unit textural feature of the lotus leave was deposited in an array on the surface. Two sets were attempted micro-pillars and micro-channels. Topographical modification induced a reduction in friction due to physical reduction of the area of true contact. In addition, the modified topography enhanced debris removal from the sliding interface, which contributed further to reduction of friction. Deterministic Micro-pillars show lower friction than deterministic micro-channels due to smaller contact areas. In the case of micro-channels, tests in orthogonal direction gave lower coefficient of friction values than those in parallel direction (which implies some dependence on the aspect ratio of the textural elements). In all, combining topographical modification to chemical modification of the surface enhances friction performance over as is silicon surfaces. The dually modified surfaces in the experiments of Yoon exhibited the lowest values of coefficient of friction and additionally show minimal observable wear. Modification of geometry contributed reduction of area whereas chemical modification reduced surface energy and whence contributed reduction of adhesion.

A spin on the approach of Yoon was attempted by Wenping et al (2007) who developed a multilayered nano-engineered coating for reduction of friction and improving wear resistance. The coating integrates a soft lubricating MoS2–PTFE layer onto hard load-supporting layers, with controlled roughness of cBN-TiN. The hard layer coating had surface features similar to that of a lotus species (C. esculenta) with alternating nano- and micro-sized "domes" and "pockets". The pockets helped retain the MoS2-particles by a basket structure formed during PTFE curing. Sliding tests demonstrated significant reduction in the friction coefficient for the multilayered coating due to lubrication being realized by continuous release of the lubricants, from the pockets.

Shafei and Alpas (2009) utilized reduction of the true area to control friction of NC-Ni films that are direct replication of lotus surfaces. Results, conceptually, confirmed the findings of Yoon and his group. The friction behavior reported was some 30% lower than that obtained for flat NC-Ni films sliding at the same conditions.

ATTACHMENT-PAD INSPIRED SURFACE TEXTURING

Regulation of friction in attachment pads is a strong function of geometry. Hexagonal texturing patterns are frequently observed in many species (frogs, bush crickets, etc.). The inner construction of the pads has a dendrite like structure that branches in the proximity of the surface, while the remaining volume is filled with a secretion liquid (Gorb S., et al., 2007). This arrangement facilitates recovery after large deformations while dissipating mechanical energy. The hexagonal surface micro-pattern may function as a friction regulating feature that prevents loss of traction due to wet hydroplaning.

Murarash et al (2011) tested an elastomeric surface textured after the hexagonal panels observed on the pads of a cricket (figure 17-b). He reported that this surface arrangement is capable of tuning elastomeric friction from 50% to 100% of an unmodified elastomeric surface through changing the textured area density and the aspect ratio of the textural elements.

The friction force, Murarash concluded, is insensitive to the area density of the hexagons on the surface. This was attributed to the nature of deformation of individual textural features when loaded. This deformation reaches a limit beyond which the flat area of the hexagons does not contribute to the true area of contact. Instead some projected area that depends on the deformation starts to engage the mating surface. This indicates that the friction force, typical for elastomeric surfaces, will change at each stage of deformation of the textural elements. As such, control of the height feature will be more dominant and influential to the friction force since it controls deformation. Again, it is the control of the true area that is the main vehicle of friction control.

Figure 16. SEM images of the papillose leaf surfaces of Nelumbo nucifera (Lotus) (a), Euphorbia myrsinites (b), Colocasia esculenta (c), and Alocasia macrorrhiza (d). Lotus has the highest density of papillae with varying heights and the smallest diameter of the papillae. The papillae of the other species have larger diameters and are covered with different wax types: wax platelets (E. myrsinitesand C. esculenta) and a wax film (A. macrorrhiza) which covers cuticular foldings.

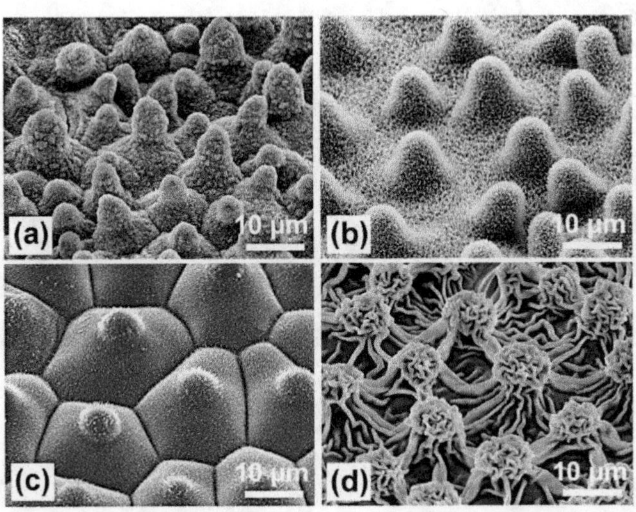

Contrary to Murarash, Huang and his group (Huang and Wang, 2013, Zhang et al. 2012, Zhang et al., 2013, Huang et al., 2012, Hanna et al., 1991, Federle et al., 2006) observed the sensitivity of friction to texture area density. They studied the performance of elastomeric patterned surfaces under wet and dry rubbing conditions. Pattern sin their studies originated from the wet patterns present on a newt. Newts can move on wet, almost vertical, glass surfaces without slipping (in a manner similar to a gecko climbing a wall). The surface micro-structure of a newt is similar to that of tree frogs toe pads, which have large hexagonal columnar epithelial cells separated by mucous-filled channels (Scholz et al., 2009). Huang and Wang fabricated spherical pins and disks made of a commercial elastomer, Polydimethyl Siloxane Pre-polymer Sylgrad-184. They implemented three types of textural arrays on the disks using lithography (figure 16). The results obtained were interesting on the counts of behavior of the COF and stability of the friction process itself

Figure 17. Some samples of Biological attachments (a) Rusty tree frog, Hypsiboasboans, on a tree trunk. FT, fluid secretion trace left on the tree surface owing to the limb sliding before friction stopped the motion. (b–d) Surface pattern in attachment pads of White's tree frog, Litoriacaerulea, great green bush cricket, Tettigoniaviridissima, and mushroom-tongued salamander, Bolitoglossaodonelli, respectively.

Reprinted ith permission from Alexey Tsipenyuk, and Michael Varenberg J. R. Soc. Interface 2014;11:20140113,e) SEM image of the elastomer surface patterned with hexagonal elements having an area density of 80% and an aspect ratio of 0.2 (Tuning elastomer friction by hexagonal surface patterning, B Murarash, Yan Itovicha, Michael Varenberg, Soft Matter, 2011,7, 5553-5557, DOI: 10.1039/C1SM00015B)

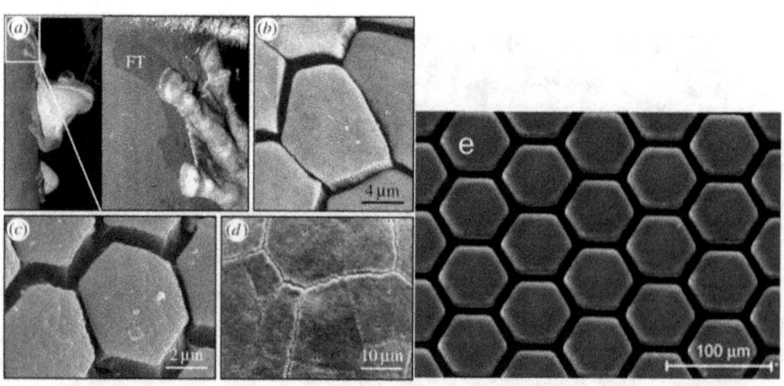

Figure 18. Configuration mode and 3D profile of several patterns modelled after tree frog

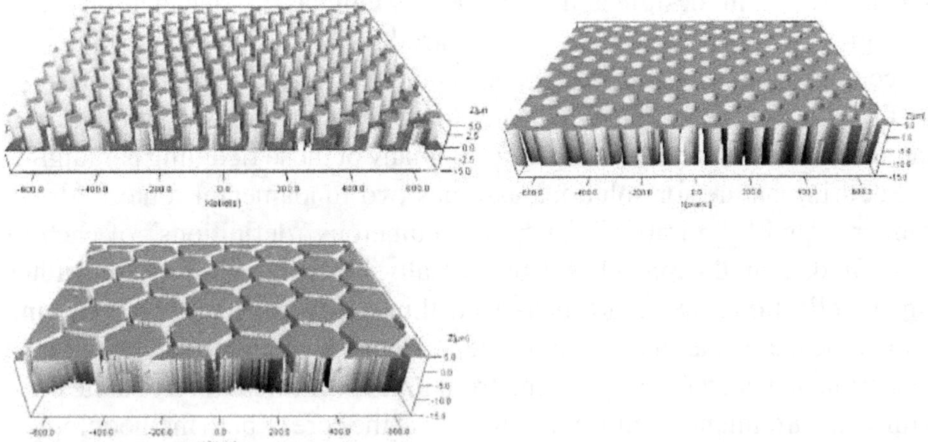

The flat (un-textured) disk in the experiments displayed a sharp transition from high COF to a lower one at a critical speed of 40 mm/s. The behavior of the textured disks, however, differed according to the texturing pattern. For dimples, regardless of the area density of features, initial increase in the COF was noted. Following that increase, and almost at the same transitional speed of 40 mm/s, gradual quasi-uniform decrease in the COF was registered. The behavior was attributed to the hydrophobic property of the PDMS.

Friction of aluminum disks with cylindrical texturing features showed a decrease with increasing the sliding speed. Area density of the features influenced the lubrication regime achieved at the different speeds (disks with higher density manifested high friction at all speeds). Friction forces increased with higher densities. Hexagonal features at higher densities caused a smooth increase in the friction force (i.e. increasing the rea density was conducive to stabilization of wet friction). This behavior was similar to the shear force velocity dependence in single frog tree pads. In all, the impact of the area density on friction was shown to depend on velocity. At lower sliding velocities, the area density has little or no-effect on friction. For high speed regime increasing the feature area density causes stabilization of friction, and the friction force shows linear dependence (an increase in density leads to a corresponding increase in the friction force).

Limbless Reptiles as Design Inspiration

Recent trends in design and engineering emphasize sustainability and ecological consciousness. This have introduced additional constraints on conception of designs and is reshaping manufacturing methods. The trend also resulted in novel paradigms that cater to design constraints. Learning from nature is a fundamental building block of many of those fledgling paradigms.

Scouring nature for solutions assumes two fundamental routes: "direct mimicry" and "inspiration". There are numerous "definitions" of each of these fundamental approaches. Conceptually, however, mimicry constitutes direct replication of a function or a part within a live form of interest. Learning is the essence of the inspiration process. Here the methodology entails the deduction of design rules stemming from functional interaction of a life form with its environment. Limbless reptiles are at the core of both methodologies. Two fields are currently of prominence: robotics and surface engineering.

Generally constructing serpentine locomotors target environments that prohibit conventional robots (i.e, those with wheels or legged appendages) from efficient operation. Serpentine, or limbless-locomotion-inspired robots are also favorable when size and shape are constrained (e.g within tight spaces, long and narrow traverses, motion on loose materials).

Limbless locomotion offers designers f robots some advantages in relation to mobility, geometry, etc. However, it also presents some limitations that confine the wide application of this class of robots. Table 4 presents a summary of the advantages of adopting limbless locomotion as a design concept in the field of robots.

Achieving true legless locomotion for field robots still represents a major challenge (even for current class of snake-inspired-robots. The difficulty stems from engineering "shoes" for the robots that are capable of manifesting true custom-anisotropic tractions (gripping) that allows traversing challenging terrains.

SURFACE ENGINEERING FOR LEGGED ROBOTS

It has been seen in previous sections that the essence of superior frictional performance of a snake is rooted in surface structure and in active control exerted by the reptile during motion. The same idea could be applied in robotics. There are two approaches that allow legged robots to traverse

Table 4. Advantages of Serpentine Robot Locomotion

Stability	A serpentine robot is inherently stable due to its inherited center of mass low potential energy.
Terrain ability	Terrain ability is the ability of a vehicle to traverse rough terrain. A serpentine mechanism can climb heights several times its own girth. This facilitates travel through challenging terrains which prohibit similarly scaled wheeled and legged machines. Moreover, serpentine robot can climb steps whose heights approach its longest linear dimension.
Traction	Traction is the force that can be applied to propel a vehicle. It is usually limited to the product of the vehicle weight and the coefficient of friction. Tractive forces can be high for natural snakes; a moving snake can exert a force up to a third of its own weight. If restrained or their locomotion is confined to certain modes, natural snakes can sustain a pull of up to four or five times its weight.
Efficiency	Snakes achieve efficiencies and performance equivalent to bio mechanisms of similar scale and mass.
Size	The small frontal area of snake mechanisms allows penetration of smaller cross-sectional areas than mass-equivalent legged or wheeled vehicles
Redundancy	Serpentine robots may employ many simple motion actuators in sequence. During operation, the loss of short segments would still permit mobility and maneuverability.
Sealing 1	The continuous unperforated surface of a serpentine device has no appendages to impede progress or to be exposed to surroundings. This allows better sealing between the mechanism internals and the environment.

difficult terrains. The first entails application of general control approaches whereas the second comprises of implementing special robot structures and materials with specific control strategies. The second approach has come to be known as morphological computation (Pfeifer, et al., 2006, 2007). The second approach, it is to be noted, is of recent origin and has not matured as of yet. However, there are several attempts for efficient implementation. The approaches are either bio-inspired or simply increasing "gripping" through using a high friction material. Kim et al. (2008) developed gecko-inspired adhesive materials and implemented them on a four-legged robot. Using the adhesive materials with a specific control mechanism for peeling toe pads of the surface allowed the robot to climb up a smooth wall. A similar strategy has been also applied to a leg-wheel hybrid robot (Daltorio, et al. 2006). Voigt et al. (2012) used foamy rubber materials as a gripper of the small robot Ratnic, with specific gripper movement control, for climbing on a pipe. These robots operate mostly on glass, smooth surfaces, or surfaces with low roughness. Extending the operational range to other surfaces of higher roughness (carpets or other felt-like or rough/dusty substrates) has been attempted as well. Spenko et al. (2008) developed special body and leg structures with compliant feet and embedded micro spines for a hexapod robot. The robot uses specific climbing control with force feedback to generate proper leg movements for climbing up different rough surfaces. Palmer et al.

(2010) developed so-called Distributed Inward Gripping (DIG) with torsion springs (i.e., passive compliance) at the legs of a six-legged robot. Using this mechanism with specific leg movement control, the robot can generate adhesive forces and walk vertically on a mesh screen. Bretl (2006) developed a four-legged robot with single peg legs wrapped in high-friction rubber and used multi-step motion planning control to enable the robot to climb vertical rock. These strategies require specialized control, structure, and material designs to handle rough surfaces.

Achieving efficient locomotion on rough surfaces is a challenging task. However, it can be achieved by inducing frictional anisotropy and mechanical interlocking between surfaces at the microscale. In other words, it can be achieved by mimicking those creatures that display anisotropic friction (such as the subject of this review: snakes). In essence, strong mechanical interlocking in one direction should enhance the grip of a robot on the surface, thereby preventing slipping or backward sliding. Meanwhile almost no mechanical interlocking in another direction should allow releasing the robot from the surface while moving forward. Implementing strong frictional isotropy should also allow the robot to grip to the surface, yet it will induce difficult release from the surface. Combined frictional anisotropy and mechanical interlocking, inspired Marvi et al. (2011) to develop active scales and their control to generate the frictional anisotropy for a snake-inspired robot (Scalybot). This combination allowed the robot to climb inclines up to 45°. Shark skin in lieu of active scales was used by Voigt et al (2012) to enhance the efficiency of legged robot locomotion on inclines. The idea in essence is not new. Shark skin was used in ancient times by fishermen to make shoes to enhance their grip on the wet wooden decks of their ships. Moreover, shark skin was also used in various tools such as in wooden rasps (McQuirty, 2011), in grips and sheath of swords, and also to sand wood and ceramics. The requirement of customized friction profile reduces the problem to proper "engineering of a surface".

POTENTIAL FOR INTEGRATED DESIGN SOLUTIONS

Engineering of a surface especially within the realm of tribology and friction control, is of diverse interpretation since it is application driven. However, common features exist between various interpretations. Namely, a properly customized surface for tribological purposes, should exhibit a dynamic surface-subsurface interaction. This is because, as mentioned in previous chapters, optimize performance of the surface in a tribological application is a function of surface features assisted by the material functionality of subsurface layers. Subsurface layers can affect the stiffness of surface features, energy storage and release while rubbing, and may also affect damage inflicted contact layers. In all, surface design for tribo-applications requires addressing performance optimization on two levels. The first is the generation of suitable roughness features for the immediate rubbing layer. Whereas the second is functionalization the properties of materials within the subsurface layers. The challenge here is to integrate both levels of design. The crucial process of integration has been a stumbling block for quite some time part of the reason is the lack of mature design paradigms. Mainly, however, the lack of an enabling technology that permits customization of thin layers within the subsurface is the major contributor.

As discussed earlier in this chapter biology can play a major role in guiding development novel surface texture-design paradigms. Functionalization and integration of subsurface layers with textural features is however advanced by the rapidly emerging additive manufacturing techniques especially three- and four-dimensional printing technology. In fact, 4-D printing is at the core of material functionalization and targeted property profiles. To this effect combining bio-inspired design and 4-D printing techniques offers a considerable potential for full customization of tribological design (among other advantages). Indeed, the past two years have witnessed the emergence of some attempts combine bio-inspiration and 3-4-D printing. Being of considerable promise, these emerging works, along with the theory behind them are seemingly transformative. Therefore, they comprise the main subject of the following chapter.

REFERENCES

Aizenberg, J., Sundar, V. C., Yablon, A. D., Weaver, J. C., & Chen, G. (2004). Biological glass fibers: Correlation between optical and structural properties. *Proceedings of the National Academy of Sciences of the United States of America*, *101*(10), 3358–3363. doi:10.1073/pnas.0307843101 PMID:14993612

Aizenberg, J., & Tkachenko, S. (2001). Article. *Nature*, *412*, 819–822. doi:10.1038/35090573 PMID:11518966

Aizenberg, J., Weaver, J. C., Thanawala, M. S., Sundar, V. C., Morse, D. E., & Fratzl, P. (2005, July 8). Skeleton of Euplectella sp.: Structural hierarchy from the nanoscale to the macroscale. *Science*, *309*(5732), 275–278. doi:10.1126cience.1112255 PMID:16002612

Anis Fatima, P. T. (n.d.). On the comparative cutting performance of nature-inspired structured cutting tool in dry cutting of AISI/SAE 4140. *Journal of Engineering Manufacture*. doi: 10.1177/0954405415617930

Ball, P. (1999). Engineering shark skin and other solutions. *Nature*, *400*(6744), 507–509. doi:10.1038/22883

Becker, N., Oroudjev, E., Mutz, S., Cleveland, J. P., Hansma, P. K., Hayashi, C. Y., ... Hansma, H. G. (2003). Molecular nanosprings in spider capture-silk threads. *Nature Materials*, *2*(4), 278–283. doi:10.1038/nmat858 PMID:12690403

Bell, F. I., McEwen, I. J., & Viney, C. (2002, March). Fibre science: Supercontraction stress in wet spider dragline. *Nature*, *416*(6876), 37. doi:10.1038/416037a PMID:11882884

Bhushan, B., & Her, E. K. (2010, February 4). Fabrication of superhydrophobic surfaces with high and low adhesion inspired from rose petal. *Langmuir*, *26*(11), 8207–8217. doi:10.1021/la904585j PMID:20131881

Bretl, T. (2006). Motion planning of multi-limbed robots subject to equilibrium constraints: The free-climbing robot problem. *The International Journal of Robotics Research*, *25*(4), 317–342. doi:10.1177/0278364906063979

Cuervo, López, Cano, Sánchez, Rudas, Estupiñán, ... Abdel-Aal. (n.d.). *Development and tribological behavior of snake-inspired Ti6Al4V surfaces with deterministic patterns*. IOP-STMP.

Daltorio, K. A. (2006). A robot that climbs walls using micro-structured polymer feet. *Proc. of Climbing and Walking Robots*, 131–138.

Emile, O., Le Floch, A., & Vollrath, F. (2006, March). Biopolymers: Shape memory in spider draglines. *Nature*, *440*(7084), 621. doi:10.1038/440621a PMID:16572162

Federle, W., Barnes, W. J., Baumgartner, W., Drechsler, P., & Smith, J. M. (2006, October 22). Wet but not slippery: Boundary friction in tree frog adhesive toe pads. *Journal of the Royal Society, Interface*, *3*(10), 689–697. doi:10.1098/rsif.2006.0135 PMID:16971337

Feng, L., Li, S. H., Li, Y. S., Li, H. J., Zhang, L. J., Zhai, J., ... Zhu, D. B. (2002). Super-Hydrophobic Surfaces: From Natural to Artificial. *Advanced Materials, 14*(24), 1857–1860. doi:10.1002/adma.200290020

Feng, L., Zhang, Y., Li, M., Zheng, Y., Shen, W., & Jiang, L. (2010). The structural color of red rose petals and their duplicates. *Langmuir, 26*(18), 14885–14888. doi:10.1021/la102406u PMID:20731398

Feng, L., Zhang, Y., Xi, J., Zhu, Y., Wang, N., Xia, F., & Jiang, L. (2008, April 15). Petal effect: A superhydrophobic state with high adhesive force. *Langmuir, 24*(8), 4114–4119. doi:10.1021/la703821h PMID:18312016

Gao, X., & Jiang, L. (2004, November). Biophysics: Water-repellent legs of water striders. *Nature, 432*(7013), 36. doi:10.1038/432036a PMID:15525973

Gao, X., Yan, X., Yao, X., Xu, L., Zhang, K., Zhang, J., ... Jiang, L. (2007). The dry-style antifogging properties of mosquito compound eyes and artificial analogues prepared by soft lithography. *Advanced Materials, 19*(17), 2213–2217. doi:10.1002/adma.200601946

Gorb, S., Varenberg, M., Peressadko, A., & Tuma, J. (2007). Biomimetic mushroom-shaped fibrillar adhesive microstructure. *J. R. Soc. Interface, 4*, 271–275. doi:) doi:10.1098/rsif.2006.0164

Greiner, C., & Schäfer, M. (2015). Bio-inspired scale-like surface textures and their tribological properties. *Bioinspiration & Biomimetics, 10*(4), 044001. doi:10.1088/1748-3190/10/4/044001 PMID:26125522

Hanna, G., Jon, W., & Barnes, W. J. (1991, January 1). Adhesion and detachment of the toe pads of tree frogs. *The Journal of Experimental Biology, 155*(1), 103–125.

Hansen, W. R., & Autumn, K. (2005). Evidence for self-cleaning in gecko setae. *Proceedings of the National Academy of Sciences of the United States of America, 102*(2), 385–389. doi:10.1073/pnas.0408304102 PMID:15630086

Huang, W., Jiang, L., Zhou, C., & Wang, X. (2012, August 31). The lubricant retaining effect of micro-dimples on the sliding surface of PDMS. *Tribology International, 52*, 87–93. doi:10.1016/j.triboint.2012.03.003

Huang, W., & Wang, X. (2013, December 1). Biomimetic design of elastomer surface pattern for friction control under wet conditions. *Bioinspiration & Biomimetics, 8*(4), 046001. doi:10.1088/1748-3182/8/4/046001 PMID:23999795

Jiang, W., & Ajay, P. (2007). Biomimetic surface structuring of coating for tribological applications. *Surface and Coatings Technology*, *201*(18), 7889–7895. doi:10.1016/j.surfcoat.2007.03.031

Kim, S. (2008). Smooth vertical surface climbing with directional adhesion. *IEEE Transactions on Robotics*, *24*(1), 65–74. doi:10.1109/TRO.2007.909786

Liu, M. J., Wang, S. T., Wei, Z. X., Song, Y. L., & Jiang, L. (2009). Bioinspired Design of a Superoleophobic and Low Adhesive Water/Solid Interface. *Advanced Materials*, *21*(6), 665–669. doi:10.1002/adma.200801782

Liu, Y., Shao, Z., & Vollrath, F. (2005, December). Relationships between supercontraction and mechanical properties of spider silk. *Nature Materials*, *4*(12), 901–905. doi:10.1038/nmat1534 PMID:16299506

Lu, Y., Hua, M., & Liu, Z. (2014, July). The Biomimetic Shark Skin Optimization Design Method for Improving Lubrication Effect of Engineering Surface, ASME. *Journal of Tribology*, *136*(3), 0317031–3170313. doi:10.1115/1.4026972 PMID:25053867

MacQuitty, M. (2011). *Shark*. Dorling Kindersley Ltd.

Martina, J. (2014). Dry friction of microstructured polymer surfaces inspired by snake skin. *Beilstein Journal of Nanotechnology*, *5*, 1091–1103. doi:10.3762/bjnano.5.122

Marvi, H., Bridges, J., Meyers, G., Russell, G., & Hu, D. L. (2011). Scalybot: A snake-inspired robot with active control of friction. Proc. of ASME Dynamic Systems and Control Conf., 443–450. doi:10.1115/DSCC2011-6174

Mayer, G. (2005, November 18). Rigid biological systems as models for synthetic composites. *Science*, *310*(5751), 1144–1147. doi:10.1126cience.1116994 PMID:16293751

Mühlberger, Rohn, Danzberger, Sonntag, Rank, Schumm, … Bergmair. (n.d.). UV-NIL fabricated bio-inspired inlays for injection molding to influence the friction behavior of ceramic surfaces. *Microelectronic Engineering, 141*, 140-144. Doi:10.1016/j.mee.2015.02.051

Murarash, B., Itovich, Y., & Varenberg, M. (2011). Yan Itovicha, Michael Varenberg Tuning elastomer friction by hexagonal surface patterning. *Soft Matter*, *7*(12), 5553–5557. doi:10.1039/c1sm00015b

Palmer, L. R., Diller, E. D., & Quinn, R. D. (2010). Toward a rapid and robust attachment strategy for vertical climbing. *Proc. of the IEEE Int. Conf. on Robotics and Automation*, 2810–2815. 10.1109/ROBOT.2010.5509680

Pérez-Rigueiro, J., Elices, M., & Guinea, G. V. (2003, June 1). Controlled supercontraction tailors the tensile behaviour of spider silk. *Polymer*, *44*(13), 3733–3736. doi:10.1016/S0032-3861(03)00245-3

Pfeifer, R., Iida, F., & Gomez, G. (2006). Morphological computation for adaptive behavior and cognition. *International Congress Series*, *1291*, 22–29. doi:10.1016/j.ics.2005.12.080

Pfeifer, R., Lungarella, M., & Iida, F. (2007). Self-organization, embodiment, and biologically inspired robotics. *Science*, *318*(5853), 1088–1093. doi:10.1126cience.1145803 PMID:18006736

Potyrailo, R. A., Ghiradella, H., Vertiatchikh, A., Dovidenko, K., Cournoyer, J. R., & Olson, E. (2007). Morpho butterfly wing scales demonstrate highly selective vapour response. *Nature Photonics*, *1*(2), 123–128. doi:10.1038/nphoton.2007.2

Sato, O., Kubo, S., & Gu, Z. Z. S. (2008). structural color films with lotus effects, superhydrophilicity, and tunable stop-bands. *Accounts of Chemical Research*, *42*(1), 1–0. doi:10.1021/ar700197v PMID:18837520

Schneider, J., Djamiykov, V., & Greiner, C. (2018). Friction reduction through biologically inspired scale-like laser surface textures. *Beilstein Journal of Nanotechnology*, *9*(1), 2561–2572. doi:10.3762/bjnano.9.238 PMID:30425902

Scholz, I., Barnes, W. J., Smith, J. M., & Baumgartner, W. (2009). Ultrastructure and physical properties of an adhesive surface, the toe pad epithelium of the tree frog, Litoriacaerulea White. *The Journal of Experimental Biology*, *212*(2), 155–162. doi:10.1242/jeb.019232 PMID:19112133

Shafiei, M., & Alpas, A. T. (2009). Nanocrystalline nickel films with lotus leaf texture for superhydrophobic and low friction surfaces App. *Surface Science*, *256*(3), 710–719. doi:10.1016/j.apsusc.2009.08.047

Shao, Z., & Vollrath, F. (1999). The effect of solvents on the contraction and mechanical properties of spider silk. *Polymer*, *40*(7), 1799–1806. doi:10.1016/S0032-3861(98)00266-3

Shao, Z., Vollrath, F., Sirichaisit, J., & Young, R. J. (1999). Analysis of spider silk in native and super contracted states using Raman spectroscopy. *Polymer, 40*(10), 2493–2500. doi:10.1016/S0032-3861(98)00475-3

Singh, R. A., & Yoon, E.-S. (2009). Biomimetics: The science of imitating nature. *Tribology & Lubrication Technology, 65*(2), 40–47.

Spenko, M., Haynes, G. C., Saunders, J. A., Cutkosky, M. R., Rizzi, A. A., Full, R. J., & Koditschek, D. E. (2008). Biologically inspired climbing with a hexapedal robot. *Journal of Field Robotics, 25*(4-5), 223–242. doi:10.1002/rob.20238

Stegmaier, T., Linke, M., & Planck, H. (2009). Bionics in textiles: flexible and translucent thermal insulations for solar thermal applications. *Philosophical Transactions of the Royal Society A: Mathematical, Physical and Engineering Sciences, 367*(1894), 1749-58.

Sun, T., Feng, L., Gao, X., & Jiang, L. (2005). Bioinspired surfaces with special wettability. *Accounts of Chemical Research, 38*(8), 644–652. doi:10.1021/ar040224c PMID:16104687

Sundar, V. C., Yablon, A. D., Grazul, J. L., Ilan, M., & Aizenberg, J. (2003). Fibre-optical features of a glass sponge. *Nature, 424*(6951), 899–900. doi:10.1038/424899a PMID:12931176

Swanson, B. O., Blackledge, T. A., & Hayashi, C. Y. (2007). Spider capture silk: Performance implications of variation in an exceptional biomaterial. *Journal of Experimental Zoology. Part A, Ecological Genetics and Physiology, 307*(11), 654–666. doi:10.1002/jez.420 PMID:17853401

Tan, T. L., Wong, D., & Lee, P. (2004). Iridescence of a shell of mollusk Haliotis Glabra. *Optics Express, 12*(20), 4847–4854. doi:10.1364/OPEX.12.004847 PMID:19484038

Voigt, D., Karguth, A., & Gorb, S. (2012). Shoe soles for the gripping robot: Searching for polymer-based materials maximizing friction. *Robotics and Autonomous Systems, 60*(8), 1046–1055. doi:10.1016/j.robot.2012.05.012

Vollrath, F., & Knight, D. P. (2001). Liquid crystal silk spinning in nature. *Nature, 410*(6828), 541–548. doi:10.1038/35069000 PMID:11279484

Vukusic, P., & Hooper, I. (2005). Directionally controlled fluorescence emission in butterflies. *Science, 310*(5751), 1151. doi:10.1126cience.1116612 PMID:16293753

Vukusic, P., & Sambles, J. R. (2003). Photonic structures in biology. *Nature*, *424*(6950), 852–855. doi:10.1038/nature01941 PMID:12917700

Work, R. W. (1985). Viscoelastic behaviour and wet supercontraction of major ampullate silk fibres of certain orb-web-building spiders (Araneae). *The Journal of Experimental Biology*, *118*(1), 379–404.

Yoon, E. S., Singh, R. A., Kong, H., Kim, B., & Kim, D. H. (2006). Article. *Tribol. Lett.*, *21*, 31–37.

Yoon, E. S., Singh, R. A., Oh, H. J., & Kong, H. (2005). The effect of contact area on nano/micro-scale friction. *Wear*, *259*(7-12), 1424–1431. doi:10.1016/j.wear.2005.01.033

Yoon, E. S., Yang, S. H., Kong, H., & Kim, K. H. (2003). Article. *Tribology Letters*, *15*(2), 145–154. doi:10.1023/A:1024409316644

Zhang, B., Huang, W., Wang, J., & Wang, X. (2013). Comparison of the effects of surface texture on the surfaces of steel and UHMWPE. *Tribology International*, *65*, 138–145. doi:10.1016/j.triboint.2013.01.004

Zhang, B., Huang, W., & Wang, X. (2012). Biomimetic surface design for ultrahigh molecular weight polyethylene to improve the tribological properties. *Proceedings of the Institution of Mechanical Engineers. Part J, Journal of Engineering Tribology*, *226*(8), 705–713. doi:10.1177/1350650112437829

Zheng, Y., Bai, H., Huang, Z., Tian, X., Nie, F. Q., Zhao, Y., ... Jiang, L. (2010). Directional water collection on wetted spider silk. *Nature*, *463*(7281), 640–643. doi:10.1038/nature08729 PMID:20130646

Zheng, Y., Gao, X., & Jiang, L. (2007). Directional adhesion of superhydrophobic butterfly wings. *Soft Matter*, *3*(2), 178–182. doi:10.1039/B612667G

Zi, J., Yu, X., Li, Y., Hu, X., Xu, C., Wang, X., ... Fu, R. (2003). Coloration strategies in peacock feathers. *Proceedings of the National Academy of Sciences of the United States of America*, *100*(22), 12576–12578. doi:10.1073/pnas.2133313100 PMID:14557541

Chapter 9
Potential of Bio-Inspiration in 3- and 4-D Printing

ABSTRACT

This chapter explores the potential of bio-inspiration in 3- and 4-D printing. The authors argue that the true potential of texturing hasn't been realized yet not because of the lack of enabling texturing technologies but because of the severe lack of detailed information about the functional details of texturing in a tribological situation, that is, how surface features, their geometry, interact with the functional gradients present within the subsurface layers to control the friction profile of a structure. The material emphasizes the potential of bio-inspired surfaces in providing a pathway for realizing true synchronization of function through a layer-by-layer customization of surface and subsurface material. In particular the chapter discusses methodologies to extract design parameters that lead to manifesting 4-D printed tribological constructs where surface and sub-surfaces respond optimally to external stimulants represented by the operation conditions of load, speed, and ambient temperature. Successful design of functional deterministic surfaces is not a product of mere biomimicry. Rather, it culminates probing the geometry, texture, form, and construction of the bio-analogue and linking these ingredients to the desired functional profile of the surface in the human engineering domain, that is, generation of bio-inspired functional surface designs stems from implementing design rules rather than replication of natural constructions. Deduction of design rules requires decoding the metrological features and the analysis of surface performance, of bio-analogues using standardized engineering

DOI: 10.4018/978-1-7998-1647-8.ch009

methods. Success in designing a bio-inspired surface, therefore, requires a trans-disciplinary approach that combines engineering, physics, and biology. These don't combine naturally since they entail different methodologies of problem solving and investigations. It is hoped that this book would bridge the gap between the disciplines in the context of biomimetic surface design and construction. Further, it is hoped that the material would equip the reader with the basic skills needed to navigate between the biological and the engineering domains.

INTRODUCTION

Tribological surfaces are multi-functional entities that exchange, channel, and respond to external operation parameters of all sorts. They also respond to internal changes of system structure as triggered by any change in environmental operational conditions (i.e. any change in load, speed, ambient temperature, or chemical components). Surfaces also play an important role in protecting the layers of system components from detrimental thermal, chemical, and mechanical effects. The multi-aspect role of surfaces is, therefore, critical in maintaining and advancing the structural, as well as, the functional health of a tribological system.

The role of the surface in preserving a rubbing system is further advanced through energy exchange between the contacts interface and the mechanically affected layers of the rubbing components. Efficiency of that exchange has been recognized lately as a direct, and crucial, contributor to preserving the structural integrity of rubbing components (Kuhlmann-Wilsdorf, 1987, Barber, 1970, Blok, 1970, Ling, 1969, Archard and Rowntree 1988). Further efficiency of energy exchange is deemed essential to reducing the energy consumed in combating friction-induced energy losses and efficiency of motion in general (Ye and Komvopoulos, 2003)., Lai, and Cheng, 1985, Tian and Kennedy, 1993, Deolalikar, et al., 2008, Lal and Mathur, 2007). As a result, modern design paradigms for tribo-systems emphasize customization of textural features of rubbing surfaces. Customization, in general, entails detailed consideration of shape and pattern of individual surface topography components. It also comprises integration of sensory elements within the structure layer of the subsurface to signal the state of the structure. Modern view of surfaces, therefore, have evolved from a "boundary" view to an "integral component" view. In this sense, modern surfaces are in effect multi-tasking-multi-objective functional engineering constructs.

In the past two decades, and due to the push for a micro-nanoscale engineering, many novel technologies have come to exist. Engineering of surface texturing on the micro and nanoscale was no exception. Many of the emerging enabling technologies allow precise control of surface features on minute scales. In the meantime, complexity of the functional demands of surfaces have also evolved. Modern, and futuristic, functional requirements of surfaces are not only complex with respect to miniaturization, but also have an added layer of complexity because a demand for synchronized multi-functions.

Surface engineering utilizes appropriate technologies to meet design constraints at minimized cost. The field is evolving as ideas, technologies, and expertise attempt to bridge the needs of stakeholders, skills of surface technologists, and visions of researchers. Bridging that gap needs forecasting future functional needs and addressing the technologies necessary to meet them. Several futuristic visions, based on surveying of the constituents of the field, and brain storming by researchers, identified pressing priorities as: self-cleaning, sensory, tribological performance, and advanced scratch resistance. A fundamental constraint on these functional demands, however, is their co-existence within a surface. Achieving this multi-functional requirement requires new insights into the design and engineering aspects of modern surface constructs. Crucial, within such a context, is to formulate surface design paradigms that facilitate management of complexity and are amenable to implementation using current emerging enabling technologies.

Demand for synchronization of function renders modern surfaces, in effect, a collection of subsystems. Each of these subsystems represents a layer of complexity that targets the fulfillment of a single functional object. Efficient management of the complex relationship that results is fundamental to effective performance of the surface and its dependent components. The goal, therefore, is to ensure that all ingredients embedded to enable fullfilment of the various demands would, effectively, function as a single unit. This, in turn, requires efficient management of complexity both at the inception and the implementation phases of the surface. Efficient management transforms the multi-systems, comprising the construct, into a homogeneous unified entity. Fulfiling this requirement depends on streamlining the different functional layers (or subsystems) so that synchronization is achieved.

In a modern surface, functional layers may be identified in relation to the performance goals of the system. In this case, the compositional components added (e.g., embedded micro-nano-sensors), or the geometrical textural patterns developed to meet a required frictional profile (e.g., hydrophobic textural element), represent a functional layer (or subsystem). Alternatively, functional layers may evolve as a result of the mechanical and structural interactions within the system (e.g., due to the different stress and deformation levels that affect the microstructure of materials).

Physically, a complying sliding solid has two main layers. The first is the surface layer which contains the topographical and roughness features. This is where the actual physical contact with the counter surface takes place. The second layer, meanwhile, is the so-called Mechanically Affected Zone (MAZ). From a mechanical function point of view a sliding solid may be dissected in two principal functional layers. The first is the interface with the contact is established and initial mechanical loading takes place. Whereas, the second is the MAZ where the subsurface contact stresses (both static and dynamic) are accommodated. Each of these functional layers may be further dissected into sub-functional layers. If, however, we consider thermal energy transport function while sliding, then we would identify two related main functional layers. The first, again is the surface layer where the thermal transport is initiated and thermal contact resistance takes place. The second would be the Heat Affected Zone (HAZ) which accommodates friction-induced thermal energy dissipation into the bulk of either contacting solids. In each of the previous examples the roughness layer has a mechanical function that intersects with the thermal function. The same applies for the MAZ and the HAZ layers. It follows that to optimize function (i.e. to minimize friction resistance and minimize wear) there is a need to synchronize the performance of the various functional layers. This synchronization should take place with respect to the interference between their functions and importance of that function to the particular friction profile of the system. Now consider the design of a modern (or futuristic) tribological system, where minimized friction and optimized wear are the fundamental, albeit not the only, goal. That is, the system has to accommodate the multi-requirements of tribological performance. In this case, the number of functional layers, and the complexity of their communication is several orders of magnitude higher than that encountered within a traditional tribo-system (where only optimized friction and damage are the goal). Within this context, one can identify as many functional subsystems as there are functional goals. Optimizing the interaction between the surface (textural or roughness) features and subsurface structure is not only crucial but demanding. Meeting such a complex requirement mandates sophisticated customization of both size and topological pattern of the surface layer. It also mandates customization of the properties and "morphing-stimulants", if any, of subsurface layers to meet design goals. Generally speaking, two generic gaols have to be met. The first is accommodation of the functional requirements of the very contacting layer. Whereas the second is customization of the properties of the subsurface layers to ensure integration of their function with the general function of the tribo-system. Constructing the surface and subsurface layers as an integral entity stands at the core of meeting these goals. However, efficient and practical implementation of surfaces of this nature has not been successful on an industrial scale, so far. Despite emergence of powerful manufacturing technologies, the major obstacle, still remains, the development of holistic design paradigms capable of tailoring surface-subsurface solutions on a commercial scale. These paradigms while currently lacking may benefit from biology. In fact, it can be shown that turning to biology may facilitate the development process and can inspire non-conventional solutions capable of meeting the requirements of modern surface functionality.

In this chapter we discuss the potential of the biological world for providing guidance to an integrated tribological solution. The discussion starts with introducing 3-and 4-D printing, their methods, and concepts. Following this brief introduction, we discuss what aspects of biology should be studied to extract printing paradigms. This is achieved through detailed analysis of the structure and function of selective legless reptile species. The discussion is facilitated through the considering surface-subsurface customization in legless squamate reptiles and how it is closely related to optimizing the function of the reptile. To this end, we detail the functional aspects of surface substructure, and how material property gradients interact with surface topographical features to maintain structural integrity of the skin and reduce friction resistance.

MULTIDIMENSIONAL (3-4-D) PRINTING

Additive Manufacturing, AM, is currently transforming production methods. Many view AM as a manifestation of the final industrial revolution, i.e., one that will forever transform the global manufacturing landscape and revolutionize all related aspects of supply chain dynamics and economics. Despite its relatively short history, in its current digitally dominated state, the technology has penetrated through almost every field of production across physical and economic scales. Additive manufacturing techniques are currently applied in various fields from toy making, to weaponry and high energy yield materials. As a result, as a technology, it is placed among an elite niche of disruptive technologies that potentially may sweep away the systems or habits it replaces because of attributes that are recognizably superior.

AM is capable of realizing objects of complex shape which are deemed unobtainable if produced by traditional techniques. These complex objects when produced by the methods of the new technology would materialize in less time, under more controlled conditions, with more tight tolerances, all while consuming less raw materials. The technology also enables fabrication of customized devices, or individual (instead of identical mass-manufactured) units. All these benefits, and a few more, take place, at least in theory, because the innovative path of building the object from design stage to acquiring final shape. Multi-dimensional printing, for example, is a flexible process. This flexibility allows producing multiple designs on the same machine (i.e., no need for change of dies or molds). This enables the manufacturer to move from mass production mode to mass customization mode with distributed manufacturing.

The main premise of AM is to enable the making of a volumetric object by adding material in a calculated and controlled manner. Addition of material, whence the designation of the technology as additive, takes place building block-by-building block. This could be a layer-by-layer, or volume-by-volume, where the thickness of the layers, rates and directions of the added material, are precisely calculated and optimized. This building scheme allows manifestation of the target object in fewer steps, at a higher speed while optimizing the materials used. Thus, compared to traditional manufacturing, Additive Manufacturing is precise, faster, and economical.

An additional fundamental benefit of Additive Manufacturing is flexibility towards raw materials. Contrary to traditional production methods, where every class of materials is handled through specialized techniques, AM, is material blind. In printing, for example, and although Plastics and polymers are very popular ink materials, other materials are currently being used. These include, ceramics, metals, living tissue, nanomaterials, and even high-energy yield materials. Although different printing methods may be applied to allow printing the variety of available stock, all methods share the same conceptual processes. Differences take place within the deposition methods, conditions, and rates of application or adding of the different material layers and technology of fusing the layers to obtain the target object. Flexibility through using a wide range of materials enables additive manufacturing to create intricate products of a near-infinite variety that are tailored to exact client specifications.

The expression Additive Manufacturing is an umbrella expression for a suite of enabling technologies. At the core of this suite is the powerful technology of multi-dimensional printing. Currently there are two interrelated printing technologies. The first is the, now familiar, 3-D printing, whereas the second is the emerging related technology of 4-D printing.

Three-dimensional (3-D) printing is a fabrication technique where manufacturing takes place through deposition of a material using a print head, nozzle, or another printer technology. This allows a *layer-by-layer* building of an object. The technology can be applied to virtually most materials polymeric or metallic. In metals printing is proving to be very beneficial and effective.

In contrast to conventional manufacturing techniques, metal-AM is a novel material processing approach for creating prototypes or parts based on incremental layer-by-layer. This layer-by-layer technique allows an unprecedented freedom in manufacturing complex structures with precision and control that is not possible through conventional manufacturing routes.

There are many economical, industrial, and environmental advantages of Metallic printing, and printing in general, over conventional manufacturing processes (some of such advantages are illustrated in figure 1).

4-D Printing

The technology of 4-D printing builds on the foundation of the now familiar 3-D printing. The 4th dimension conceptually stands for enabling objects to morph in space along a pre-conceived path in functional space. Complimentary implementation of the fourth dimension entails pre-control of property gradients in an object to ensure some behavior of interest. Control of properties, therefore, is the enabling tool that sets the path for the evolution of the objects and allow for their morphing under a triggering condition (temperature-stress-pressure, etc.,). Initially 4-D printing was defined in terms of time evolution of 3-D printed structures. Accordingly, the shape, property, or functionality of a printed object can change as a function of time pending that an appropriate material is used. In metals, this definition is extended to reflect the possibility of customizing the properties of each layer within the build to obtain a deliberate functional gradient of properties. Thus, because printing is a layer-by-layer process, it is possible, at least in principle to control all aspects of each layer so that a gradient in properties may take place. This definition, therefore, promotes 4-D printing as a feasible means to achieve self-assembly, multi-functionality, and self-repair regardless of the printed material. It also accommodates the possibility of deliberate design of a given thickness, or volume, of an object to reflect a predetermined behavior (which represents the core of surface-subsurface customization).

Realization of the technological potential of 4D printing stands on two foundations: the so-called smart materials, and availability of appropriate stimulus to trigger morphing of shape or customized functionality. Thus, it is the deliberate application of a suitable material-stimulus combination that leads to successful implementation of 4D the printing of an object.

Figure 1. Summary advantages of additive manufacturing technology and its impact on modern manufacturing.

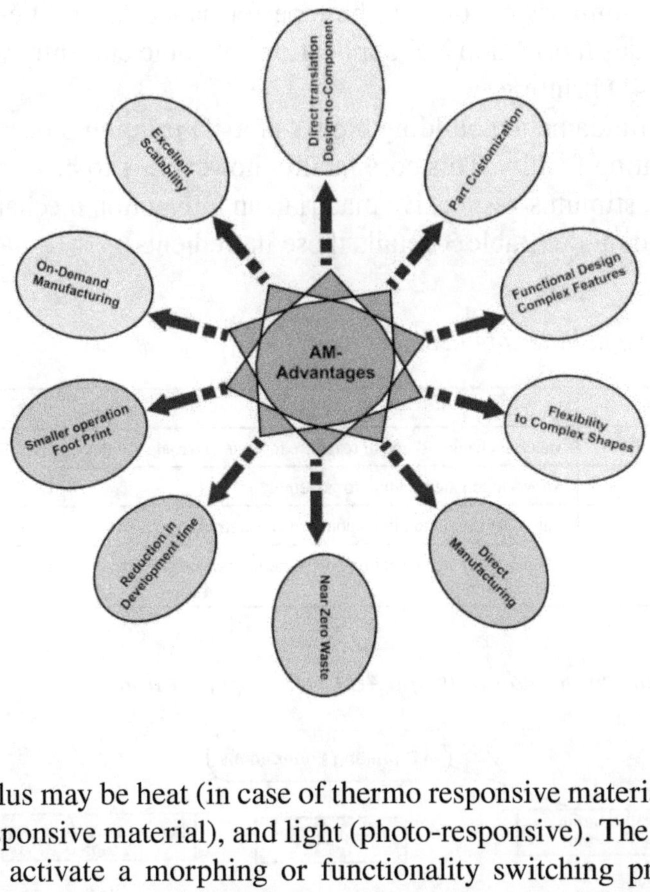

A stimulus may be heat (in case of thermo responsive material), solution (chemo-responsive material), and light (photo-responsive). The stimuli will apply may activate a morphing or functionality switching process. This switching process may be reversible or non-reversible depending on the specific object to be printed and its envisioned operational needs. Applying the right stimulus can induce the ability of a printed object to switch its geometric configuration for example (e.g. including surface morphology) from one configuration to another in a fully controlled manner. Pairing the material to a suitable stimulus requires in-depth knowledge of response patterns of materials at all conditions. Thus successful 4 D printing depends on finding a printable material that exhibits the desired response when stimulated by a predesigned condition. That is, finding material that will exhibit and intrinsic response under the prevailing condition within the intended application.

The intrinsic characteristic (response), then, can be exploited to the desired performance of an object or a layer within the object. Many intrinsic responses may be envisioned. These include: self-healing, self-sensing, self-actuating, and shape or property changing. In this case the materials intrinsic responses provide an active intelligent element of performance in the product. Table 1 provides a summary of some of these performance capabilities. Whereas, table 2 provides fabrication and applications of some currently widely used materials in 4D printing.

The core fundamental building blocks of a 4D printing process, figure 2, is a 3-D printing facility. This core facility however, is to be complimented by: stimulus, stimulus-responsive material, an interaction mechanism, and a mathematical model (table 3 details these ingredients).

Table 1. The capabilities of intelligent materials

Function	Description	Status
Shape memory	Material changes shape in response to some external stimulus	Established
Self-assembly	Allows automated folding for assembly	Established
Self-actuating	Automated actuation in response to an external stimulus	Research stage
Self-sensing	Allows automated detection and sometimes quantification of external stimuli	Established, but is still the focus of research

Figure 2. Technical foundations of a 4-D printing operation

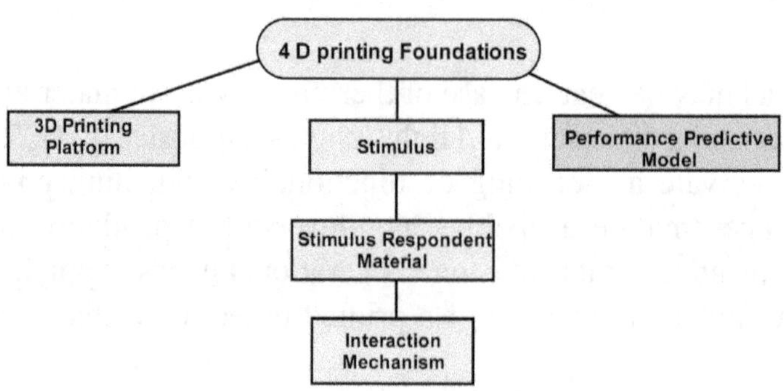

Table 2. Fabrication methods of intelligent materials and their field applications

Intelligent Materials	Fabrication Methods	Applications
Shape memory materials		
Thermo-responsive	Bonded together by the organic polymer point by point	Aerospace space-deployable structures (Fulcher, et al., 2010), Biomedicine stimuli-sensitive implant materials (Lendlein, and Kelch, 2005), textiles - smart clothing (Mondal and Hu 2006), and automobiles - fastening system
Athermally induced	Photo-triggered – temporary shape is fixed by covalently crosslinking cinnamic groups under UV light of $\lambda > 260$ nm and the shape recovery is triggered by exposure to UV-light irradiations of $\lambda < 260$ nm and is water sensitive When the polymer system is exposed to water, deformation of the polymer composite into a temporary shape takes place. The shape fixing is done by subsequent drying	Biomedicine, endovascular thrombectomy device (Wu, et al., 2010)
Self-assembly, self-actuating and self-sensing material	Allows automated folding for assembly	Biological applications, Linear and rotary actuators, Touch screens, Underwater robots (Won, et al., 2016), and biological imaging
Electroactive Polymer		
IPMC	The Nafion solution and alcohol-water solution is mixed as precursor materials, and mixed liquid of Ag fine particles and Nafion solution is used as electrode materials of IPMC	Three-layered and five-layered IPMC structures
Bucky Gel	The electrode-base materials-electrode structure is solidified point by point accumulatively	A hand shape Bucky Gel
DE	Polyacrylic acid material is cured by UV light, then the surface of polyacrylic acid matrix material is coated with flexible electrode material	In the initial stage of development

FUNCTIONAL ADAPTATION AND PROPERTY GRADIENTS IN SNAKES

Relationship between scale counts and the allometric variables of length and mass are closely correlated in snakes. The number of scales on a given snake species has a unique range. This range is one trait used to identify individuals. While the range of scale counts remains constant for the species, variation among sexes of the same family do exist. Differences in scale counts also exist between same sexes of the same species due to change of habitat (across continents or due to variation in local food supply). An important variation that takes place in scales is change in size to accommodate growth. Figure 3 explores the general trend that emerges upon comparing variation in ventral scale parameters (number and area) to length and mass. Figure 3-a, plots the snout to vent length SVL versus the median ventral scale counts of approximately seven hundred species. No special statistical treatment other than averaging is performed to account for differences in families or genera. The plot implies that, in general, the number of ventral scales directly varies in proportion to the SVL of the individual. The plot includes members of the Pythonide family, which comprise some of the heaviest and longest snakes. From a structural point of view the data shows that the number of ventrals offsets the adverse effect of length on facility of locomotion. Increased length implies increased mass as well. The heavier the snake there is a need to divide its mass on a higher number of sections to reduce the resistance to motion. Sequential distribution of mass helps in that regards. The same trend emerges when plotting the masses against the number of scales.

Table 3. The core fundamental building blocks of a 4D printing process

Building Block	Definition
3D printing facility	Usually, a 4D printed structure is created by combining several materials in the appropriate distribution into a single, one-time printed structure (Raviv et al. 2014). The differences in material properties, such as swelling ratio and thermal expansion coefficient, will lead to the desired shape-shifting behavior. Therefore, 3D printing is necessary for the fabrication of multi-material structures with simple geometry.
Stimulus	Stimulus is required to trigger the alterations of shape/ property/functionality of a 4D printed structure. The stimuli that researchers have used in 4D printing thus far include water (Tibbits, 2014, Tibbits, et al., 2014, Gladman, et al., 2016, Ge 2013, Pei, 2014), heat (Ge, et al., 2014, Zhang, et al., 2016, Kuksenok et al., 2016, Bakarich, et al.,2015, Roy, et al., 2010, Stuart, et al., 2010), a combination of heat and light (Sun, et al., 2012), and a combination of water and heat (Ge, et al., 2014, Meng, et al.,2013). The selection of the stimulus depends on the requirements of the specific application, which also determines the types of smart materials employed in the 4D printed structure.
Smart or stimulus-responsive material	stimulus-responsive material is one of the most critical components of 4D printing. Stimulus-responsive materials can be classified into several sub-categories, as shown in Fig. 4. The capability of this group of materials is defined by the following characteristics: self-sensing, decision making, responsiveness, shape memory, self-adaptability, multi functionality (Young, 2016), and self-repair. Several review studies on stimulus-responsive materials have been provided by Roy et al. (Lindel et al., 1993), Stuart et al. (Arnold and Bennett, 1988), Sun et al. (2012), and Meng et al. (2013).
Interaction mechanism	In some cases, the desired shape of a 4D printed structure is not directly achieved by simply exposing the smart materials to the stimulus. The stimulus needs to be applied in a certain sequence under an appropriate amount of time, which is referred to as the interaction mechanism in this review paper. For example, one of the main interaction mechanisms is constrained-thermo-mechanics. In this mechanism, the stimulus is heat and the smart material has the shape memory effect. It contains a 4-step cycle. First, the structure is deformed by an external load at a high temperature; second, the temperature is lowered while the external load is maintained; third, the structure is unloaded at the low temperature and the desired shape is achieved; fourth, the original shape can be recovered by reheating the structure.
Mathematical Modeling	Math is necessary for 4D printing in order to design the material distribution and structure needed to achieve the desired change in shape, property, or functionality. Theoretical and numerical models need to be developed to establish the connections between four core elements: material structure, desired final shape, material properties, and stimulus properties. These will be discussed in additional detail in the following sections.

Figure 3. Variation in ventral scale parameters (scale counts, area, and loading) against the allometric parameters mass and length.

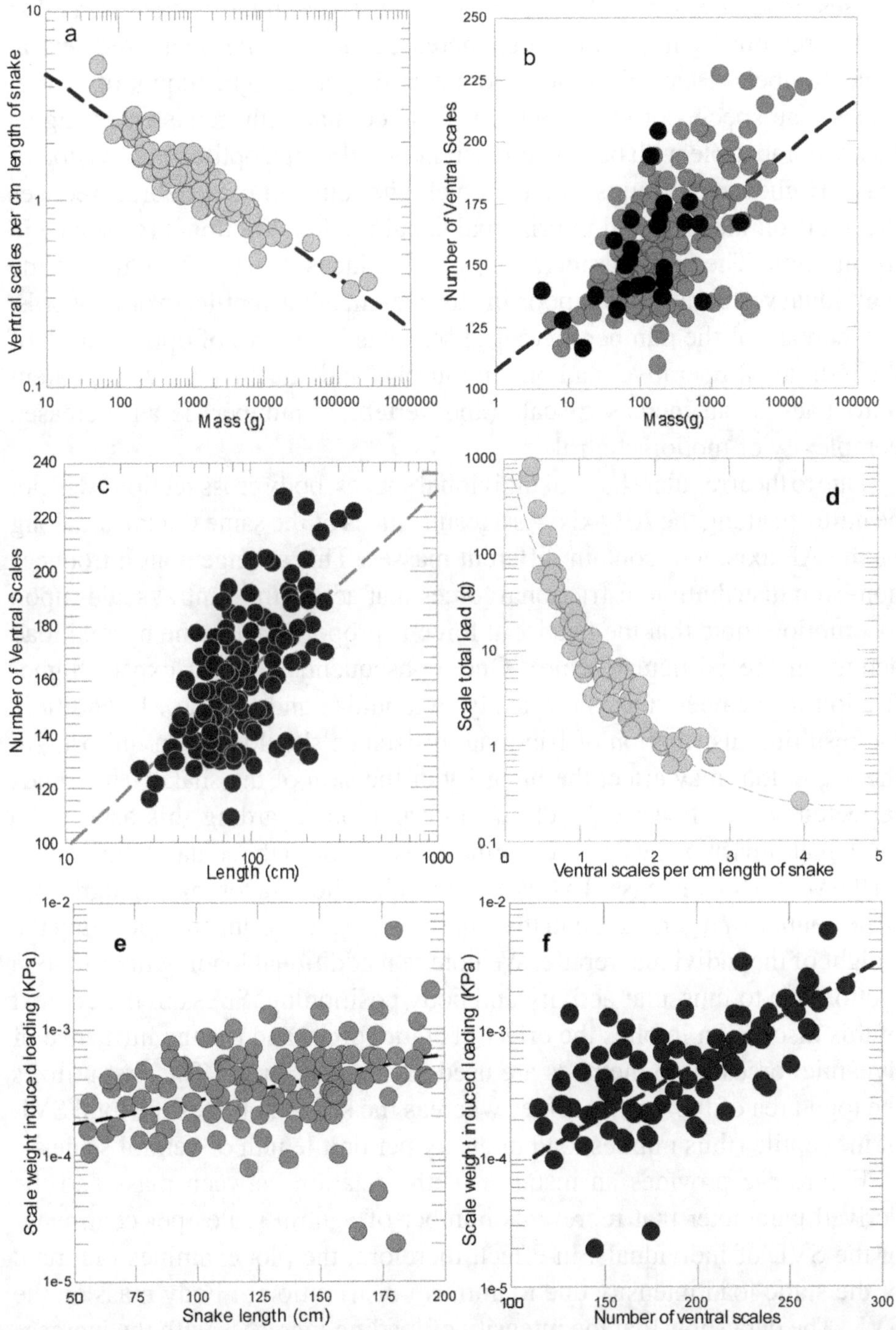

This is shown in figure 3-b, which examines the relationship between mass and number of ventrals. The black circles shown in the figure are actual recorded masses of some species. Again, the trend reflected in the plot supports that of the previous figure. Namely that increased mass implies increased ventral scale numbers. Heavier species, moreover, display a higher count of ventral scales than species of lesser mass. It is noted that both trends emerging so far are compatible with optimized locomotion through optimized division of mass. Higher scale counts, and thereby higher number of vertebrae, reduces the loads on muscles and in principal should reduce frictional resistance in locomotion. The high number, moreover, reduces the static loading on the individual vertebra due to supporting the weight of the reptile. In other words, the increase of the number of ventral scales is indicative of optimization of the individual operative load on the muscles and attachments. An apparent draw back of an increased scale (and vertebrate number) is an increased complexity of motion control.

Due to the irregular shape of individual snakes, body cross sections will not be uniform along the AP-axis. As a result, slices of the same thickness, along on the AP axis, will contain different masses. This arrangement introduces non-even distribution of frictional forces that act on the ventral scales upon locomotion (note that the frictional force is proportional to the normal load acting on the particular scale). One, consequently, should expect higher friction for heavier snakes (typically Pythonidae and Boidae). In addition, the resulting distribution of frictional resistance should be non-uniform, to the extent that may affect the integrity of the skin of the snake. The trends reflected so far, however, yield no information regarding this aspect. To investigate this hypothesis, we examine the variation of the actual static loading on the ventrals with mass. To this end, we mine the data for mass, length, and scale counts for a general trend that characterizes stress due to supporting the weight of the individual reptile. We note that additional loading may arise in motion due to muscular activity and body positioning. Stress in the context of this discussion implies the effect of static forces and not the influence of dynamic forces. Two methods are used in devising stress. The first utilizes the total area of the ventral scales whereas the second utilizes the total SVL of the reptile (thus implies load intensity per unit length of ventral scales).

Figure 3-c provides an instance of the relation between mass and the derived parameter that represents number of ventral scales per centimeter of the SVL of individuals. In effect, therefore, the plot examines the trend of the static load intensity due to uniform distribution of body mass on the SVL. The data show that the intensity of loading increases with the increase of body mass. A similar trend is reflected in figure 3-d which depicts the variation in the linear count of ventral scales with static loading. The plot shows that scales of shorter scale length (along the AP-axis) undergo lower load intensity due to supporting the weight of the snake.

An additional indicator of the sense of ventral scale number comes from examining weight induced loading intensity. The data are plotted in figure 3-e. The stress data, however, were computed from actual mass data (not model predictions). Weight induced loading intensity resulted from dividing the weight of the individual species by the total area of the ventral scales (again the area resulted from actual measurements). The plot reflects an increase in loading as a result of increasing the SVL length of the reptile a trend which is supported by the general attitude of the data plotted in figure 3-f.

The overall picture emerging from the preceding plots imply a *controlled* distribution of ventral scale counts among the various species. The data show that the number of ventral scales is a strong function of the allometric traits (length and mass) rather than the taxonomic classification. It was shown that variation in number takes place among sexes and among clades. This variation, however, is not as strong as that displayed relative to the overall variation in mass and length. From a structural point of view, the data reflect a universal trend of division of mass among a higher number of linkages as a mean of alleviating the load intensity acting on individual vertebrae.

The correlation between number of vertebrae and efficiency of locomotion is not surprising. Locomotion depends, in part, on the geometry and extension range of vertebrae (refer to chapter--). The ventral scales, however, connect to the ribs of the snake by the costo-cutaneous musculature (Lindel et al., 1993). The vertebrae, ribs, muscles and ventral scales are all part of a functional complex involved in locomotion (Arnold and Bennett). Among snakes, the numbers of ventral scales to the number of vertebrae vary in a one-to one ratio (Kerfoot, 1970) with the exception of some primitive snakes. It follows that the number of ventral scales in a species influences locomotion and may determine whether movement of the reptile is optimized. The plots also implied that the division of weight among linkages has consequences with respect to friction forces. It seems that the number of ventral scales plays a role in controlling the thresholds of friction resistance to motion and the actual frictional forces supported by individual scales. Limiting the frictional forces controls the shear stress acting on the scales and thereby contributes to preserving the integrity of the skin. Fine-tuning of the friction force however is also a function of the microstructure of the individual scales (Alibardi, 2002). The *tuning* takes place through limiting the actual area of contact between the reptile and the substratum whence reducing or increasing the frictional resistance to motion. This highlights the importance of studying the microstructure of the scales with special emphasis on the structural micro-composition of the skin of the ventral scales.

Skin Structure

The skin of a snake is mainly composed of keratins. Two types of keratins are present within the skin: *α-keratin* and *β-keratin* figure 4. The first type, α-keratin is a polypeptide chain that forms a right-handed α-helix (Alibardi, 2002, 2007). Two of these polypeptide chains twist together to form a left-handed helical structure known as a coiled coil. These coiled coil dimers, approximately 45 nm long, are bonded together with disulfide bonds, utilizing the many cysteine amino acids found in α-keratins (Alibardi, 1998). The dimers then align, their termini bonding with the termini of other dimers, and two of these new chains bond length-wise, all through disulfide bonds, to form a protofilament. (Alibardi and Thompson 2002)

Figure 4. Generalized structure of α- and β- keratin filaments

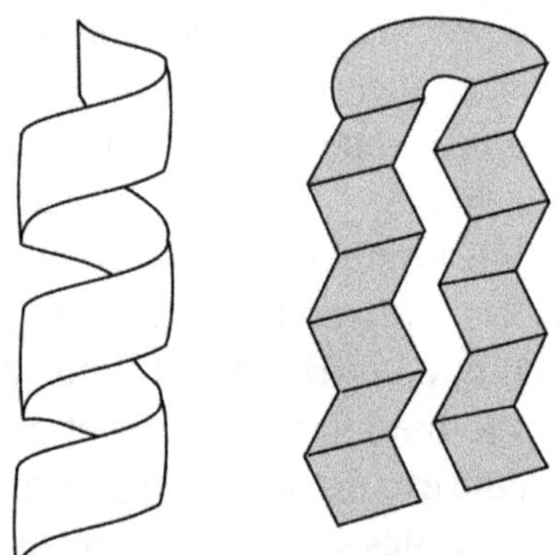

Two protofilaments aggregate to form a protofibril, and four protofibrils polymerize to form the intermediate filament (IF). The IF is the basic subunit of α-keratins. These IFs are able to condense into a super-coil formation of about 7 nm in diameter, and can be type I, acidic, or type II, basic. The IFs are finally embedded in a keratin matrix that either is high in cysteine or glycine, tyrosine and phenylalanine residues. The different types, alignments, and matrices of these IFs account for the large variation in α-keratin structures found in mammals (Fraser and Macrae 1973). Alpha-keratin is structuraly stable. When exposed to mechanical stress, α-keratin structures can retain their shape and therefore can protect what they surround.(Klein, et al., 2010) Under high tension, alpha-keratin can even change into beta-keratin, a stronger keratin formation that has a secondary structure of beta-pleated sheets.(Klein and Gorb 2012) Alpha-keratin tissues also show signs of viscoelasticity, allowing them to both be able to stretch and absorb impact to a degree, though they are not impervious to fracture. Alpha-keratin strength is also affected by water content in the intermediate filament matrix; higher water content decreases the strength and stiffness of the keratin cell due to their effect on the various hydrogen bonds in the alpha-keratin network (Alibardi 1998).

β-keratins add much more rigidity to reptilian skin than alpha-keratins alone do to mammalian skin. β-keratins are impregnated into the stratum corneum of the reptilian skin, providing waterproofing and the prevention of desiccation. The epidermal layer in a snake entails seven layers. These are organized in plies of cells with tight packing. The epidermal layer (sub-layers included) encases the body of the retile to form an outer shield. Figure 5 details the seven layers present within the epidermis. Described from the inside of the skin, the first layer is the "stratum germinativum". This is the deepest layer of the skin. It contains rapidly dividing cells. Six additional sub-layers, again from the inside of the skin, follow. Together the six sub-layers form a so-called "epidermal generation" (old and new skin layers). Thus, stacked above the stratum germinativum, there exists: the clear layer and the lacunar layer. The lacunar layer matures in the old skin layer as the new skin is growing beneath. Following there is the α–layer, the mesos layer and the β-layer. The mesos layer is similar to the human stratum corneum (Lillywhite and Maderson, 1982) and contains several layers of flat and extremely thin cells surrounded by intercellular lipids (Maderson, 1965). These three layers consist of cells that consistently keratinize with the production of two types of keratin: α (hair-like) and β (feather like).

Keratinisation continuously transforms these cells into a hard-protective layer. Finally, there is the "oberhautchen" layer, which forms the toughest outer most layer of keratinized dead skin cells. The oberhautchen layer contains the fine surface structure known as the micro-ornamentation (Maderson 1966). Before the molt, a new layer of epidermis forms under the currently existing one; the two layers are zipped together by a spinulae structure (Maderson 1967). During the molt, the reptile sheds the outer (older) layer of the epidermis. The principle constituents of snakeskin are keratin fibres (Maderson

1967). Keratinized-cells constitute the outer part of the skin. The process of keratinization consists in synthesizing keratins that will potentially form the keratin fibers. The keratinization brings an increase of keratin production from the cells that start to begin platter before dying (Maderson 1985).

Two types of keratins form the epidermis: the α-keratin (which in a snake is acid or neutral) and the β keratin (which in a snake is basic). The β-layer consists mainly of β-keratin; this type of keratin is not present in other layers of the skin. Alpha α-keratin constitutes most of the epidermal layers and it contributes to the mechanical properties of skin cells (Irish et al., 1988).

The oberhautchen consists mainly of β-keratin. The presence of the two different types of keratinaceous protein α and β distinguishes the reptilian epidermis from its' mammalian counterpart (Lillywhite and Maderson, 1982). The shed epidermis of snakeskin consists of four layers: the outermost Oberhautchen, the β-layer (mainly protein), the mesos layer (lipid-rich), and the inner α-layer (mainly protein). The oberhautchen consists of a particular type of β-cells that play a major role in the shedding process. Together, this layer and the β-layer, both containing β-keratins, are considered as a unique β-layer in the mature epidermis. The oberhautchen layer is the outermost ply within the epidermal layers. It contains the micro-textural ornamentation. It is also the layer which in direct contact with the surroundings. That is, it is the most active layer of the skin in the sense that it simultaneously accommodates contact and frictional effects.

Figure 5. General structure of the epidermis of a squamate reptile. The figure details the "outer" generation layer (the layer about to be shed); and the "inner" generation layer (the new replacement skin layer).

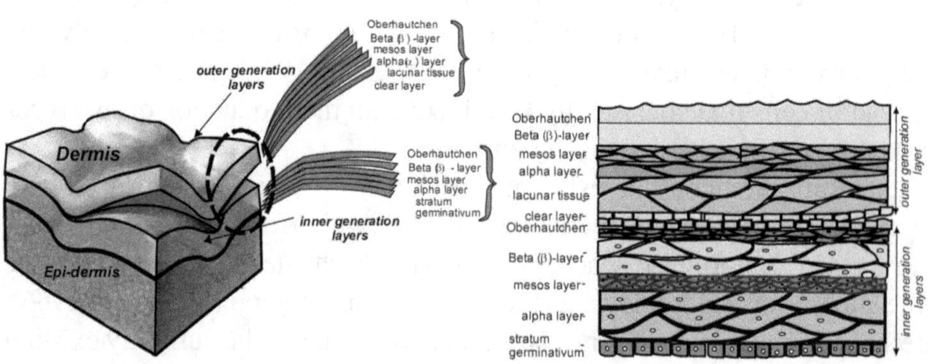

From a tribological point of view the six layers of the skin work as a unit that accommodates contact stresses as well as frictional tractions. That is the function of the skin layers is closely related to locomotion and its efficiency. The body of the snake, as mentioned, in chapter 7, has different local and global frictional requirements. These are governed by economy of effort. Since the snake moves on different substrates of different composition and tpography the skin structure has to be adaptable and optimized for a range of tribological conditions. Accomodation of the variable needs, both local and global, have to be satisfied by the snake structure since it operates within a variable, mostly hostile, triboloical environment. The complex functional and optimization requirements are satisfied through customization of the structural and mechanical properties of the skin layers. Further exploration of the structural layout of the skin layers reveals intricate multiscale customization. Close examination of the skin layers at at the the micron and submicron scale reveals functional property gradients. It also indicates a link between tribological performance and the building-block strucres of the skin layers.

Structural Customization in Skin Substructure

The starting point is examination of functional customization in relation to particular species. Here we focus on two species: *P. regius* and *B. Constrictor* (figure 6). Table 4 presents a summary of species taxonomy and major dimensional features. Both snakes are massive they also share the same rectilinear locomotion mode. Figure 7 illustrates the general appearance of their ventral scale structure.

Table 4. Summary of species taxonomy and major dimensional features

	B. constrictor	*P. regius*
Family	Boidae	Pythonidae
Subfamily		Python
Genus	Boa	P. regius
Length (cm)	186	150
Number of ventral scales	239	208
Ratio of length to maximum diameter	8.7	10.2
Mass (Kg)	12.50	1.3

Figure 6. General appearance of (a) P. regius and (b) B. constrictor.

Figure 7. SEM-Micrographs of Ventral scale ornamentation in ventral scales of a P. regius (a, b) and B. Cpnstrictor (c,d) at different magnifications.

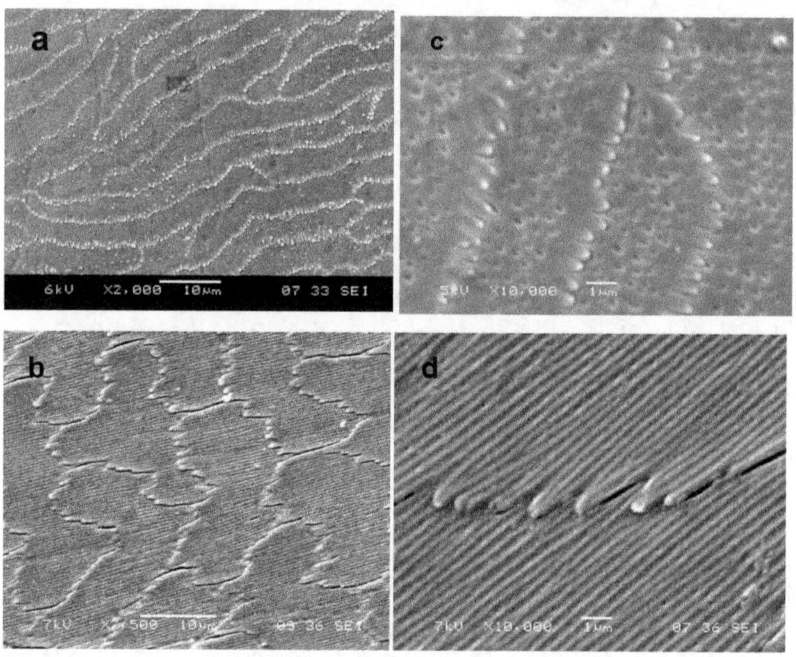

The general TEM aspect of the APR and RTB skins are shown in figure 7. In both species analyzed the skin is composed of 6 layers, whose main characteristic are as follows:

- Oberhautchen layer (O): It is the outermost layer of the skin, with a thickness between 110 and 150 nm. It is richer in sulphur than the other layers and it is reinforced with nanoparticles that contribute to increase elastic modulus and hardness (Figures 8 to 11). From the tribological point of view, this layer acts as a stiff, wear-resistant "pad" supported by a more compliant substrate. The literature reviewed so far do not give detailed information about the chemistry or mechanical properties of the Oberhautchen layer, and the reinforcement particles are generically called "inclusions". Klein and Gorb, for example, found these so-called inclusions in *G. Colubrinus* snake, but not in three other species considered in their study (Klein, et al., 2010). Analysis of the nanoparticles within this layer using atomic-resolution STEM reveals that these paticles are most likely of the $CaCO_3$-type whihc forms in the skin due to calcification processes. Noticed also is an increase in the content of sulphurwithin a layer close to the surface (aboutt 120-150 nm from the surface). Sulphur contributes to the increase in hardness and affects the stiffness of the skin due to the strengthening effect of disulfide bonds (Klein and Gorb 2012). Disulfide bonds may have direct influence on the friction behavior of the skin due to their possible role in the ordering of C=C, C–H and C=O bonds within the very thin lipid layer as reported in the case of *L. Californiae* (Klein and Gorb 2012). Possible evidence of such ordering is noted (Figures.9-b and d*)*, which show a noticable increase in contrast within a layer of approximate thickness 5 nm immediately adjacent to the surface (note the annular dark-field images). The bright areas in the images are formed by strongly-diffracted electron beams deviated, with respect to the direct beam, due to interaction with ordered structures. The structures, in the present case, correspond to the calcium-rich particles within the thin layer immediately adjacent to the surface.

Figure 8. Brigh Field TEM images showing the whole thickness of the skin of APR and RTB skins. O: Oberhautchen, m: mesos, l:lacunar tissue, c:clear.

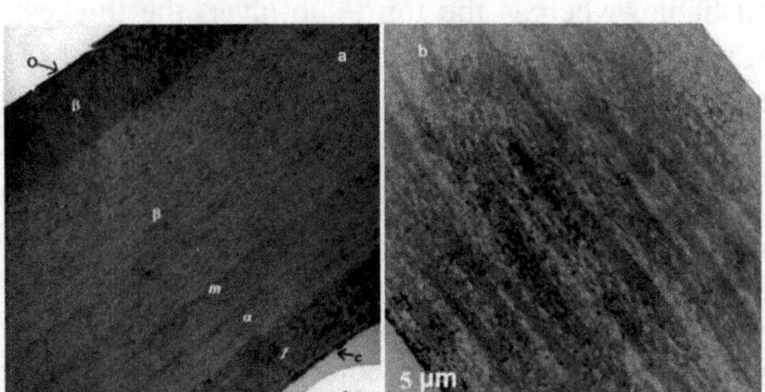

Figure 9. TEM/STEM images of the Oberhautchen layer of APR and RTB skins. The dark-field contrast is expected to be correlated with structure ordering since brighter areas correspond to strongly diffracted intensities. BF (Bright-Field), HAADF (High-Angle Annular Dark-Field), MAADF (Medium-angle Annular Dark-Field).

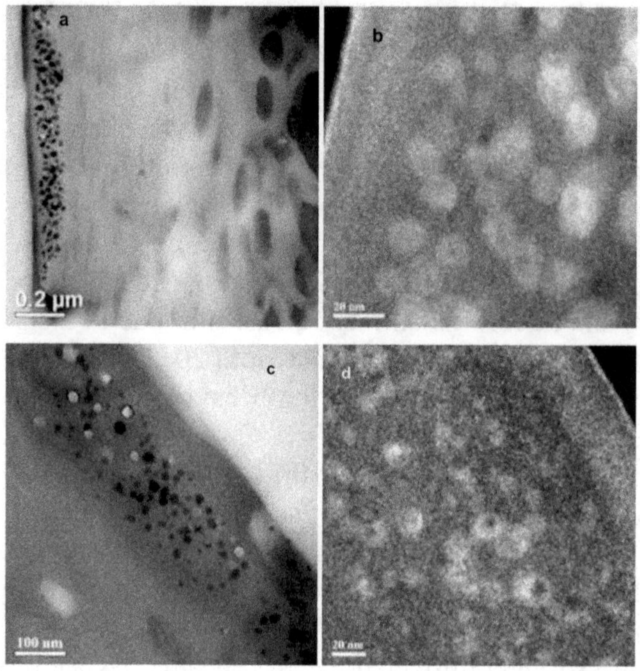

- *β-layer* (figure 11): This is the thickest layer within the shedskin. It comprises three sub-layers that are differentiated according to function and morphology of the β-keratin cells. The first sub-layer next to the Oberhautchen layer is denser and individual cells can hardly be differentiated. The following sub-layer is composed of oval cells clearly distinguishable, whereas the third sub-layer, the thickest, comprises aligned, well-packed cells. No particles or fibers are observed. Figure 9.11 depicts the different sub-layers of β layer in both snakes. During maturation of the epidermis, the relatively thin Oberhautchen fuses together with the thick β-layer resulting in a homogeneous stratum without cell boundaries (i.e., the *β*-dense layer mentioned above). In the lowermost region of the β-layer, cell boundaries are still recognizable (Maderson, 1965). When combined together these two layers compose the outer surface of the snake epidermis, which interacts with the environment.

Figure 10. Relative variation of the concentration of Ca, C, S and in the Oberhautchen layer. RTB sample. Images in HAADF-STEM mode.

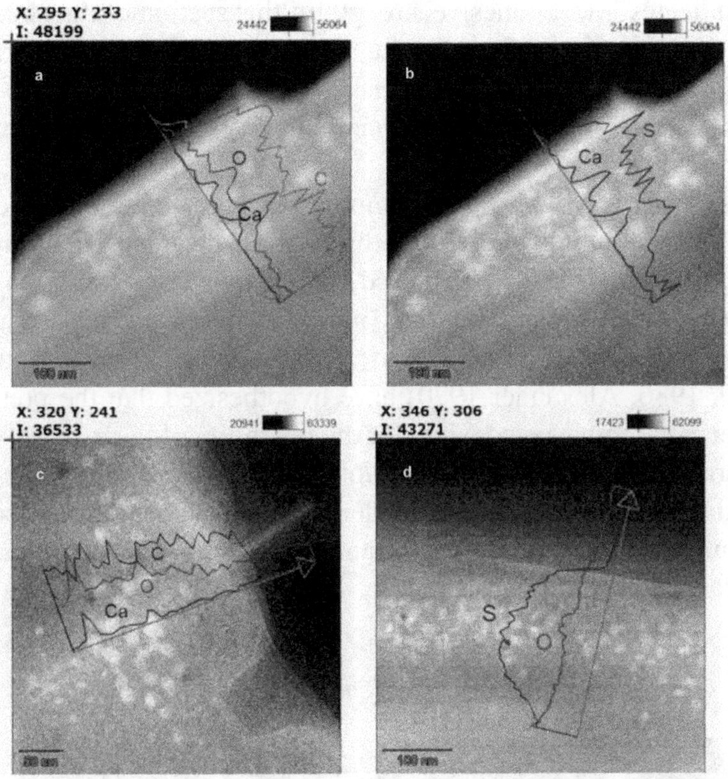

Figure 11. Structure of Oberhautchen (O) and sub-layers of β layer in APR and RTB skin. BF-TEM.

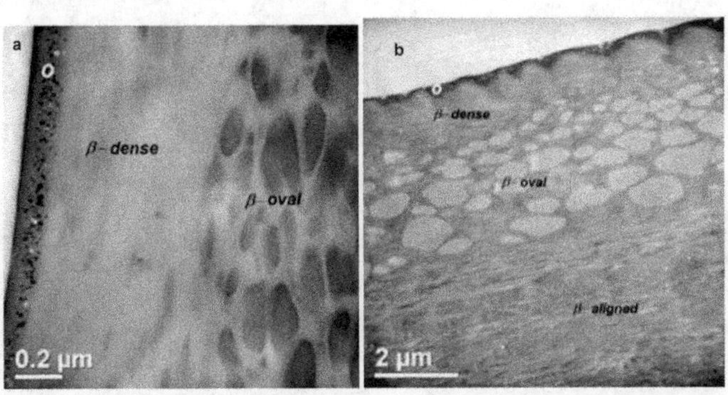

- *Mesos layer* (*m*): This is a thin layer that separates the β-keratin cells from the α-keratin cells. Analysis of cryo-fractured surfaces in figure 12 indicates lower cohesive strength in this section of the skin, with a delamination-type failure that suggests the existence of high residual stresses at the α/*m* and β/*m* interfaces. Also, the rupture of individual fibers can be observed predominantly in this layer due to its less-dense cell packing.
- *Layer:* This is a dense layer composed of aligned, elongated cells. Its thickness is roughly one fifth of that of the β-layer and according to the observations of the cryo-fractured surface its cohesive strength appears higher than neighboring layers (mesos layer and lacunar tissue). See figure 8. Several investigations (Maderson, 1972, 1985, Landmann, 1979, 1986, Alexander 1970) have hypothesized that the presumably softer α-keratin is responsible for the flexibility of the α-layer, whereas β-keratin makes the β-layer inflexible (Maderson, 1972, 1985, Landmann, 1979, 1986, Alexander 1970). Note also the difference in geometry of the two keratins (figure 2) which may also contribute to the relevant softness observed.

Figure 12. Cryo-fracture surfaces of APR and RTB samples. SEM.

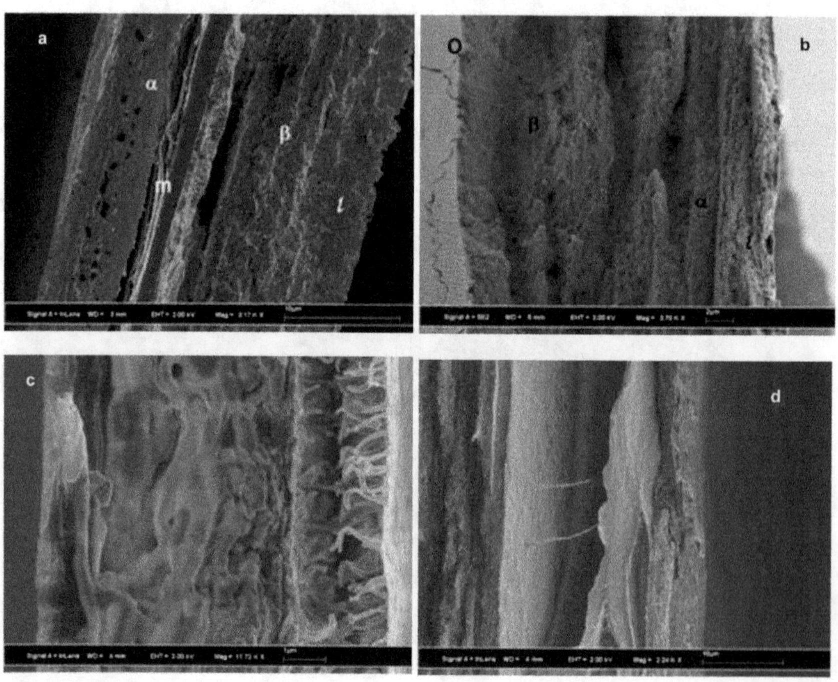

- *Lacunar tissue*: this is a densely packed layer, comparable to α and β layers, although it is more homogenous than reported in the literature for other snake species (Klein, et al., 2010). The cells are mostly elongated and aligned, and the thickness is similar to that of the α layer in both the APR and RTB skins (see figures 12-a, b and 13).
- *Clear layer:* this is the most internal layer of the skin; it is considerably thinner than all the other layers with the exception of the Oberhautchen layer. In general, several individual cells can be distinguished in the TEM images, as shown in figure 14.

Figure 13. Cryo-fractured surface of APR and RTB skin samples. The texture of the Oberhautchen layer can be seen, as well as the differences in the fracture mode of α and β layers.

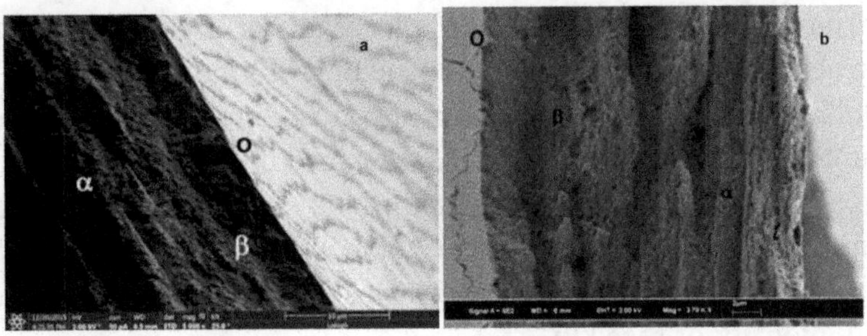

Figure 14. Cryo-fractured surface of APR and RTB skin samples showing the texture of the Clear layer.

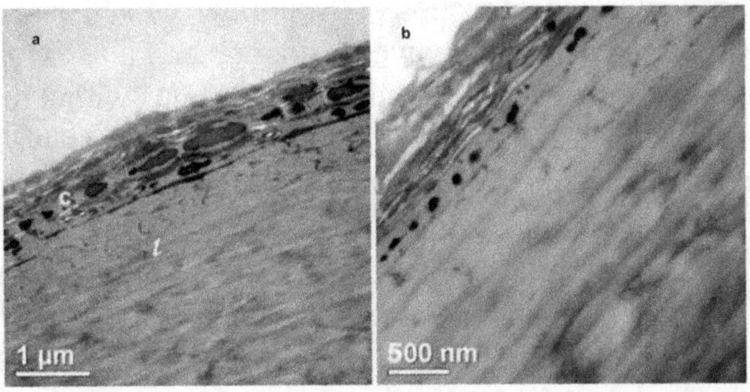

The mesos-layer consists of 2-10 layers of flattened cells containing α–filaments, which do not form a matrix keratin pattern. The α-layer consists of several keratinized cell layers (number of layers can vary according to species and seasonal influence, perhaps even in one species) and contains α-keratin (mammalian-type) with a coiled-coil conformation rather than β-keratin (bird-type) with a pleated sheet conformation present in the β-layer (Klein, et al., 2010, Irish et al., 1988, Raviv and Zhao, 2015). The lacunar tissue and the clear layer make up the inner most part of the epidermis. The clear layer lies directly above the stratum germativum. The lacunar tissue consists of 1-4 cell layers. It consists of one layer of flattened cells. The lacunar and the clear layers are similar and their cells resemble immature α-cells.

The functional structure of the skin layers appears to be closely related to locomotion and its efficiency. The objectives of the functional-layout of the skin layer, and the consequential property gradients, is to optimize the frictional environment dominant during locomotion. Two prevailing themes appear to characterize the organization and order of the skin layers. The first theme, addresses efficient transmission of contact forces originating from the muscular layer. The second theme, addresses the dissipation of the irregular ground reaction forces that transmits to the body due to moving on substrates of nonuniform topography. These two themes, synchronize during motion to economize the effort invested in moving the reptile and to enhance the structural integrity of the skin.

The reptile is likely to move on rough substrates such as: soil, forest bottom, tree trunks, etc. the surfaces, which are deemed tribological hostile, are not perfectly smooth. They are very rough and of a regular topography. This irregularity affects the direction of the normal, as well as the frictional, reaction forces acting on the body while motion. Irregular topography will also close stresses to concentrate around certain points within the body that are in direct contact with the peaks of the substrate topography. This in turn can contribute to injury of the skin which renders the snake vulnerable to predators. Thereby, stress concentration, is in essence detrimental to structural integrity of the contacting layer of the skin. Stress concentration may also increase the energy consumed in propelling the mass of the snake. It will be noticed that these two design constraints are also common to the design of tribological surfaces and rubbing components in general. Addressing these two factors affects the cost of operation and the sustainable usable lifecycle of the component.

To understand the first theme, it is beneficial to first discuss the general build of the body of a snake in relation it's skeleton. A snake's body is built around a vertebrate skeleton. The body of the reptile comprises a series of vertebrae joined buy a network of muscles and tissue (figure 15, 16). The body may be resembled chain of links connected through hinges. These hinges provide range of motion that depends on the size and dimensions of the particular species. The dimensions and flexibility of the hinges are a function of motion of the particular species. A snake has to overcome the friction with the locomotion medium (the substratum). The reptile also has to generate sufficient propulsive forces in the direction of the trunkwhile minimizing slippage in the normal direction. Further, there is a need to minimize the loss to muscular efforts due to friction.

Figure 15. General view of a skeleton of a snake showing the ribs and the verbrae. Picture is licensed under the Creative Commons Attribution (https://en.wikipedia.org/wiki/Snake_skeleton)

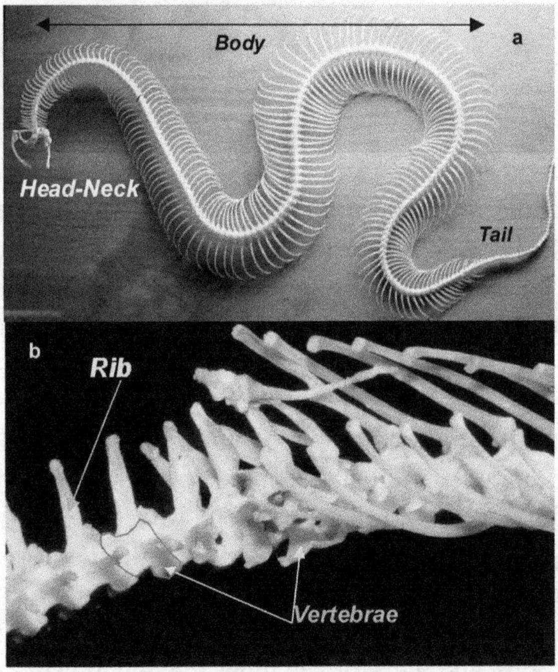

Figure 16. Model for musculo-skeletal structure of a snake and its relation to forces

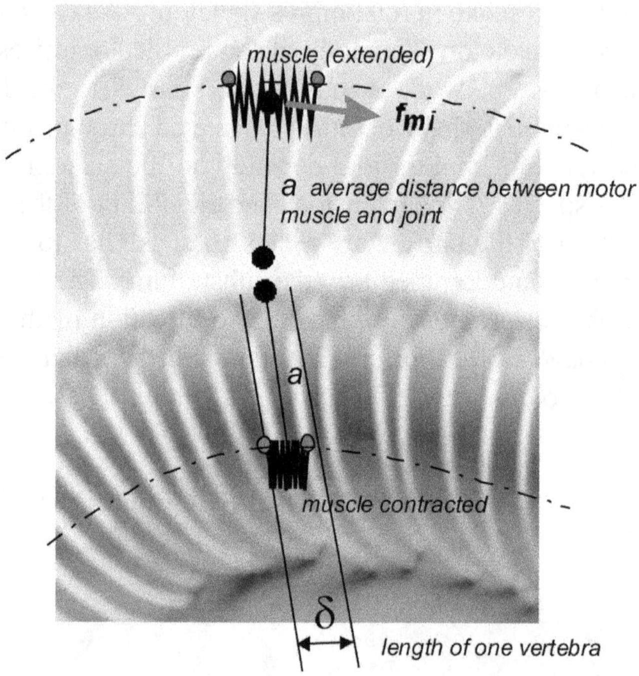

That is, to maximize the portion of the muscular effort that is utilized to propel the mass of the body forward. This condition, necessitates minimization of friction -induced losses. That is the amount of muscular effort dedicated to combat friction. These, nontrivial, requirements are met through a combination of flexibility in body form, optimized body friction surface topological features, and compliant subsurface structural build-up. This last trait of the body facilitates interaction between surface features and the muscle layers within the body (which is the origin of locomotion and contact forces). Forces necessary for locomotion originate from the muscles, which are connected to the vertebrae. The snake controls its friction profile by changing the tension of the muscles (compression or relaxation of a muscle or group of muscles individually or collectively). In doing so, it controls the contact forces and the body contour. Changing the body contour, practically changes the local curvature of the body and therefore changes the real area of contact between the particular body segments and the sliding substrate. Here, it is noted that the accommodation of friction abides by the so-called Amonton's laws of friction. Thus, in essence, the snake controls its local friction profile by simultaneous manipulation of the contact force and the real area of contact.

This process requires a segmented body profile to prevent excessive bending stresses from acting on passive parts of the body. Segmentation of the body thus, the body confines the structural of deformation to the active segments engaged in contacting the substrate. Segmentation of the body, therefore, shields neighboring passive body zones from structural deformation and associated stresses.

Referring to figures 7, through 9, it is noticed that there exist three layers of beta β– keratins. These layers cushion the layer containing the surface features (Oberhautchen). The first layer is the so-called β-dense layer which is compacted layer. This is followed by the β– oval layer which extends parallel to the contour of the body (whence the cross-section appears oval shaped). The layered structure of the skin contributes to accommodating the generation of contacting forces The final layer is the β-aligned layer which is laid almost parallel to the length of the reptile (i.e. along the snout to vent axis of the body).

The shape of individual β-keratin fibers, in the order and orientation observed, serves to enhance the structural rigidity of the contact layer within the skin. That is, the orientation of the β-structures within the skin first strengthens resistance to tear and the deformation of the MAZ layer within the body. Secondly, it works to accommodate, and dissipate part of the work of the friction tractions affecting the contacting layer. It is beneficial to note here that due to the irregular shape of the sliding substrates. The direction of the friction forces may not necessarily align with the major Anterior-Posterior axis of the reptile

Misalignment of reaction forces may lead to more local deformation and may sacrifice the local structural integrity of the skin. It also may lead to intense release of friction work locally which may raise the local energy density, and again, may cause damage. The apparent "flexible mesh" layout of the β-layers seems to act first and damp some of the potential detrimental effects of deformation and, second, to prevent concentration of local friction energy by dissipating it within the mashed structure.

Further damping of forces, and deformation, appears to take place through fiber attachments of the mesos layer. Fibers within this layer appear to be flexible and relatively easier to break or detach. This provides a dissipation mechanism to excessive stresses and work of deformation. It also provides flexibility to accommodate sudden changes in local deformation and stretching or compacting of the layers. That is, it acts as a buffer between the force-generation layer (α-layer) and the MAZ layer (β-layer). Therefore, it partially accommodates the deformation resulting from the generation of the contacting forces by the muscles.

The contacting forces may not be exactly of equal direction or magnitude to the reaction forces originating from the substrate. Such a situation can lead to an unbalanced stress profile on both sides of the skin and hence can promote cracking or internal injury. In other words, the mesos layer acts as a shock absorber that regulates any difference between the internally generated contact forces and externally acting attractive forces.

Accommodation of contact force generation is another design constraint that must be addressed by the skin layers. From an energetic point of view, it is not economical for the snake to precisely customize the contact forces generated by the muscles. This is because the likelihood of wasting more energy in the regulation process. For economy of effort purposes, there is a need to harvest part of the work performed by the muscles to generate contact forces. This takes place through the α-keratin fibers which are helical in shape and resemble coil springs. Their shape implies that the fibers can compress and extend in response to muscular forces, along with any vertical deformation resulting from sliding over an even topography. That is, the helical spring form of these keratins accommodates shape changes and potentially may harvest the portion of the energy (work) of that deformation. The helical shape of the fibers is additionally beneficial in isolating local deformation. It provides flexibility and prevents global spreading of deformation that may cause unnecessary injury to the rest of the body. Similar to the beta layer the α-layer is separated from the musculature layer by a damping zone (i.e. the lacunar layer). Again, is noticed that the fibril interface of the layers facilitates attachment and detachment thus allowing controled dissipation of potentially excess work.

MECHANICAL PROPERTY CUSTOMIZATION

Elastic Modulus

The major requirements of locomotion is to generate a large enough force to move the body in the desired direction. In case of the snake, the whole body does not have to move at the same time. In fact, part of the body remains instantaneously stagnant to provide for balance and stability. Since the mass of the body is not evenly distributed along the active length, then the force required to move each of the linkage (vertebrae) within the body is not constant, nor uniform. The force generated varies in proportion to the mass distribution. A consequence of this distribution is that the local functional requirements and mechanical property gradients will also vary relative to the body mass distribution. This is satisfied through variation of the thicknesses of the various layers that make up the skin both in depth and along the body length.

Figure 17 depicts the variation of the six layers of the skin along the snout-two-vent length of a *P. regius*. Figure 17-a detailed the thickness of the β-layer (with all its component-layers). Here it is noted that the layer of larger thickness is that of the β-aligned structures. This layer, which is aligned with SVL axis is almost of uniform thickness along the body. The β-oval layer which is oriented approximately along the lateral axis of the body is next in thickness (approximately one fourth of the thickness of the β-aligned layer). In contrast to the β-aligned layer, the thickness of the overlay or is directly proportion to the mass distribution along the SVL of the reptile. The maximum thickness of that layer is roughly located within the section with the maximum body mass of the snake (approximately at mid length). Finally, that is the β-dense layer which manifests the minimum thickness among the three layers. This layer, which is fused with the Oberhauchen layer is of uniform thickness along the body length. Thickness distribution of the β - oval layer appears to influence the thickness distribution of the layers containing b-keratins. When all the thicknesses are summed the total is directly proportional to the mass distribution of the snake along the main body axis.

Thickness distribution of the layers containing α-keratins is detailed in figure 17-b. The α-layer is the thickest as indicated in the plot. However, in contrast to the β-layer it's thickness is inversely distributed with respect to the mass distribution along the body. Thus, the heaviest portions of the body manifest the least thickness. Next in order of magnitiude in layer thickness is the mesos-layer. Here the thickness linearly decreases with the increase in body length. The parts close to the head are thicker than those toward the event. The clear layer, which also contains some α-keratin fibers, is of the least thickness compared to other layers. The plot indicates also that the clear layer is of nonuniform thickness, although that the increase in thickness along the body is almost uniform. A sharp increase however is manifested within the fourth quarter (i.e the trailing segment) of the body. Such a trend, again, is opposite to the distribution of the body mass along the length. For comparison, the plot includes a thickness distribution of the Oberhauchen layer. The plot shows that the Oberhautchen layer is of minimum thickness compared to the α-layer. However, with respect to the thickness distribution, the thickness of the Oberhautchen layer manifests an opposite trend to that observed for the α-layer. In particular, the thickness is directly proportional to the distribution of the body mass.

A comparison of the thickness distribution of the β-total layer (i.e. some of the thicknesses of all three layers containing β- keratin) and the α-layer is given in figure 17-c. The plot further compares the thickness of these two layers and the total thickness of the skin. The thickness of the β-total layer is almost 5 times the thickness of the α-layer. Therefore, the β-layer constitutes the main bulk of the skin. Both layers show a contrasting distribution of thickness with respect to the distribution of the body mass along the length. The β-layer is directly proportional to the mass distribution along the SVL. Thus, the maximum thickness is almost located where the maximum mass is located. In addition, because the dominance of the β-layer the overall thickness of the skin (i.e. the sum of the six compositional layers) also follow the distribution of mass along the SVL.

Figure 17. Variation of the thickness of the different layers of the APR skin as a function of the normalized length of the snake.

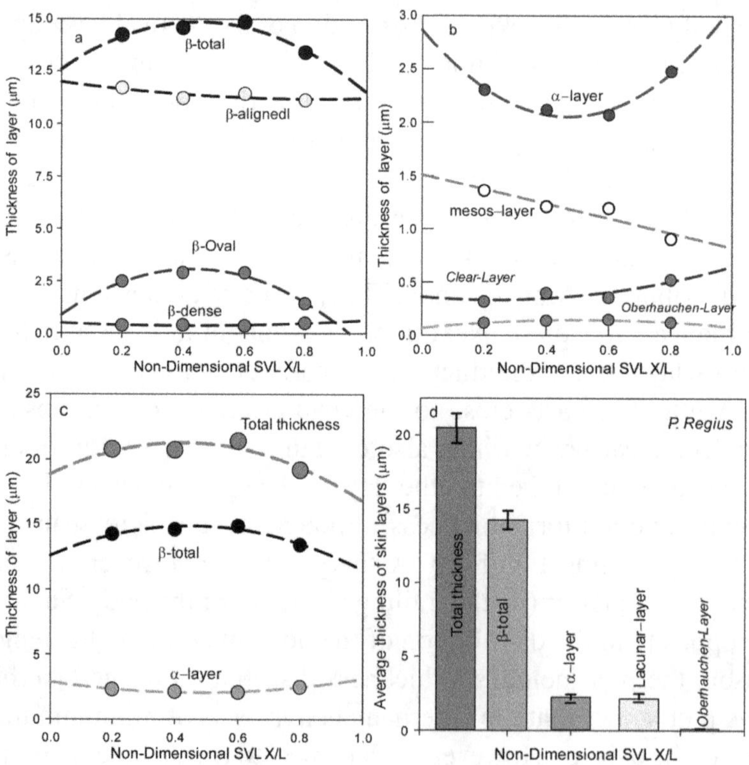

The relative order of magnitude of the sickness of the different layers is depicted in figure 17-d. The figure compares the average thickness of the main layers relative to the total thickness of the skin. It is interesting that the average thickness of the α-layer is of the same order of magnitude of the lacunar layer specially that both contain α-keratin fibers (although apparently of different properties).

A more revealing comparison is obtained when the thickness distribution of the two general functional layers is attempted. This is shown in figure 18-a. Here the thickness of all the β-layers and all the layers containing α-filaments are plotted against the mass distribution. Thus, in effect the plot compares the thickness of the layers that accommodate the force generation, the layers that accommodate external actions, and distribution of the body mass along the SVL. The α-layers in the figure, include the main α-layer, mesos, lacunar, and clear layers. The β-layer meanwhile includes the three β-layers discussed earlier, in addition to the Oberhauchen layer which contains the ornamentation and surface features. The plot shows that while the β-layer follows the mass distribution of the body along the SVL, the α-layers manifest a distribution that slightly decreases towards the vent of the reptile. This may be mainly attributed to the influence of the mesos layer and its geometrical traits. In a living snake, each scale has two sides. Each side is exposed to a mechanical environment.

The side containing the outer layer of the skin (i.e. Oberhautchen where the roughness features are present) is exposed to the surrounding environment. It directly interacts with the locomotion substrate. The Outer Scale Layer, OSL, comprises the β-keratin layers. Mechanically it is subject to frictional tractions, and action forces due to interacting of the local mass of the active body segment with the topography of the substrate. Due to the local topography, and in response to the contacting forces generated by the muscles, the direction of the normal reaction of the ground may not be perfectly normal to the axis of the body. As such, from OSL perspective, forces acting on the body may cause compressive stresses, as well as shear stresses. This condition has its impact on the rigidity requirements of the skin and manifests in the hardness and modulus of elasticity values of the skin layers.

The Inner Side Layer, *ISL*, is also in contact with the mechanical environment. However, this environment is contained within the body, within the stratum germativum layer of the skin, as it mechanically manifests contact and locomotion force generation layer (s). Again, efficiency of force transmission, energy harvesting, and dissipation of potentially damaging transient strains, are satisfied through mechanical properties and perspective variation in the gradients. Thus, for force regulation purposes and structural integrity, the skin of the reptile may be considered as a closed system. This system interacts with two neighboring systems. The first, interacts with the ISL, and generates contact and locomotion forces. Whereas, the second system, locomotion environment, generates reaction forces from outside the body. Thus, in effect, the skin entertains information related to forces, stresses, and associated strains both internal and external. For energy conservation and structural integrity these forces have to be balanced and the gradients have to be minimized. Optimizing the formations so that internal injury may be prevented, therefore, is the major constraint on the mechanical properties. Thus, it is anticipated that these properties vary in relation to the particular loads on the particular side of the active ventral scale and the relevant system in direct contact with that side.

The ISL, should efficiently transfer the contact forces generated by the muscle system. As such, the rigidity requirements should be relatively relaxed compared to the OSL of the active ventral scale. This is because the OSL accommodates external reactions and should be able to damp the formations. It also should be of higher hardness so it would resist scratch and tear. The internal layers should also be of a lower modulus of elasticity to cushion the formations and provide a softer layer that absorbs sudden changes in topography of the sliding layer.

Figure 18. Comparison of the thickness distribution of the two general functional layers in a B. Constrictor

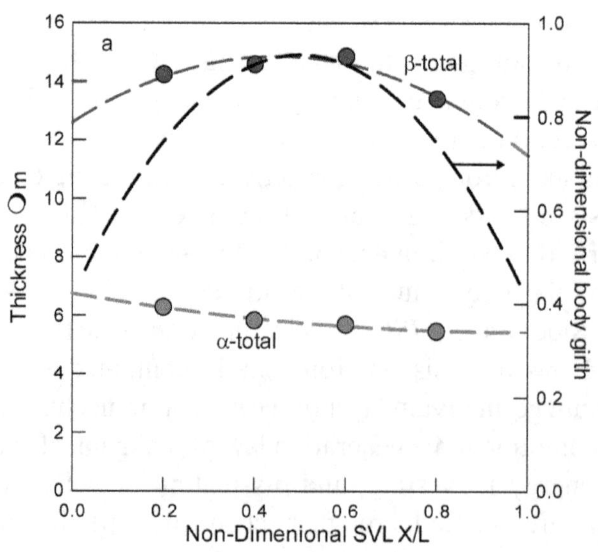

Results of nano-indentation tests on the skin of several species have confirmed the proposed order of magnitude of mechanical properties. Table 5 (Klein and Gorb 2012) provide the results of indentation experiments performed on the skin of a Kenyan sand boa (G. Colubrinus). The data examine the variation in the hardness and the modulus of elasticity on the ISL, and OSL of the snake at three depths intervals (0.4-0.5 μm, 1.0-1.1 μm, and 1.6-1.7 μm). The data show that the ISL is softer than the OSL (H_{OSL}/ H_{ISL} »1.67). This implies that the OSL is more resistant to penetration and to deformation than the ISL (i.e. the OSL is more rigid than the ISL of the ventral scales).

Table 5. Comparison of mechanical properties of ventral scales in G. Colubrinus

	OSL (β-Layer Side)		ISL (Clear Layer Side)	
Depth μm	**Elastic Modulus (GPA)**	**SD**	**Elastic Modulus (GPA)**	**SD**
0.4-0.5	5.02	0.59	3.3	0.36
1.0-1.1	4.73	0.23	3.43	0.56
1.6-1.7	4.46	0.18	3.48	0.36
Depth μm	**Hardness (GPA)**	**SD**	**Hardness (GPA)**	**SD**
0.4-0.5	0.31	0.07	0.15	0.07
1.0-1.1	0.32	0.04	0.17	0.04
1.6-1.7	0.32	0.03	0.15	0.03

(Klein and Gorb 2012)

The gradients of the modulus of elasticity in case of P. regius is shown in figure 19. The data ploted pertain to nano-indentations at a series of depths starting at the OSL within two spots within one ventral scale. The depths examined include the thickness of the Oberhauchen layer, β-dense, β-Oval, and roughly the upper third of the β-aligned layer. It can be seen that the Oberhautchen layer has noticeably different properties when compared to the β layer, and that the β layer itself presents a gradual variation as a function of the sub-layer considered.

It is noticed that the very outer layer of the skin (the Oberhauchen) is the most rigid layer. Past this layer a considerable steep gradient of the elastic modulus takes place within the β-dense layer. (dE/dh» -3.5 GPa/μm). A gradual transition takes place within the β-oval layer when the slope is not as pronounced as that of the β-dense layer (. (dE/dh» -0.6 GPa/μm). The modulus of elasticity of the OSL becomes quasi-constant within the. β-aligned layer would hardly any change in the modulus of elasticity is noticed. The ISL of the ventral scales manifests a similar trend (Klein and Gorb 2012) although the values of the modulus at the edge layer of the scale is likely to be smaller than that of the Oberhauchen layer (as mentioned earlier).

Landmann (53) suggested that the β-layers main function is to protect the epidermis against abrasion. Alibardi and Toni (40) showed that the α- and β-keratin differ in chemical structure and therefore presumably also in material properties (55). The β-keratin molecule is stiff and inelastic, which presumably makes materials containing it, more robust and therefore protecting the underlying layers against damage, whereas the α-keratin presumably provides flexibility to the material of the skin (53, 55).

Variation in the modulus of elasticity also takes place along the length of the reptile figure 20-a. The figure plots the variation in the elastic modulus of the ventral skin for the Western diamond backed (C. Atrox). The curves were evaluated for four sections within each of the ventral scales as shown in figure 20-b. The distribution of the elastic modulus is inversely proportional to the mass distribution of the snake.

Figure 19. Elastic modulus and hardness of RTB sample. 1 and 2 labels correspond to independent tests performed on different areas of the same sample.

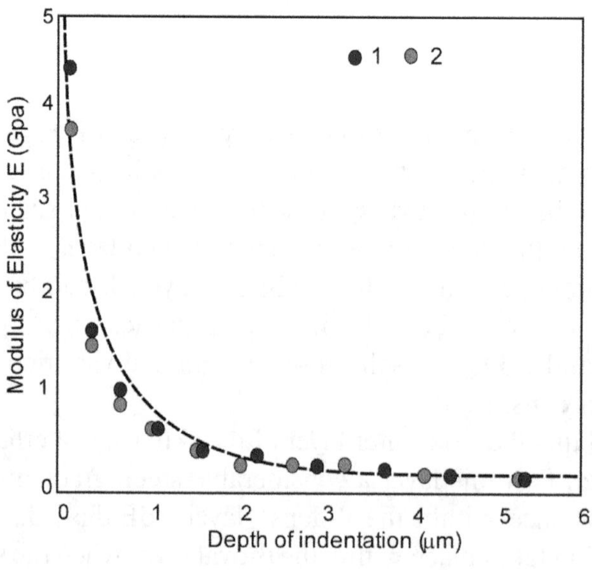

Figure 20. Distribution of the Elastic Modulus along the non-dimensional SVL of a C. Atrox evaluated for four positions on each scale (b)

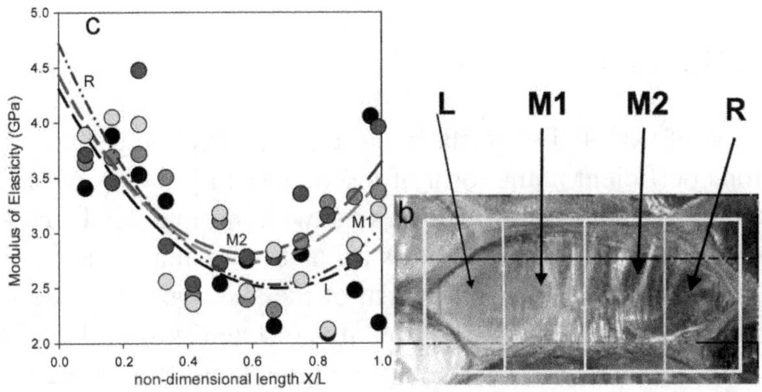

The values show a rapid decrease almost until the middle of the length of the snake (X/L=0.5). Past this point, the elastic modulus shows a sharp increase. Values do not show appreciable variation with respect to position on the scale approximately until X/L=0.4. Past that section, clear differences between values on the right-hand side and the left-hand side of the scale are distinguished (compare the curves labeled R, M2 to those labeled L, M1).

Variation in the modulus of elasticity and hardness reflect on the dynamics of contact between the active segments of the snake trunk and the locomotion substrate. In this case, the elastic modulus is key to the interaction between the roughness features, contained within the Oberhautchen layer, and the subsurface layers. The interaction between roughness features and subsurface is characterized through the so-called contact stiffness. Contact stiffness influences dynamic compliance of the snake body during motion. Compliance of the body is a function of the forces generated by the muscles and is directly related to the size of the real area of contact between the active segments of the body and the substrate. Therefore, in effect, the contact stiffness directly influences friction tractions acting on the ventral scales and thereby the energy consumed in overcoming friction.

FUNCTIONAL INTERACTION BETWEEN SURFACE AND SUB-SURFACE LAYERS

Contact Stiffness

Deformation of skin and subsurface tissue during friction can contribute to the friction coefficient in the form of visco-elastic hysteresis or ploughing. The contribution to hysteresis increases with normal load and contact pressure. Contact pressure, in turn, is a function of the surface roughness of the skin. So in order to check which of the skin types is more likely to entertain larger hysteresis contribution, it is of interest to evaluate, at least, qualitatively, which skin type develops a higher contact pressure under the action of the same nominal normal load. For such a purpose we invoke the classical Greenwood-Williamson model.

Greenwood and Williamson (1966) developed a model for rough surfaces in contact. The model assumes that the surface roughness has a constant radius (Rs) and that the heights of the roughness elements follow a Gaussian distribution. For elastic deformation of the roughness elements the contact load over a circular contact spot (similar to that established due to the contact between the acoustic probe and the skin) s given by:

$$P\left(d\right)=\frac{4}{3}\eta A_{n}E^{*}\sqrt{R_{s}}\int_{d}^{\infty}\left(Z-d\right)^{\frac{3}{2}}\phi\left(Z\right)dZ \tag{1}$$

For Gaussian distribution, there is no closed form solution for equation (1). Shi and Polycarpou (2005), showed that an approximation to the Gaussian distribution assumed in equation (1) is an exponential roughness height distribution of the form,

$$\phi=C\exp\left[-\frac{\lambda d}{R_{q}}\right] \tag{2}$$

Adopting the distribution of equation two simplifies the integration and therefore allows a closed form solution of equation one. Accordingly, for an exponential height distribution and elastic contact conditions, the contact pressure will be given by,

$$P\left(d\right)=\frac{C\sqrt{\pi}}{\lambda^{\frac{5}{2}}}\beta A_{n}E^{*}\left(\frac{R_{q}}{R_{s}}\right)^{\frac{1}{2}}\exp\left(-\frac{\lambda d}{R_{q}}\right) \tag{3}$$

Equation three emphasizes the influence of the roughness of the skin surface on the magnitude and distribution of the pressure acting on the skin in dry friction. Equation three may be rearranged to explicitly express the contact pressure in terms of standard surface parameters. Thus, for Hertzian contact conditions we write,

$$\frac{P\left(d\right)}{A_{n}}\propto\delta E^{*}\left(R_{s}R_{q}^{3}\right)^{\frac{1}{2}}\exp\left(-\frac{3d}{R_{q}}\right) \tag{4}$$

Where: δ is the density of the roughness features. Equation (4) yields the contact stiffness as

$$\left|k_{c}(d)\right|=\left|\frac{\lambda}{R_{q}}P(d)\right| \tag{5}$$

Figure 21 depicts the distribution of the contact stiffness along the SVL of C. Atrox. Again, the figure includes four plots with the labeling corresponding to that depicted in figure 20-b.

The general trend of the data manifests an increase in the stiffness up to the middle section of the body. The values do not show considerable difference with respect to position on the scale. Past the midsection threshold differences in values are distinguishable. The stiffness drops with length and data for the left-hand portion of the scales show independent values. It is also noted that the heavier sectionswithin the body display higher contact stiffness valleys (which in the meantime are the same zones showing the lowest modulus of elasticity).

Figure 21. Distribution of the contact stiffness along the non-dimensional SVL of C. Atrox

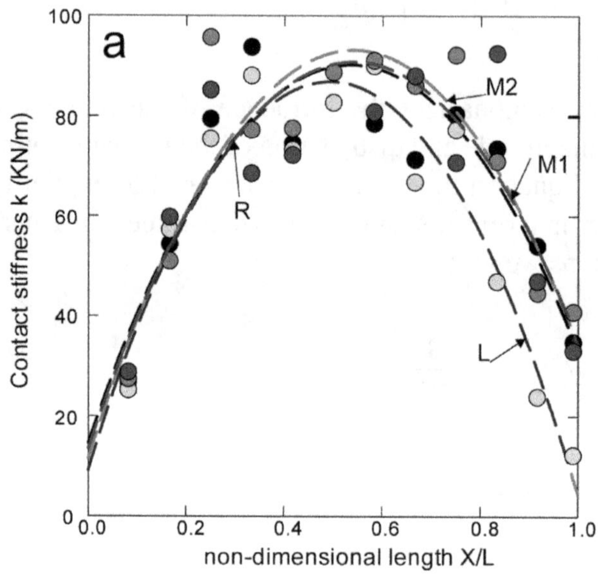

Consequence for Friction

Closer consideration of equations 4 and 5 indicates that the contact stress P(a)/ A_n in equation 4 is a function of the parameters. The first is the composite modulus of elasticity E*whereas the second is the roughness parameters R_s and R_q. The first term, E*, is mainly dominated by the modulus of elasticity of the skin (see figure 20). This modulus varies in the reverse to the contact stiffness (compare figures 20-a and figure 21). Such a contrasting behavior may, therefore, be attributed to the influence of the roughness parameters. To that effect, the compliance of the heaviest part of the body would be minimized due to the influence of the roughness parameters. This causes a reduced friction force at the heavier sections (figure 22). Consequently, the heavy parts of the body would need a relatively smaller force to establish contact and combat friction resistance. This, in turn, dedicates larger portion of the work performed by the muscles to moving the body and therefore contributes to economy of effort.

Figure 22. Variation of the COF for C. Atrox along the non-dimensional SVL in relation to the contact stiffness distribution along the SVL of body. The figure depicts the average values for forward motion (F) and backward motion (B)

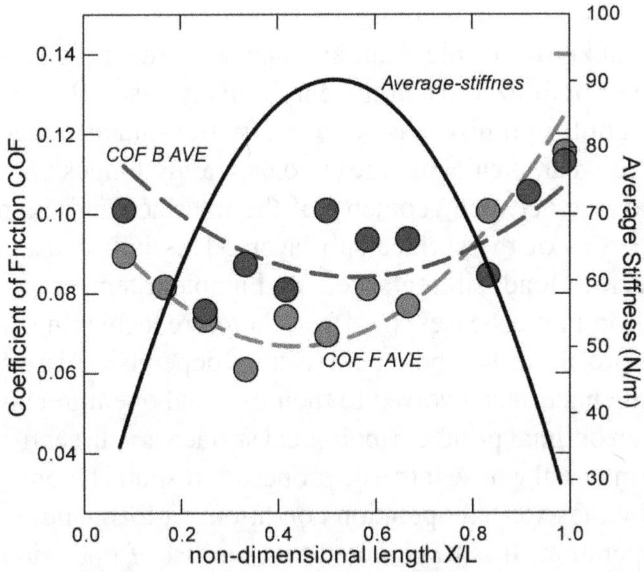

DESIGN FOR 4-D PRINTING

The design process for 4-D printing is more demanding than that of 3-D. The added fourth dimension, whether it is time or evolution of functionality, impacts design requirements conventionally, i.e. when time represents the fourth dimension, finding a suitable material is of prime importance. In such a case the material first has to be suitable for processing in a layer-by-layer fashion, and second has to be compatible with the time morphing criteria. If evolution of functionality represents the fourth dimension, then a suitable, printable material, is also a fundamental requirement. However, in this case more information about the behavior of the material during and after printing has to be fully characterized. Characterization of behavior in the second case entails complete prior knowledge of the properties of the layers to be deposited and the path of their evolution in response to a) the physical parameters of the printing process; and b) how to evolve post printing during operation. In all, it is essential to have prior knowledge of the printability of the material to be used (which is the fundamental requirement for conventional 3-the processing). The second requirement depends on the envisioned morphing of the printed object. In what follows, the printability requirement won't be discussed as it falls out of the scope of this chapter. The focus would be on the second condition with emphasis on the design criteria for a 4-D printed tribological object. That is an object that entails complying surfaces which engage in friction.

In the context of tribological applications, the required functional profile is one of optimized friction-wear. In many cases, this translates into minimized friction, optimize dissipation of friction-induced thermal energy, and minimized wear. Achieving these goals mainly hinges on reaching the functional balance between geometry of the interface and the mechanical-thermal properties of the surface sub layers. This is because the external conditions such as load, sliding speed, and ambient temperature, are fixed before operation in most cases. In effect, therefore, achieving and optimize performance profile in a tribological system depends on streamlining the response of the materials involved to their external operation conditions. In other words, an optimal printed tribological surface is a direct function of the choice of the material whose intrinsic properties respond favorably to stimuli represented by the external operation conditions of load, sliding speed, and ambient temperature. It is to be noted that the list of operation conditions may be extended as well to include additional stimulants as appropriate. As such, we may view optimized function of a printed tribological object as direct consequence of the response of the material used in the build to external conditions. Here, the material response is the objective of morphing of function. Whereas the external conditions, known a priori in most cases, represent the stimulant. Here the fourth dimension would be the morphing of function due to the effects of the stimulants (which are represented by the load and the sliding speed).

The design constraints for the tribological system would be predetermined performance metrics (e.g. limits of wear and coefficient of friction under an operation nominal load and sliding speed). The design challenge would be to find a material that meets the pre-described constraints. That material may be an alloy, and element, or a combination of alloys that when printed in a layer by layer fashion, then stimulated, would function as a single entity that meet the performance requirements. Therefore, theoretically at least, the 4D printing process would materialize a tribological system of which layers are functionally customized through imposing precalculated material property gradients. The layers, a gain, would collectively interact and respond to load and speed stimulation as a single unit. This response, in effect, would control the frictional response of the system and consequently yield the required wear profile.

Customization and control of properties in a layer-by-layer fashion in multidimensional printing is not quite perfected yet. The topic still represents a technical gap within the AM field. However, there are considerable efforts taking place to address this gap (Raviv and Zhao, 2015). These efforts are likely to mature in the very near future and considerable control of properties would be practically achieved. As such, the real challenge is to understand the role of layering and imposing property gradients within the subsurface zones in facilitating tribological response. Understanding the functionality of layers, on a fundamental level, is still lacking in tribological research. This is despite almost a century since inauguration of tribology as a scientific discipline. The problem is further complicated by the additional need to understand the nature of interaction between surface topography and subsurface properties, in a dynamic situation. To date, the majority of the work considers the initiation of rubbing or steady state friction (i.e. the so-called steady-state attrition stage) without bridging the gap with the running in period. Lack of fundamental understanding also extends to knowledge of evolution of the friction-where the process in terms of dynamic property changes and function of surface-subsurface layers during rubbing. This is particularly where biology can be valuable.

The philosophy of design generation in nature centers around function. Underpinning this paradigm is the general principle of "separation of function". Separation of function enhances survivability of species in a fiercely competitive world that promotes economy of living material. Natural designs, also, portray an integrative relation between form and function. The pursuit of a single purpose requires careful considerations of the imperatives of balancing economy of design (with regards to materials and effort) and efficiency of function.

The philosophy of design generation in nature centers around function. The underpinning paradigm is harmony between form and function. Such a paradigm manifests in a multitude of schemes that aim to enhance economy of effort and survivability of species. On a more detailed level there are many strategies that manifest while studying the logic and means of mitigation of destructive influences due to mechanical loading at interfaces (both static and dynamic). These strategies mitigate stress concentration, for example, across interfaces through chemical concentration gradients, or microstructural features. In squids, the beaks show stiffness gradient from stiff compliant. Cortical bone manifests the change in porosity. An additional technique to enhance force transmission while reducing interfacial stress when joining too hard materials is the insertion of a flexible connecting layer, or bearing tissue.

Mitigation of friction, associated stresses, and enhancement of surface integrity are enabled in snakes through material functional grading. Material property gradients in a layer-by-layer fashion words optimizing the frictional response of the skin both from the mechanical and structural perspective.

Resistance to normal elastic contact in case of an elastically graded material, when the Poisson's ratio is almost constant, comprises three cases (figure 23) when the young modulus increases exponentially towards the depths (α<0) figure 23a, the indentation response is stiffer. The stiffness of the response reaches a vertical asymptote as the permissible limit of the exponential increase in the elastic modulus is reached. When the material properties are genius and the modulus of elasticity is constant throughout the depth material, then the response increases sharply and is not bounded. When, however, the material is engineered such that the modulus of elasticity decreases exponentially with the increased depth, as observed snake structure, then the response is compliant. The elastically softer material beneath the indented surface promotes a compliant response when compared with homogeneous material in this case the indentation load carrying capacity vanishes and asymptotic limit as the modulus of elasticity decreases exponentially beneath the surface.

Figure 23. Indentation response of an elastically graded surface. (a) Schematic illustration of an exponential increase (α > 0) or decrease (α <0) in Young's modulus as a function of depth z beneath the indented surface; (b) Schematic illustration of the variation of normalized indentation load as a function of normalized depth of penetration for the homogeneous and graded surfaces. For the graded surfaces, the variation of P with the penetration depth h is a complex function of the surface and indenter properties, the indenter radius, and the gradient in elastic modulus α.

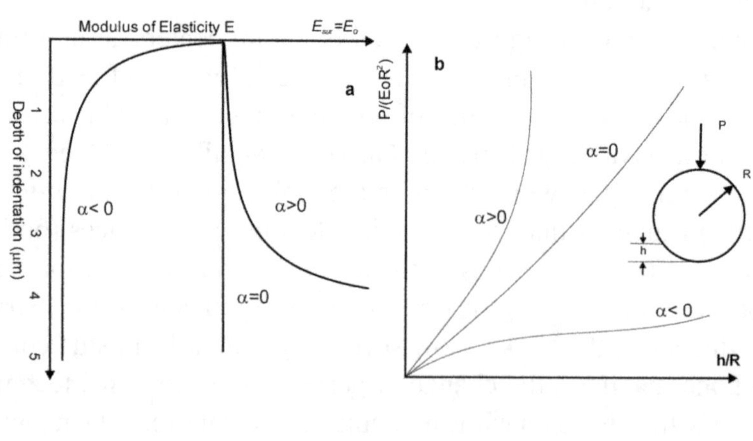

Taking an in-depth look at the examples presented in the previous sections, we note that active control of friction is enabled deliberate distribution of material property gradients. Linking the gradients to mass distribution is key in this context. This is because moving the mass of the reptile is the primary objective of the muscular effort performed by the muscles responsible. Efficiency of motion in this context demands minimizing the effort lost friction. It is here where the material gradients become effective. Consider the distribution of the coefficient of friction along they SVL of the reptile, figure 22, in relation to the mass distribution it is noted that the coefficient of friction is minimal at the section with the heaviest mass. The result of this distribution is that the friction force, which represents the shear acting on the body, would be almost across the body. This leads to a quasi-uniform shear stress on the ventral skin. Reaching this state of sheer loading effect depends on controlling the frictional response of the body.

Performance of the surface in a tribological application is the resultant of the performance of the textural layer (i.e. roughness patterns) and the performance of the subsurface layers. A surface of superior performance is in reality a" system" of optimized roughness and an optimal complementary subsurface layer. In other words, it is the optimized interaction of material properties and the topological pattern of the surface that result in a successful system in tribology.

There are two fundamental laws in friction. These are known as the Amonotons laws. These pertain to control of friction. They imply that regulation of friction may be achieved through controlling the external normal force which keeps the surface in contact, or by controlling the true area of contact between the sliding surface and its substrate. Here we should note that the true area of contact depends on the mechanical properties of the materials and on the roughness (i.e. the textural pattern) of the surface.

A snake moves by purposeful mitigation of friction. In so doing, snake follows the two laws of Amonotons. In addition, it relies on the optimization of surface patterns and subsurface layer performance. Unique to a snake, however, is that it applies both laws at once. It's contact force originates from its muscles. So, if they can relax or contract the muscle to provide a force to comply the part of the body to the ground. The act of relaxing and contracting affects the local curvature of the body, whence the value of the true area of contact with the ground. At this stage, the roughness (micron -sized so-called fibrils within the ventral scales) takes over. Their job is to fine-tune the area of contact and whence fine-tune friction regulation. If one adds to this process the observations that the underlying skin, and its mechanical properties, work to absorb or dissipate friction energy, then we conclude that a snake in reality is a synchronized tribological system. In addition, upon measuring the surface parameters and their distribution compared to the mass distribution along the body one observes a close correlation.

It would be noted that the contact stiffness, figure 21, is maximum at the heaviest parts. This leads to a stiff response by the skin leading to minimizing the real area of contact between the moving section and the substrate. This, in turn, reduces the friction force. Distribution of the modulus of elasticity along the SVL is inversely proportional to the mass distribution. Thus, heavier body sections show a lower modulus of elasticity. Heavier sections, in theory, appear to be more compliant, and less massive sections appear to be stiffer. Considering that the mass distribution in conjunction to the distribution of ventral scale areas, could give a measure of natural loading. That is the force of contact are arising from body mass. Plotting this as a parameter of loading against the distribution of the modulus of elasticity, figure 24, shows direct correlation. The distribution of that newly defined loading parameter along the SVL is directly proportional to the distribution of the modulus of elasticity. Further, profiting this parameter in relation to the coefficient of friction along the in longitudinal axis of the body shows direct proportionality as well.

Figure 24. Correlation of the COF in C. Atrox with the empirical loading parameter (girth/ scale area). Data are plotted in relation to variation of the parameters along the SVL of the reptile.

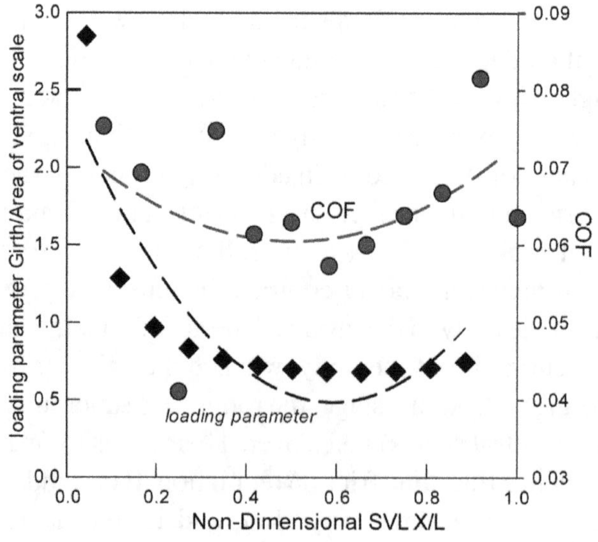

Preserving the structural integrity of the subsurface in the reptile is closely correlated to subsurface material property gradients. The modulus of elasticity within the contact layer of the skin varies with depth. This variation starts at the very outer layer of the active zone of the skin (i.e. the outer side of the surface which is in direct contact with the substrate). Two gradients are noticed, the first is within the first 2 μm of the skin where an increase of the modulus of elasticity is noticed.

Figure 25. Variation of the Effective modulusof elasticity for 2 snakes G. Colubrinus and L. Californaie as a function of depth within the skin, data extracted from table 3 (45)

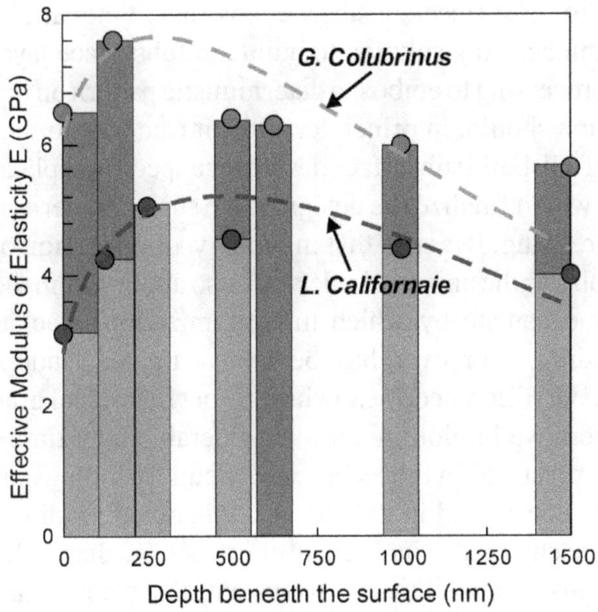

This increase is due to the presence of the inclusions (figure 9 a and c). The thickness of the layer containing the inclusions is considerably small compared to the overall thickness of the skin. Despite its small thickness its properties (hardness and modulus of elasticity) are considerably higher than the rest the layers within the depth of the β-layer (figure 25). The increase in the modulus of elasticity acts to direct any possible cracks towards the surface layer. It further leads to the suppression of crack penetration arising around any protrusion present on the contacting substrate. The mechanism responsible for this act centers around reduction of the stresses around the surface features (which is the initiator of any crack taking place during sliding).

One of the key aspects of textured surfaces in nature is built-in redundancy (bot at the surface architecture level and the subsurface level). Redundancy is also present in conventional surfaces however not to the extent present in natural constructs. Building redundancy in engineering surfaces hinges on the ability of building the active zone in the material layer-by-layer. That is, in general, being able to control the micro-structure of the subsurface, and its properties, and in the same time being able to customize the geometry and topology of the textural features on the surface. Such a concept lends itself, by default, too additive manufacturing techniques. In the next decade, or so, viable printing of metals will advance drastically. This would allow true customization of surface-subsurface constructs. One can always envision an additive manufacturing solution to build the subsurface layers and a laser texturing postprocessing to emboss a deterministic pattern on the printed part. This combination should, in principle, allow it true tribological-system-by-design. Printing will build subsurface design for a specific application, whereas laser texturing would finalize the design of the surface deterministic pattern.

From the foregoing, it is seen that in biology, optimization of tribological performance follows the universal rules that also apply within the engineering domain. The mechanism by which this optimization is achieved, namely functional grading, moreover has been investigated, independently, by engineers. A major difference, meanwhile, is that functional grading has been present and operative in biology for a considerable long time before it was discovered and practiced in the technical domain. In biology, and especially in the examples considered in the chapter, functional grading is present on a fine scale. Executing it tribological construct on the scale observed in biology is beyond the capabilities most conventional manufacturing methods. Additive manufacturing techniques may, however, present a viable mean which enables a tribological construct that mimics many bio-analogues. A key advantage of AM is the layer-by-layer building scheme. This method of constructing the target object allows for customizing both the scale and the property gradients. It further affords customization of the thickness of each layer, the direction of laying the material, and the scale of the particular layers in proportion to load and anticipated stress distribution. In so doing, printing can activate the fourth dimension that arises because of the response of the material and property gradients to the external sliding parameters (i.e. the stimulants in terms of 4-D printing).

Physical and functional response of the deposited layers, and consequently optimization of performance, principally depend on processing parameters during fabrication. Final performance, and its optimized evolutionary path, further depend on the response of the material used in printing to the physical parameters of processing. Thus, it is precise control of the local microstructure and chemistry of the material used, while processing, that realize the target functional response. Control of microstructure makes it possible to manipulate properties, on the local and global scale. In a manner that achieves the objective response to external stimulation. That is, control of the evolution of layer properties is key to realization of a dynamic 4-D printing paradigm.

To manifest a feasible 4-D printing functional grading in a metallic construct is essential to explore the printability of the material. Printability in this context implies identification of feasible processing regions within the so-called metal-process space. These regions are zones of which response to manufacturing operation physical parameters would result in property gradients that aligns with the required stimulant response. Thus a "printable zone" within the material-process space it is all for which applying a certain processing parameter set would result in a defect free build. Exploring the printability of many metals and alloys fit for establishing functional profile is currently a technical gap within the AM community.

A new material to be used as a feedstock in 4-D printing (or 3-D) application undergoes lengthy, and costly, experimental procedures to identify behavior under manufacturing process parameters once the complete mapping of the structure-process parameter name is established the material can be used on a larger scale. To date, little effort has been directed to gate printability of metallic functional materials. However, the gap is closing rapidly and many are directing the attention to complement the knowledge base. To this end, it is anticipated that in-process control of layer behavior, at the desired level, would let you in a period of two or three years at the most. This will shift bio-inspiration of tribological constructs the paradigm domain practice domain.

REFERENCES

Alexander, N. J. (1970). Comparison of α and β keratin in reptiles. *Z Zellforsch, 110*, 153–165. PMID:5490547

Alibardi, L. (1998). Differentiation of the epidermis during scale formation in embryos of lizard. *Journal of Anatomy*, *192*(2), 173–186. doi:10.1046/j.1469-7580.1998.19220173.x PMID:9643418

Alibardi, L., & Thompson, M. B. (2002). Keratinization and ultrastructure of the epidermis of late embryonic stages in the alligator (Alligator mississippiensis). *Journal of Anatomy*, *201*(1), 71–84. doi:10.1046/j.1469-7580.2002.00075.x PMID:12171478

Alibardi, L., & Toni, M. (2007). Alpha and beta keratins of snake epidermis. *Zoology (Jena, Germany)*, *110*(1), 41–47. doi:10.1016/j.zool.2006.07.001 PMID:17169542

Archard, J. F., & Rowntree, R. A. (1988). The temperature of rubbing bodies; part 2, the distribution of temperatures. *Wear*, *128*(1), 1–7. doi:10.1016/0043-1648(88)90249-9

Arnold, S., & Bennett, A. (1988). Behavioral variation in natural populations. V. Morphological correlations of locomotion in the garter snake. Thamnophis radix. *Biological Journal of the Linnean Society. Linnean Society of London*, *34*(2), 175–190. doi:10.1111/j.1095-8312.1988.tb01955.x

Bakarich, S. E., Gorkin, R., & Spinks, G. M. (2015). 4D printing with mechanically robust, thermally actuating hydrogels. *Macromolecular Rapid Communications*, *36*(12), 1211–1217. doi:10.1002/marc.201500079 PMID:25864515

Barber, J. R. (1970). The conduction of heat from sliding solids. *International Journal of Heat and Mass Transfer*, *13*(5), 857–869. doi:10.1016/0017-9310(70)90131-6

Blok, H. (1970). The postulate about the constancy of scoring temperature. *NASA Special Publication*, *237*, 153.

Deolalikar, N., Sadeghi, F., & Marble, S. (2008). Numerical modeling of mixed lubrication and flash temperature in EHL elliptical contacts. *Journal of Tribology*, *130*(1), 011004. doi:10.1115/1.2805429

Fraser, R. D. B., & Macrae, T. P. (1973). Conformation in Fibrous Proteins and Related Synthetic Polypeptides. Academic Press.

Fulcher, J. T., Lu, Y. C., Tandon, G. P., & Foster, D. C. (2010). Thermomechanical characterization of environmentally conditioned shape memory polymer using nanoindentation. *Proceedings of the Society for Photo-Instrumentation Engineers*, *76*, 4–40. doi:10.1117/12.846974

Ge, Q., Dunn, C. K., Qi, H. J., & Dunn, M. L. (2014). Active origami by 4D printing. *Smart Materials and Structures*, *23*(9), 094007. doi:10.1088/0964-1726/23/9/094007

Ge, Q., Qi, H. J., & Dunn, M. L. (2013). Active materials by four-dimension printing. *Applied Physics Letters*, *103*(13), 131901. doi:10.1063/1.4819837

Gladman, Matsumoto, Nuzzo, Mahadevan, & Lewis. (2016). Biomimetic 4D printing. *Nat. Mater.*

Greenwood, J. A., & Williamson, J. P. (1966). Contact of nominally flat surfaces. *Proceedings of the Royal Society of London. Series A, Mathematical and Physical Sciences*, *295*(1442), 300–319. doi:10.1098/rspa.1966.0242

Irish, F. J., Williams, E. E., & Seling, E. (1988). Scanning electron microscopy of changes in the epidermal structure occurring during the shedding cycle in squamate reptiles. *Journal of Morphology*, *197*(1), 105–126. doi:10.1002/jmor.1051970108 PMID:29890790

Jamal, M., Kadam, S. S., Xiao, R., Jivan, F., Onn, T. M., Fernandes, R., ... Gracias, D. H. (2013). Bio-origami hydrogel scaffolds composed of photo-crosslinked PEG bilayers. *Advanced Healthcare Materials*, *2*(8), 1142–1150. doi:10.1002/adhm.201200458 PMID:23386382

Kerfoot, W. (1970). The effect of functional changes upon the variability of lizard and snake body scale numbers. *Copeia*, *1970*(2), 252–260. doi:10.2307/1441647

Khoo, Z. X., Teoh, J. E. M., Liu, Y., Chua, C. K., Yang, S., An, J., ... Yeong, W. Y. (2015). 3D printing of smart materials: A review on recent progresses in 4D printing. *Virtual and Physical Prototyping*, *10*(3), 103–122. doi:10.1080/17452759.2015.1097054

Klein, M.-C. G., Deuschle, J., & Gorb, S. N. (2010). Material properties of the skin of the Kenyan sand boa Gongylophis colubrinus (Squamata, Boidae). *J. of Comp. Phys. A: Neuroethology, Sensory, Neural, and Behav. Phys.*, *196*, 659–668.

Klein, M.-C. G., & Gorb, S. N. (2012). Epidermis architecture and material properties of the skin of four snake species. *Journal of the Royal Society, Interface*, *9*(76), 3140–3155. doi:10.1098/rsif.2012.0479 PMID:22896567

Kuhlmann-Wilsdorf, D. (1987). Temperatures at interfacial contact spots: Dependence on velocity and on role reversal of two materials in sliding contact. *ASME J. Trib.*, *109*(2), 321–329. doi:10.1115/1.3261361

Kuksenok, O., & Balazs, A. C. (2016). Stimuli-responsive behavior of composites integrating thermo-responsive gels with photo-responsive fibers. *Materials Horizons*, *3*(1), 53–62. doi:10.1039/C5MH00212E

Lai, W. T., & Cheng, H. S. (1985). Temperature analysis in lubricated simple sliding rough contacts. *ASLE Transactions*, *28*(3), 303–312. doi:10.1080/05698198508981625

Lal, B., & Mathur, G. N. (2007). Tribo-investigation of PEEK-short carbon fibre composites in severe testing conditions. *Journal of Synthetic Lubrication*, *24*(3), 149-66.

Landmann, L. (1979). Keratin formation and barrier mechanisms in the epidermis of Natrix natrix (Reptilia: Serpentes): an ultrastructural study. *Journal of Morphology*, *162*(1), 93–126. doi:10.1002/jmor.1051620107 PMID:30200689

Landmann, L. (1986). Biology of the integument. In The skin of reptiles. Springer.

Lendlein, A., & Kelch, S. (2005). Shape-memory polymers as stimuli-sensitive implant materials. *Clinical Hemorheology and Microcirculation*, *32*, 105–116. PMID:15764819

Lillywhite, H. B., & Maderson, P. F. A. (1982). Biology of Reptilia. Physiology C. Academic Press Inc Ltd.

Lindel, L., Forsman, A., & Merila, J. (1993). Variation in number of ventral scales in snakes: Effects on body size, growth rate and survival in the adder, Vipera berus. *Journal of Zoology (London, England)*, *230*(1), 101–115. doi:10.1111/j.1469-7998.1993.tb02675.x

Ling, F. F. (1969). On temperature transients in sliding interface. *ASME J. Lubr. Tech.*, *1*(91), 397–405. doi:10.1115/1.3554950

Maderson, P. F. (1985). Chapter. In C. Gans, F. Billett, & P. F. Maderson (Eds.), Biology of Reptilia (pp. 525–598). New York: Wiley.

Maderson, P. F. A. (1965). Histological changes in the epidermis of snakes during the sloughing cycle. *Journal of Zoology*, *146*(1), 98–113.

Maderson, P. F. A. (1966). Histological changes in the epidermis of the Tokay (Gekko gecko) during the sloughing cycle. *Journal of Morphology*, *119*(1), 39–50. doi:10.1002/jmor.1051190105

Maderson, P. F. A. (1972). When? Why? and How? Some speculations on the evolution of the vertebrate integument. *American Zoologist*, *12*(1), 159–171. doi:10.1093/icb/12.1.159

Maderson, P. F. A., & Licht, P. (1967). The epidermal morphology and sloughing frequency in normal and prolactin injected Anolis caroli;zensis (Iguanidae, Lacertilia). *Journal of Morphology*, *123*(2), 157–172. doi:10.1002/jmor.1051230205 PMID:6073684

Mao, Y., Yu, K., Isakov, M. S., Wu, J., Dunn, M. L., & Qi, H. J. (2015). Sequential self-folding structures by 3D printed digital shape memory polymers. *Scientific Reports*, *5*(1), 5. doi:10.1038rep13616 PMID:26346202

Meng, H., & Li, G. (2013). A review of stimuli-responsive shape memory polymer composites. *Polymer*, *54*(9), 2199–2221. doi:10.1016/j.polymer.2013.02.023

Mondal, S., & Hu, J. (2006). Temperature stimulating shape memory polyurethane for smart clothing. *Indian Journal of Fibre and Textile Research*, *31*, 66–71.

Pei, E. (2014). 4D printing: Dawn of an emerging technology cycle. *Assembly Automation*, *34*(4), 310–314. doi:10.1108/AA-07-2014-062

Raviv, D., Zhao, W., McKnelly, C., Papadopoulou, A., Kadambi, A., Shi, B., ... Olguin, C. (2014). Active printed materials for complex self evolving deformations. *Scientific Reports*, 4. PMID:25522053

Raviv, D., Zhao, W., McKnelly, C., Papadopoulou, A., Kadambi, A., Shi, B., ... Tibbits, S. (2015). Active Printed Materials for Complex Self-Evolving Deformations. *Scientific Reports*, *4*(1), 7422. doi:10.1038rep07422 PMID:25522053

Roy, D., Cambre, J. N., & Sumerlin, B. S. (2010). Future perspectives and recent advances in stimuli-responsive materials. *Progress in Polymer Science, 35*(1-2), 278–301. doi:10.1016/j.progpolymsci.2009.10.008

Shi, X., & Polycarpou, A. A. (2005) Measurement and modeling of normal contact stiffness and contact damping at the meso scale. *J. Vib. Acoust., 127*(1), 52-60.

Stuart, M. A. C., Huck, W. T., Genzer, J., Müller, M., Ober, C., Stamm, M., ... Urban, M. (2010). Emerging applications of stimuli-responsive polymer materials. *Nature Materials, 9*(2), 101–113. doi:10.1038/nmat2614 PMID:20094081

Sun, L., Huang, W. M., Ding, Z., Zhao, Y., Wang, C. C., Purnawali, H., & Tang, C. (2012). Stimulus responsive shape memory materials: A review. *Materials & Design, 33*, 577–640. doi:10.1016/j.matdes.2011.04.065

Tian, X., & Kennedy, F. E. (1993). Temperature rise at the sliding contact interface for a coated semi-infinite body. *ASME J. Trib., 115*(1), 1–15. doi:10.1115/1.2920976

Tibbits, McKnelly, Olguin, Dikovsky, & Hirsch. (2014). *4D printing and universal transformation*. Academic Press.

Tibbits, S. (2014). 4D printing: Multi-material shape change. *Archit. Des., 84*, 116–121.

Won, S., Kyung, T., & Sung, H. (2016). Review of biomimetic underwater robots using smart actuators. *International Journal of Precision Engineering and Manufacturing, 13*, 1281–1292.

Wu, J., Yuan, C., Ding, Z., Isakov, M., Mao, Y., Wang, T., ... Qi, H. J. (2016). Multi-shape active composites by 3D printing of digital shape memory polymers. *Scientific Reports, 6*(1), 6. doi:10.1038rep24224 PMID:27071543

Wu, L., Jin, C., & Sun, X. (2010). Synthesis, properties, and light-induced shape memory effect of multiblock polyesterurethanes containing biodegradable segments and pendant cinnamamide groups. *Biomacromolecules, 12*(1), 235–241. doi:10.1021/bm1012162 PMID:21125994

Ye, N., & Komvopoulos, K. (2003). Three-dimensional finite element analysis of elastic-plastic layered media under thermomechanical surface loading. *ASME J. of Trib., 125*(1), 52-9.

Yu, K., Dunn, M. L., & Qi, H. J. (2015). Digital manufacture of shape changing components. *Extreme Mechanics Letters*, *4*, 9–17. doi:10.1016/j.eml.2015.07.005

Zhang, Q., Zhang, K., & Hu, G. (2016). Smart three-dimensional lightweight structure triggered from a thin composite sheet via 3D printing technique. *Scientific Reports*, *6*(1), 6. doi:10.1038rep22431 PMID:26926357

Related Readings

To continue IGI Global's long-standing tradition of advancing innovation through emerging research, please find below a compiled list of recommended IGI Global book chapters and journal articles in the areas of green cities, environmental management, and sustainable urban development. These related readings will provide additional information and guidance to further enrich your knowledge and assist you with your own research.

Abdel-Aal, H. A. (2018). Review of Friction and Surface Properties of Snakeskin. In M. Habib (Ed.), *Handbook of Research on Biomimetics and Biomedical Robotics* (pp. 276–315). Hershey, PA: IGI Global. doi:10.4018/978-1-5225-2993-4.ch012

Adebiyi, I. D., Popoola, P. A., & Pityana, S. (2017). Mitigation of Wear Damage by Laser Surface Alloying Technique. In I. Management Association (Ed.), Materials Science and Engineering: Concepts, Methodologies, Tools, and Applications (pp. 1384-1403). Hershey, PA: IGI Global. doi:10.4018/978-1-5225-1798-6.ch055

Akpinar, E., & Ersin, S. (2017). A Comparative Investigation of Structural Performance of Typical and Non-Ducitle Public RC Buildings Strengthened Using Friction Dampers and RC Walls. In V. Plevris, G. Kremmyda, & Y. Fahjan (Eds.), *Performance-Based Seismic Design of Concrete Structures and Infrastructures* (pp. 112–127). Hershey, PA: IGI Global. doi:10.4018/978-1-5225-2089-4.ch005

Almeida, L. E., Martins, A. F., Gomes, S. R., & Cunha, F. A. (2018). Thermal Decomposition Kinetics Studies of HTPB/Al/AP Solid Propellants Formulated With Iron Oxide Burning Rate Catalyst in Nano and Micro Scale. In R. Goncalves, J. Atilio Fritz Fidel Rocco, & K. Iha (Eds.), *Energetic Materials Research, Applications, and New Technologies* (pp. 211–233). Hershey, PA: IGI Global. doi:10.4018/978-1-5225-2903-3.ch010

Amuda, M. O., & Akinlabi, E. T. (2020). Trend and Development in Laser Surface Modification for Enhanced Materials Properties. In I. Management Association (Ed.), Additive Manufacturing: Breakthroughs in Research and Practice (pp. 271-295). Hershey, PA: IGI Global. doi:10.4018/978-1-5225-9624-0.ch011

Amuda, M. O., Lawal, T. F., & Akinlabi, E. T. (2017). Research Progress on Rheological Behavior of AA7075 Aluminum Alloy During Hot Deformation. *International Journal of Materials Forming and Machining Processes*, *4*(1), 53–96. doi:10.4018/IJMFMP.2017010104

Armghan, A., Hu, X., & Javed, M. Y. (2018). Design and Simulations of Negative Refractive Index Metamaterial (NIR) SRR and CSRR Structures. In *Modeling and Simulations for Metamaterials: Emerging Research and Opportunities* (pp. 141–147). Hershey, PA: IGI Global. doi:10.4018/978-1-5225-4180-6.ch006

Armghan, A., Hu, X., & Javed, M. Y. (2018). Fundamentals of Metamaterials. In *Modeling and Simulations for Metamaterials: Emerging Research and Opportunities* (pp. 76–89). Hershey, PA: IGI Global. doi:10.4018/978-1-5225-4180-6.ch003

Assovskiy, I. G. (2018). Laser Ignition of Metalized Solid Propellants. In R. Goncalves, J. Atilio Fritz Fidel Rocco, & K. Iha (Eds.), *Energetic Materials Research, Applications, and New Technologies* (pp. 79–99). Hershey, PA: IGI Global. doi:10.4018/978-1-5225-2903-3.ch004

Azadi, M. (2018). Deposition of TiN, TiC, and DLC Coatings by PACVD. In A. Pakseresht (Ed.), *Production, Properties, and Applications of High Temperature Coatings* (pp. 381–402). Hershey, PA: IGI Global. doi:10.4018/978-1-5225-4194-3.ch014

B. J. M. (2018). Studies on Metal Oxide Nanoparticle Doped PVP Polymer Nanocomposites. In R. Somashekar, & T. Urs G. (Eds.), Emergent Research on Polymeric and Composite Materials (pp. 209-224). Hershey, PA: IGI Global. doi:10.4018/978-1-5225-3023-7.ch009

Balachandran, P. V., & Rondinelli, J. M. (2017). Informatics-Based Approaches for Accelerated Discovery of Functional Materials. In I. Management Association (Ed.), Materials Science and Engineering: Concepts, Methodologies, Tools, and Applications (pp. 153-184). Hershey, PA: IGI Global. doi:10.4018/978-1-5225-1798-6.ch007

Balla, V. K., Das, M., Datta, S., & Kundu, B. (2018). Articulating Biomaterials: Surface Engineering, Tribology, and Biocompatibility. In I. Management Association (Ed.), Biomedical Engineering: Concepts, Methodologies, Tools, and Applications (pp. 859-910). Hershey, PA: IGI Global. doi:10.4018/978-1-5225-3158-6.ch038

Banerjee, S., Poria, S., Sutradhar, G., & Sahoo, P. (2020). Optimization of Tribological Behavior of Mg-Wc Nanocomposites at Elevated Temperature. *International Journal of Surface Engineering and Interdisciplinary Materials Science*, *8*(1), 25–43. doi:10.4018/IJSEIMS.2020010103

Barua, R., Datta, S., Datta, P., & Chowdhury, A. R. (2019). Scaffolds and Tissue Engineering Applications by 3D Bio-Printing Process: A New Approach. In K. Kumar & J. Davim (Eds.), *Design, Development, and Optimization of Bio-Mechatronic Engineering Products* (pp. 78–99). Hershey, PA: IGI Global. doi:10.4018/978-1-5225-8235-9.ch004

Bhaskar, S. V., & Kudal, H. N. (2017). Effect of TiCN and AlCrN Coating on Tribological Behaviour of Plasma-nitrided AISI 4140 Steel. *International Journal of Surface Engineering and Interdisciplinary Materials Science*, *5*(2), 1–17. doi:10.4018/IJSEIMS.2017070101

Bhuyan, D. (2018). Designing of a Twin Tube Shock Absorber: A Study in Reverse Engineering. In K. Kumar & J. Davim (Eds.), *Design and Optimization of Mechanical Engineering Products* (pp. 83–104). Hershey, PA: IGI Global. doi:10.4018/978-1-5225-3401-3.ch005

Binda, R. V., Rocha, R. J., & Almeida, L. E. (2018). Study of Factors Influencing the Life Predictions of Solid Rocket Motor. In R. Goncalves, J. Atilio Fritz Fidel Rocco, & K. Iha (Eds.), *Energetic Materials Research, Applications, and New Technologies* (pp. 234–245). Hershey, PA: IGI Global. doi:10.4018/978-1-5225-2903-3.ch011

Bolar, G., & Joshi, S. N. (2018). Experimental Study on Surface Integrity, Dimensional Accuracy, and Micro-Hardness in Thin-Wall Machining of Aluminum Alloy. *International Journal of Materials Forming and Machining Processes*, *5*(2), 13–31. doi:10.4018/IJMFMP.2018070102

Borah, R., & Kumar, A. (2019). Enhanced Cellular Activity on Conducting Polymer: Chitosan Nanocomposite Scaffolds for Tissue Engineering Application. In N. Ramdani (Ed.), *Polymer Nanocomposites for Advanced Engineering and Military Applications* (pp. 150–189). Hershey, PA: IGI Global. doi:10.4018/978-1-5225-7838-3.ch006

Boschetto, A., & Bottini, L. (2017). Surface Characterization in Fused Deposition Modeling. In I. Management Association (Ed.), 3D Printing: Breakthroughs in Research and Practice (pp. 22-47). Hershey, PA: IGI Global. doi:10.4018/978-1-5225-1677-4.ch002

Chen, L. (2017). Thermal-Mechanical Effects and Near-Critical Fluid Dynamic Behaviors in Micro-Scale. In L. Chen & Y. Iwamoto (Eds.), *Advanced Applications of Supercritical Fluids in Energy Systems* (pp. 55–101). Hershey, PA: IGI Global. doi:10.4018/978-1-5225-2047-4.ch003

Chowdhury, S., Goyal, P., Hariharan, R., & Samui, P. (2017). Determination of Stability of Rock Slope Using Intelligent Pattern Recognition Techniques. In I. Management Association (Ed.), Artificial Intelligence: Concepts, Methodologies, Tools, and Applications (pp. 558-579). Hershey, PA: IGI Global. doi:10.4018/978-1-5225-1759-7.ch024

Costa, F. S., & Gonzáles, C. A. (2018). Chapman-Jouguet Combustion Waves in Van Der Waals and Noble-Abel Gases. In R. Goncalves, J. Atilio Fritz Fidel Rocco, & K. Iha (Eds.), *Energetic Materials Research, Applications, and New Technologies* (pp. 51–78). Hershey, PA: IGI Global. doi:10.4018/978-1-5225-2903-3.ch003 da Silva, G. (2018). Green Propellants. In R. Goncalves, J. Atilio Fritz Fidel Rocco, & K. Iha (Eds.), *Energetic Materials Research, Applications, and New Technologies* (pp. 170–184). Hershey, PA: IGI Global. doi:10.4018/978-1-5225-2903-3.ch008

Das, M. K., Barman, T. K., Kumar, K., & Sahoo, P. (2017). Effect of Process Parameters on MRR and Surface Roughness in ECM of EN 31 Tool Steel Using WPCA. *International Journal of Materials Forming and Machining Processes, 4*(2), 45–63. doi:10.4018/IJMFMP.2017070104

Das, S. K., Roy, S., & Sahoo, P. (2020). Corrosion and Wear Behavior of Electroless Nickel Coatings. In P. Sahoo (Ed.), *Handbook of Research on Developments and Trends in Industrial and Materials Engineering* (pp. 210–227). Hershey, PA: IGI Global. doi:10.4018/978-1-7998-1831-1.ch010

DeLuca, L. T., Bohn, M. A., Gettwert, V., Weiser, V., & Tagliabue, C. (2018). Innovative Solid Rocket Propellant Formulations for Space Propulsion. In R. Goncalves, J. Atilio Fritz Fidel Rocco, & K. Iha (Eds.), *Energetic Materials Research, Applications, and New Technologies* (pp. 1–24). Hershey, PA: IGI Global. doi:10.4018/978-1-5225-2903-3.ch001

Dixit, A. (2018). Application of Silica-Gel-Reinforced Aluminium Composite on the Piston of Internal Combustion Engine: Comparative Study of Silica-Gel-Reinforced Aluminium Composite Piston With Aluminium Alloy Piston. In K. Kumar & J. Davim (Eds.), *Composites and Advanced Materials for Industrial Applications* (pp. 63–98). Hershey, PA: IGI Global. doi:10.4018/978-1-5225-5216-1.ch004

Du, B. (2020). Flashover and Surface Charge in GIL Insulator. In *Electrical Insulation Breakdown and Its Theory, Process, and Prevention: Emerging Research and Opportunities* (pp. 46–72). Hershey, PA: IGI Global. doi:10.4018/978-1-5225-8885-6.ch002

Du, B. (2020). Surface Charge Property of SiR/SiC Composites with Field-Dependent Conductivity. In *Electrical Insulation Breakdown and Its Theory, Process, and Prevention: Emerging Research and Opportunities* (pp. 219–254). Hershey, PA: IGI Global. doi:10.4018/978-1-5225-8885-6.ch008

Duari, S., Mukhopadhyay, A., Barman, T. K., & Sahoo, P. (2018). Tribological Performance Optimization of Electroless Nickel Coatings Under Lubricated Condition. In K. Kumar & J. Davim (Eds.), *Design and Optimization of Mechanical Engineering Products* (pp. 250–280). Hershey, PA: IGI Global. doi:10.4018/978-1-5225-3401-3.ch013

Erinosho, M. F., Akinlabi, E. T., & Pityana, S. (2020). Enhancement of Surface Integrity of Titanium Alloy With Copper by Means of Laser Metal Deposition Process. In I. Management Association (Ed.), Additive Manufacturing: Breakthroughs in Research and Practice (pp. 245-270). Hershey, PA: IGI Global. doi:10.4018/978-1-5225-9624-0.ch010

Faria, P. C. (2018). An Analysis of the Initiation Process of Electroexplosive Devices. In R. Goncalves, J. Atilio Fritz Fidel Rocco, & K. Iha (Eds.), *Energetic Materials Research, Applications, and New Technologies* (pp. 322–333). Hershey, PA: IGI Global. doi:10.4018/978-1-5225-2903-3.ch015

Farkas, J., Csanády, E., & Csóka, L. (2018). Effects of Tool Wear on Surface Roughness and Cutting Force in Thermoplastics Turning. *International Journal of Materials Forming and Machining Processes*, *5*(1), 1–11. doi:10.4018/IJMFMP.2018010101

Gomes, S. R., & Rocco, L. J. (2018). Design and Testing of a Lab-Scale Hybrid Rocket Motor to Evaluate Swirl Injection Characteristics. In R. Goncalves, J. Atilio Fritz Fidel Rocco, & K. Iha (Eds.), *Energetic Materials Research, Applications, and New Technologies* (pp. 185–210). Hershey, PA: IGI Global. doi:10.4018/978-1-5225-2903-3.ch009

Goncalves, R. F., Iha, K., & Rocco, J. A. (2018). Enhancement of Energetic Materials Combustion Process. In R. Goncalves, J. Atilio Fritz Fidel Rocco, & K. Iha (Eds.), *Energetic Materials Research, Applications, and New Technologies* (pp. 25–50). Hershey, PA: IGI Global. doi:10.4018/978-1-5225-2903-3.ch002

Gupta, R., Kumar, K., & Sharma, N. (2018). Multi-Performance Optimization in Friction Stir Welding of Aluminum Alloy Using Response Surface Methodology. In D. Kim, S. Sekhar Roy, T. Länsivaara, R. Deo, & P. Samui (Eds.), *Handbook of Research on Predictive Modeling and Optimization Methods in Science and Engineering* (pp. 240–263). Hershey, PA: IGI Global. doi:10.4018/978-1-5225-4766-2.ch011

H, S., & Alabur, M. (2018). Structure-Property Relations in Polymers and Polymer Blends Using X-Ray Diffraction Technique. In R. Somashekar, & T. Urs G. (Eds.), *Emergent Research on Polymeric and Composite Materials* (pp. 122-138). Hershey, PA: IGI Global. doi:10.4018/978-1-5225-3023-7.ch005

Haidar, S., Sekh, M., Ghose, J., & Sutradhar, G. (2017). Frictional Behavior of Aluminium MMC Foam Synthesized Using Dual Foaming Agent. *International Journal of Surface Engineering and Interdisciplinary Materials Science*, *5*(2), 18–32. doi:10.4018/IJSEIMS.2017070102

Hassan, M. Z., Roslan, S. A., Rasid, A. Z., Mohamed Yusoff, M. Z., Sapuan, S. M., & Muhammad-Sukki, F. (2020). Optimizing the Mercerisation Effect on the Mode I Fracture Toughness of Bambusa Vulgaris Bamboo Using Surface Response Method. In M. Mastura, & S. Sapuan (Eds.), Implementation and Evaluation of Green Materials in Technology Development: Emerging Research and Opportunities (pp. 112-129). Hershey, PA: IGI Global. doi:10.4018/978-1-7998-1374-3.ch006

He (Herman) Tang. (2017). Comprehensive Considerations on Material Selection for Lightweighting Vehicle Bodies Based on Material Costs and Assembly Joining Technologies. *International Journal of Manufacturing, Materials, and Mechanical Engineering*, *7*(4), 1–14. doi:10.4018/IJMMME.2017100101

Henao, J., & Sotelo, O. (2018). Surface Engineering at High Temperature: Thermal Cycling and Corrosion Resistance. In A. Pakseresht (Ed.), *Production, Properties, and Applications of High Temperature Coatings* (pp. 131–159). Hershey, PA: IGI Global. doi:10.4018/978-1-5225-4194-3.ch006

Ilori, O. O., Adetan, D. A., & Umoru, L. E. (2017). Effect of Cutting Parameters on the Surface Residual Stress of Face-Milled Pearlitic Ductile Iron. *International Journal of Materials Forming and Machining Processes*, *4*(1), 38–52. doi:10.4018/IJMFMP.2017010103

Iosif, K. Z. (2020). Performance Improvement of Mechanical Components by Precision Coating. In J. Zbitou, C. Pruncu, & A. Errkik (Eds.), *Handbook of Research on Recent Developments in Electrical and Mechanical Engineering* (pp. 430–454). Hershey, PA: IGI Global. doi:10.4018/978-1-7998-0117-7.ch016

Jajimoggala, S. (2019). Optimization of Hot Extrusion Process Parameters Using Taguchi Based Grey Relation Analysis: An Experimental Approach. *International Journal of Materials Forming and Machining Processes*, *6*(1), 1–18. doi:10.4018/IJMFMP.2019010101

Jayalakshmi, S., & Singh, R. A. (2017). Processing Routes, Mechanical, and Tribological Properties of Light Metal Matrix Nanocomposites. In I. Management Association (Ed.), Materials Science and Engineering: Concepts, Methodologies, Tools, and Applications (pp. 991-1037). Hershey, PA: IGI Global. doi:10.4018/978-1-5225-1798-6.ch040

Kadda, M., & Nadia, B. (2020). Elastic Behavior of the Plain Journal Bearing Coated With a Textured Surface and a Non-Textured Surface: Plain Journal Bearing at Textured Surface Behavior. *International Journal of Surface Engineering and Interdisciplinary Materials Science*, *8*(1), 55–77. doi:10.4018/IJSEIMS.2020010105

Kakaravada, I., Mahamani, A., & Pandurangadu, V. (2018). Investigation on Cutting Force, Flank Wear, and Surface Roughness in Machining of the A356-TiB2/TiC in-situ Composites. *International Journal of Materials Forming and Machining Processes*, *5*(2), 45–77. doi:10.4018/IJMFMP.2018070104

Koopaee, M. J., Gilani, C., Scott, C., & Chen, X. (2018). Bio-Inspired Snake Robots: Design, Modelling, and Control. In M. Habib (Ed.), *Handbook of Research on Biomimetics and Biomedical Robotics* (pp. 246–275). Hershey, PA: IGI Global. doi:10.4018/978-1-5225-2993-4.ch011

Kumar, G. R., Rajyalakshmi, G., & Manupati, V. K. (2017). Surface Micro Patterning of Aluminium Reinforced Composite through Laser Peening. *International Journal of Manufacturing, Materials, and Mechanical Engineering*, *7*(4), 15–27. doi:10.4018/IJMMME.2017100102

Kumar, K., & Davim, J. P. (2018). Comparative Study of Mechanical and Tribological Behaviour of Thermoplast-Based Composites: With Inorganic and Natural Particulate Fillers. In R. Somashekar, & T. Urs G. (Eds.), Emergent Research on Polymeric and Composite Materials (pp. 78-98). Hershey, PA: IGI Global. doi:10.4018/978-1-5225-3023-7.ch003

Kumari, N., & Kumar, K. (2018). Efficacy of Composites for Fabrication of Orthotic Calipers. In R. Somashekar, & T. Urs G. (Eds.), Emergent Research on Polymeric and Composite Materials (pp. 267-286). Hershey, PA: IGI Global. doi:10.4018/978-1-5225-3023-7.ch012

Li, C., & Ali, H. M. (2020). Comparative Study on Tribological Properties of Nanofluids in Friction-Wear Experiments and Grinding Processing. In *Enhanced Heat Transfer Mechanism of Nanofluid MQL Cooling Grinding* (pp. 298–316). Hershey, PA: IGI Global. doi:10.4018/978-1-7998-1546-4.ch013

Li, C., & Ali, H. M. (2020). Experimental Evaluation of the Lubrication Properties of the Wheel/Workpiece Interface in MQL Grinding Using Vegetable Oils. In *Enhanced Heat Transfer Mechanism of Nanofluid MQL Cooling Grinding* (pp. 249–274). Hershey, PA: IGI Global. doi:10.4018/978-1-7998-1546-4.ch011

Li, C., & Ali, H. M. (2020). Experimental Evaluation on the Effect of Nanofluids Physical Properties With Different Concentrations on Grinding Temperature. In *Enhanced Heat Transfer Mechanism of Nanofluid MQL Cooling Grinding* (pp. 203–225). Hershey, PA: IGI Global. doi:10.4018/978-1-7998-1546-4.ch009

Li, C., & Ali, H. M. (2020). Experimental Evaluation on Tribological Performance of the Wheel/Workpiece Interface in NMQL Grinding With Different Concentrations of Al2o3 Nanofluids. In *Enhanced Heat Transfer Mechanism of Nanofluid MQL Cooling Grinding* (pp. 317–336). Hershey, PA: IGI Global. doi:10.4018/978-1-7998-1546-4.ch014

Li, C., & Ali, H. M. (2020). Experimental Research on Grinding Temperature With Different Workpiece Materials. In *Enhanced Heat Transfer Mechanism of Nanofluid MQL Cooling Grinding* (pp. 226–248). Hershey, PA: IGI Global. doi:10.4018/978-1-7998-1546-4.ch010

Li, C., & Ali, H. M. (2020). Experimental Research on Heat Transfer Performance in MQL Grinding With Different Nanofluids. In *Enhanced Heat Transfer Mechanism of Nanofluid MQL Cooling Grinding* (pp. 182–202). Hershey, PA: IGI Global. doi:10.4018/978-1-7998-1546-4.ch008

Li, C., & Ali, H. M. (2020). Experimental Research on Minimum Quantity Lubrication Surface Grinding With Different Cooling and Lubrication Conditions. In *Enhanced Heat Transfer Mechanism of Nanofluid MQL Cooling Grinding* (pp. 132–159). Hershey, PA: IGI Global. doi:10.4018/978-1-7998-1546-4.ch006

Li, C., & Ali, H. M. (2020). Experimental Study of Lubricating Property at Grinding Wheel/Workpiece Interface Under NMQL Grinding. In *Enhanced Heat Transfer Mechanism of Nanofluid MQL Cooling Grinding* (pp. 275–297). Hershey, PA: IGI Global. doi:10.4018/978-1-7998-1546-4.ch012

Li, C., & Ali, H. M. (2020). Machining Mechanism of Minimum Quantity Lubrication Grinding. In *Enhanced Heat Transfer Mechanism of Nanofluid MQL Cooling Grinding* (pp. 43–75). Hershey, PA: IGI Global. doi:10.4018/978-1-7998-1546-4.ch003

Li, C., & Ali, H. M. (2020). Modeling and Simulation of Surface Topography in Single Abrasive Grain Grinding. In *Enhanced Heat Transfer Mechanism of Nanofluid MQL Cooling Grinding* (pp. 358–380). Hershey, PA: IGI Global. doi:10.4018/978-1-7998-1546-4.ch016

Li, C., & Ali, H. M. (2020). Modeling and Simulation of the Surface Topography Generation With Engineered Grinding Wheel. In *Enhanced Heat Transfer Mechanism of Nanofluid MQL Cooling Grinding* (pp. 381–403). Hershey, PA: IGI Global. doi:10.4018/978-1-7998-1546-4.ch017

Li, C., & Ali, H. M. (2020). Optimization Design of Process Parameters for Different Workpiece Materials in NMQL Grinding With Different Vegetable Oils. In *Enhanced Heat Transfer Mechanism of Nanofluid MQL Cooling Grinding* (pp. 337–357). Hershey, PA: IGI Global. doi:10.4018/978-1-7998-1546-4.ch015

Li, Y., Wang, J., Wang, S., Li, D., Song, S., Zhang, P., ... Yuan, H. (2019). Immiscible Two-Phase Parallel Microflow and Its Applications in Fabricating Micro- and Nanomaterials. In H. Santana, J. Lameu da Silva Jr, & O. Taranto (Eds.), *Process Analysis, Design, and Intensification in Microfluidics and Chemical Engineering* (pp. 136–166). Hershey, PA: IGI Global. doi:10.4018/978-1-5225-7138-4.ch005

Ma, W., Shang, J., Lu, J., & Meng, J. (2017). Material and Tribology Issues of Self-Lubricating Copper Matrix Composite. In I. Management Association (Ed.), Materials Science and Engineering: Concepts, Methodologies, Tools, and Applications (pp. 414-438). Hershey, PA: IGI Global. doi:10.4018/978-1-5225-1798-6.ch017

Mahamood, R. M., Akinlabi, E. T., Shukla, M., & Pityana, S. (2020). Improving Surface Integrity Using Laser Metal Deposition Process. In I. Management Association (Ed.), Additive Manufacturing: Breakthroughs in Research and Practice (pp. 220-244). Hershey, PA: IGI Global. doi:10.4018/978-1-5225-9624-0.ch009

Mahamood, R. M., Shukla, M., & Pityana, S. (2017). Laser Additive Manufacturing in Surface Modification of Metals. In I. Management Association (Ed.), 3D Printing: Breakthroughs in Research and Practice (pp. 183-203). Hershey, PA: IGI Global. doi:10.4018/978-1-5225-1677-4.ch010

Maity, K., & Pradhan, S. (2017). Study of Chip Morphology, Flank Wear on Different Machinability Conditions of Titanium Alloy (Ti-6Al-4V) Using Response Surface Methodology Approach. *International Journal of Materials Forming and Machining Processes*, *4*(1), 19–37. doi:10.4018/IJMFMP.2017010102

Mattos, E. D., & Dutra, R. D. (2018). Characterization and Quantification of Polymer Content in Plastic Explosives Using FT-IR Techniques. In R. Goncalves, J. Atilio Fritz Fidel Rocco, & K. Iha (Eds.), *Energetic Materials Research, Applications, and New Technologies* (pp. 288–321). Hershey, PA: IGI Global. doi:10.4018/978-1-5225-2903-3.ch014

Mejia, G. L. (2018). Solid Rocket Motor Internal Ballistics Simulation Considering Complex 3D Propellant Grain Geometries. In R. Goncalves, J. Atilio Fritz Fidel Rocco, & K. Iha (Eds.), *Energetic Materials Research, Applications, and New Technologies* (pp. 146–169). Hershey, PA: IGI Global. doi:10.4018/978-1-5225-2903-3.ch007

Mendonça, F. B., & Urgessa, G. S. (2018). Pre-Test and Analysis of a Reinforced Concrete Slab Subjected to Blast From a Non-Confined Explosive. In R. Goncalves, J. Atilio Fritz Fidel Rocco, & K. Iha (Eds.), *Energetic Materials Research, Applications, and New Technologies* (pp. 272–287). Hershey, PA: IGI Global. doi:10.4018/978-1-5225-2903-3.ch013

Meshginqalam, B., Ahmadi, M. T., Tousi, H. T., Sabatyan, A., & Centeno, A. (2017). Surface Plasmon Resonance-Based Sensor Modeling. In M. Ahmadi, R. Ismail, & S. Anwar (Eds.), *Handbook of Research on Nanoelectronic Sensor Modeling and Applications* (pp. 361–394). Hershey, PA: IGI Global. doi:10.4018/978-1-5225-0736-9.ch014

Mohamad, J. (2017). Investigating Bauschinger Effect and Plastic Hardening Characteristics of Sheet Metal under Cyclic Loading. *International Journal of Materials Forming and Machining Processes*, *4*(2), 1–14. doi:10.4018/IJMFMP.2017070101

Mostafa, A. O. (2019). Mechanical Properties and Wear Behavior of Aluminum Grain Refined by Ti and Ti+B. *International Journal of Surface Engineering and Interdisciplinary Materials Science*, *7*(1), 1–19. doi:10.4018/IJSEIMS.2019010101

Mukherjee, A., & Mitra, P. (2018). Physical Characteristics of CBD Synthesized CdZnS Thin Films: CdZnS Thin Film Preparation Techniques, Properties, and Characterization. In R. Somashekar, & T. Urs G. (Eds.), Emergent Research on Polymeric and Composite Materials (pp. 186-208). Hershey, PA: IGI Global. doi:10.4018/978-1-5225-3023-7.ch008

Mukhopadhyay, A., Barman, M., & Sahoo, P. (2019). Optimization of Fractal Dimension of Turned AISI 1040 Steel Surface Considering Different Cutting Conditions: Fractal Dimension of Turned Steel Surface. *International Journal of Surface Engineering and Interdisciplinary Materials Science*, *7*(2), 19–33. doi:10.4018/IJSEIMS.2019070102

Mukhopadhyay, A., Barman, T. K., & Sahoo, P. (2018). Electroless Nickel Coatings for High Temperature Applications. In K. Kumar & J. Davim (Eds.), *Composites and Advanced Materials for Industrial Applications* (pp. 297–331). Hershey, PA: IGI Global. doi:10.4018/978-1-5225-5216-1.ch013

Mukhopadhyay, A., Duari, S., Barman, T. K., & Sahoo, P. (2017). Tribology of Electroless Ni-P Coating Under Lubricated Condition: An RSM and GA Approach. *International Journal of Surface Engineering and Interdisciplinary Materials Science*, *5*(1), 37–57. doi:10.4018/IJSEIMS.2017010103

Osorno, D. M., & Castro, C. (2018). Cellulose Application in Food Industry: A Review. In R. Somashekar, & T. Urs G. (Eds.), Emergent Research on Polymeric and Composite Materials (pp. 38-77). Hershey, PA: IGI Global. doi:10.4018/978-1-5225-3023-7.ch002

Parida, A. K. (2019). Finite Element Analysis of Tool Wear in Hot Machining Process: Hot Machining. In K. Kumar, N. Kumari, & J. Davim (Eds.), *Non-Conventional Machining in Modern Manufacturing Systems* (pp. 232–248). Hershey, PA: IGI Global. doi:10.4018/978-1-5225-6161-3.ch011

Patel, M. K., & Ghanshyam, C. (2020). Fundamentals of Electrostatic Spraying: Basic Concepts and Engineering Practices. In I. Management Association (Ed.), Environmental and Agricultural Informatics: Concepts, Methodologies, Tools, and Applications (pp. 79-107). Hershey, PA: IGI Global. doi:10.4018/978-1-5225-9621-9.ch005

Paul, S. R., Nayak, S. K., Anis, A., & Pal, K. (2018). Methods of Preparation of Nanoparticle Formulations for the Treatment of COPD: A Short Review. In R. Somashekar, & T. Urs G. (Eds.), Emergent Research on Polymeric and Composite Materials (pp. 247-266). Hershey, PA: IGI Global. doi:10.4018/978-1-5225-3023-7.ch011

Pradhan, M. K. (2020). Tribological Behavior, Machinability, and Optimization of EDM of AA-2014 Hybrid Composite Reinforced With SiC and Glass Particulates. In P. Sahoo (Ed.), *Handbook of Research on Developments and Trends in Industrial and Materials Engineering* (pp. 228–269). Hershey, PA: IGI Global. doi:10.4018/978-1-7998-1831-1.ch011

Ramalingam, A., & Muthuvel, S. (2017). Identification of Optimal Condition for Solid State Welding of Al/SiCp Composite using Taguchi Principles Integrated TOPSIS (TPIT) Method. *International Journal of Manufacturing, Materials, and Mechanical Engineering, 7*(2), 19–37. doi:10.4018/IJMMME.2017040102

Ramdani, N., & Azibi, M. (2018). Properties Enhancement of Polymer/Ceramic Nanocomposites: Polymer Nanocomposites. In R. Somashekar, & T. Urs G. (Eds.), Emergent Research on Polymeric and Composite Materials (pp. 99-121). Hershey, PA: IGI Global. doi:10.4018/978-1-5225-3023-7.ch004

Real, O. R., Castro-Toscano, M. J., Rodríguez-Quiñonez, J. C., Serginyenko, O., Hernández-Balbuena, D., Rivas-Lopez, M., ... Lindner, L. (2019). Surface Measurement Techniques in Machine Vision: Operation, Applications, and Trends. In M. Rivas-Lopez, O. Sergiyenko, W. Flores-Fuentes, & J. Rodríguez-Quiñonez (Eds.), *Optoelectronics in Machine Vision-Based Theories and Applications* (pp. 79–104). Hershey, PA: IGI Global. doi:10.4018/978-1-5225-5751-7.ch004

Rocco, L. Jr. (2018). Disintegration of Liquid Sheet Produced by Swirl Injector. In R. Goncalves, J. Atilio Fritz Fidel Rocco, & K. Iha (Eds.), *Energetic Materials Research, Applications, and New Technologies* (pp. 133–145). Hershey, PA: IGI Global. doi:10.4018/978-1-5225-2903-3.ch006

Rocha, R. J. (2018). Accelerated Aging Tests of HTPB-Based Propellants. In R. Goncalves, J. Atilio Fritz Fidel Rocco, & K. Iha (Eds.), *Energetic Materials Research, Applications, and New Technologies* (pp. 246–271). Hershey, PA: IGI Global. doi:10.4018/978-1-5225-2903-3.ch012

Rudresh, B. M., Ravikumar, B. N., & Madhu, D. (2018). Tribological Response of Polyamide 66 and Polytetrafluroethylene (PA66/PTFE) Blends in Two Body Abrasion Through Multipass Condition. *International Journal of Surface Engineering and Interdisciplinary Materials Science*, *6*(1), 1–16. doi:10.4018/IJSEIMS.2018010101

S., R., & Siddaramaiah. (2018). Effect of Cenosphere on the Performance of Polyurethane/Polystyrene Interpenetrating Polymer Network Green Composites. In R. Somashekar, & T. Urs G. (Eds.), *Emergent Research on Polymeric and Composite Materials* (pp. 158-185). Hershey, PA: IGI Global. doi:10.4018/978-1-5225-3023-7.ch007

Sadri, E., & Ashrafizadeh, F. (2018). High Temperature Nanocomposite Coatings by Plasma Spraying for Friction and Wear Applications. In A. Pakseresht (Ed.), *Production, Properties, and Applications of High Temperature Coatings* (pp. 216–245). Hershey, PA: IGI Global. doi:10.4018/978-1-5225-4194-3.ch009

Sagiri, S. S., Nayak, S. K., Lakshmi, S., & Pal, K. (2018). Preparation and Characterization of Biopolymeric Nanoparticles as Drug Delivery Vehicles: A Comparative Study. In R. Somashekar, & T. Urs G. (Eds.), Emergent Research on Polymeric and Composite Materials (pp. 225-246). Hershey, PA: IGI Global. doi:10.4018/978-1-5225-3023-7.ch010

Sahoo, P., & Roy, S. (2017). Tribological Behavior of Electroless Ni-P, Ni-P-W and Ni-P-Cu Coatings: A Comparison. *International Journal of Surface Engineering and Interdisciplinary Materials Science*, *5*(1), 1–15. doi:10.4018/IJSEIMS.2017010101

Sahu, A. K., & Mahapatra, S. S. (2019). Optimization of Surface Roughness Parameters by Different Multi-Response Optimization Techniques During Electro-Discharge Machining of Titanium Alloy. In K. Kumar, N. Kumari, & J. Davim (Eds.), *Non-Conventional Machining in Modern Manufacturing Systems* (pp. 82–108). Hershey, PA: IGI Global. doi:10.4018/978-1-5225-6161-3.ch004

Samanta, E., Sekh, A. A., & Khatua, D. (2020). An Expert System-Based Automation in Indian Traction System for a One Way Single Platform Station by Introducing PLC. In S. Bhattacharya (Ed.), *Advancements in Instrumentation and Control in Applied System Applications* (pp. 213–230). Hershey, PA: IGI Global. doi:10.4018/978-1-7998-2584-5.ch013

Savonov, R. I. (2018). Dynamics of the Internal Flow in Swirl Atomizers by CFD Simulations. In R. Goncalves, J. Atilio Fritz Fidel Rocco, & K. Iha (Eds.), *Energetic Materials Research, Applications, and New Technologies* (pp. 100–132). Hershey, PA: IGI Global. doi:10.4018/978-1-5225-2903-3.ch005

Sawant, S. (2018). Deep Learning and Biomedical Engineering. In U. Kose, G. Guraksin, & O. Deperlioglu (Eds.), *Nature-Inspired Intelligent Techniques for Solving Biomedical Engineering Problems* (pp. 283–296). Hershey, PA: IGI Global. doi:10.4018/978-1-5225-4769-3.ch014

Sharma, S., Cheema, G. S., Singh, S. S., & Verma, D. (2018). Improvement of Mechanical Properties and Morphological Studies of Friction Stir Processed Composites: A Short Review. In K. Kumar & J. Davim (Eds.), *Design and Optimization of Mechanical Engineering Products* (pp. 152–179). Hershey, PA: IGI Global. doi:10.4018/978-1-5225-3401-3.ch008

Sheikholeslami, M. (2019). Influence of Melting Surface on Nanofluid Convective Heat Transfer. In *Applications of Nanofluid Transportation and Heat Transfer Simulation* (pp. 642–689). Hershey, PA: IGI Global. doi:10.4018/978-1-5225-7595-5.ch010

Shunmugavel, M., Polishetty, A., Goldberg, M., Singh, R. P., & Littlefair, G. (2017). Tool Wear and Surface Integrity Analysis of Machined Heat Treated Selective Laser Melted Ti-6Al-4V. In I. Management Association (Ed.), 3D Printing: Breakthroughs in Research and Practice (pp. 228-240). Hershey, PA: IGI Global. doi:10.4018/978-1-5225-1677-4.ch012

Singh, S., & Pal, K. (2018). Advanced Numerical and Experimental Methods Used in Material Science for Evaluating Mechanical and Damping Nature of Composite Materials. In A. Kumar, P. Patil, & Y. Prajapati (Eds.), *Advanced Numerical Simulations in Mechanical Engineering* (pp. 63–75). Hershey, PA: IGI Global. doi:10.4018/978-1-5225-3722-9.ch004

Subramani, N. K. K. N., S., Shivanna, S., B. M., J. R., & Hatna, S. (2018). Advances in Polymer-Based Piezoelectric Systems. In R. Somashekar, & T. Urs G. (Eds.), Emergent Research on Polymeric and Composite Materials (pp. 139-157). Hershey, PA: IGI Global. doi:10.4018/978-1-5225-3023-7.ch006

Tesi, A., Galvanetto, E., Annicchiarico, C., & Capitani, R. (2017). Tribologic Analysis, Wear Evolution and Torque Trend Estimation of an LSD Clutch Pack. *International Journal of Surface Engineering and Interdisciplinary Materials Science*, *5*(1), 16–36. doi:10.4018/IJSEIMS.2017010102

Thejas Urs, G., & Somashekar, R. (2018). Fibre Diffraction and Whole Powder Pattern Fitting in Polymers. In R. Somashekar, & T. Urs G. (Eds.), Emergent Research on Polymeric and Composite Materials (pp. 1-37). Hershey, PA: IGI Global. doi:10.4018/978-1-5225-3023-7.ch001

Vargas-Bernal, R. (2017). Advances in Functional Nanocoatings Applied in the Aerospace Industry. In I. Management Association (Ed.), Materials Science and Engineering: Concepts, Methodologies, Tools, and Applications (pp. 1722-1744). Hershey, PA: IGI Global. doi:10.4018/978-1-5225-1798-6.ch071

Venkateswarlu, G., Davidson, M., Tagore, G., & Sammaiah, P. (2017). Influence of Process Parameters on Microstructure of Friction Stir Processed Mg AZ31 Alloy. In I. Management Association (Ed.), Materials Science and Engineering: Concepts, Methodologies, Tools, and Applications (pp. 1293-1305). Hershey, PA: IGI Global. doi:10.4018/978-1-5225-1798-6.ch051

Wiśniewski, J. H. (2017). Decreasing Wear of Large Vertical Axis Wind Turbines by Employing a Multi-Level Turbine Concept. In I. Management Association (Ed.), Renewable and Alternative Energy: Concepts, Methodologies, Tools, and Applications (pp. 1015-1028). Hershey, PA: IGI Global. doi:10.4018/978-1-5225-1671-2.ch032

Yahayaa, M. S., Raof, N. A., Ibrahim, Z., Ahmad, A., & Gomes, C. (2019). Modifications Required for Palm Oil to be Qualified as a Mechanical Lubricant. *International Journal of Manufacturing, Materials, and Mechanical Engineering*, *9*(1), 50–66. doi:10.4018/IJMMME.2019010104

Yan, X., Zang, J., Xiong, T., Sui, X., Huang, Y., & Xiao, Z. (2017). The Basic Thermal Hydraulic Issues of Applying Supercritical Fluid to Nuclear Reactors. In L. Chen & Y. Iwamoto (Eds.), *Advanced Applications of Supercritical Fluids in Energy Systems* (pp. 481–553). Hershey, PA: IGI Global. doi:10.4018/978-1-5225-2047-4.ch015

Yang, F., Jin, P. J., Cebelak, M., Ran, B., & Walton, C. M. (2018). The Application of Venue-Side Location-Based Social Networking (VS-LBSN) Data in Dynamic Origin-Destination Estimation. In I. Management Association (Ed.), Intelligent Transportation and Planning: Breakthroughs in Research and Practice (pp. 355-375). Hershey, PA: IGI Global. doi:10.4018/978-1-5225-5210-9.ch016

Yuvaraj, N., & Pradeep Kumar, M. (2019). Performance and Surface Evaluation Characteristics on Cryogenic-Assisted Abrasive Water Jet Machining of AISI D2 Steel. In K. Kumar, N. Kumari, & J. Davim (Eds.), *Non-Conventional Machining in Modern Manufacturing Systems* (pp. 202–231). Hershey, PA: IGI Global. doi:10.4018/978-1-5225-6161-3.ch010

About the Author

Hisham Abdel-Aa, obtained his first degree in Mechanical Engineering from Alexandria University, Alexandria Egypt. He received his Doctoral degree in Precision Engineering from the University of North Carolina in Charlotte in 1998. Since his graduation he has been active in research and teaching in the US and abroad. He occupied several invited positions in Europe, South America and Asia. Including a Research Professorship at the Laboratoire de Materiux et Procedes at Arts et Métiers ParisTech in France where he founded the first research group for Bio-inspired sustainable surfaces for manufacturing. Dr. Abdel-Aal, has lectured extensively on the subjects of tribology, Bio-inspired surfaces, and thermodynamics of wear, and is the author or co-author of over 150 research articles in books, journals, and conference proceedings the subjects of Tribology, surface engineering, Bio-inspired Surface engineering, and bio-mimetics.

Index

www.ingramcontent.com/pod-product-compliance
Lightning Source LLC
LaVergne TN
LVHW080602300926
849162LV00005B/1023